IC TIMER HANDBOOK

. . . with 100 projects & experiments

Other TAB books by the author:

No. 787 *OP AMP Circuit Design & Applications*
No. 901 *CET License Handbook–2nd Edition*
No. 930 *Servicing Medical & Bioelectronic Equipment*
No. 1012 *How to Design and Build Electronic Instrumentation*
No. 1070 *Digital Interfacing with an Analog World*
No. 1152 *Antenna Data Reference Manual–including dimension tables*
No. 1182 *The Complete Handbook of Radio Receivers*
No. 1194 *How to Troubleshoot & Repair Amateur Radio Equipment*
No. 1224 *The Complete Handbook of Radio Transmitters*
No. 1230 *The Complete Handbook of Amplifiers, Oscillators and Multivibrators*
No. 1250 *Digital Electronics Troubleshooting*
No. 1271 *Microcomputer Interfacing Handbook: A/D & D/A*
No. 1273 *Amateur Radio Novice Class License Study Guide–3rd Edition*
No. 1289 *Amateur Radio License Study Guide . . . for all classes*

No. 1290
$15.95

IC TIMER HANDBOOK

. . . with 100 projects & experiments

BY JOSEPH J. CARR

FIRST EDITION

FIRST PRINTING

Printed in the United States of America

Library of Congress Cataloging in Publication Data

Carr, Joseph J.
IC timer handbook—with 100 projects & experiments.

Includes index.
1. Timing circuits. 2. Integrated circuits.
I. Title.
TK7868.T5C37 621.381'73 81-9209
ISBN 0-8306-0007-8 AACR2
ISBN 0-8306-1290-4 (pbk.)

Contents

Introduction

Timer circuits have been with us since the advent of electronics. These useful circuits may be monostable multivibrators (i.e., one-shots), which produce single pulses of fixed duration, or astable multivibrators that produce wavetrains of constant frequency. The latter form of circuit is used as the clock circuit in digital projects and computers.

The main goal of this book is to present enough timer theory that you can make changes in the projects presented. It is rare that you will find the *exact* circuit needed in any textbook such as this, so it is necessary to gain skill in designing and modifying circuits for your own use. Because of this need, we are presenting theory descriptions and the design equations and/or criteria for each class of timer.

The first chapter of this book will introduce you to ordinary bipolar integrated circuits, especially the operational amplifier. The op-amp, ordinarily thought of as a linear device, is useful in several different types of monostable and astable multivibrator.

We will also study some of the timer circuits that use discrete electronic components (e.g., transistors instead of integrated circuits). These circuits are still used to some extent, but have been eclipsed by integrated circuit devices. The theory is presented, however, in order to give you some insight into the operation of timer circuits, many of which are used in the design of IC timers.

The TTL and CMOS lines of IC digital logic devices, and general digital electronics theory (Chapter 4) are also presented. The ordinary digital logic elements are often configured into timer circuits, even though they are not specifically designed for timer application.

The ubiquitous 555 integrated circuit timer is covered in detail. This handy little device is used so widely in electronics that it may well be the most popular integrated circuit timer on the market! The 555 is very easy to apply, and has enough functions brought to external pins to make it a constant source of surprise applications. The 555 lacks the sensitivity to power supply voltage variations that plagues certain discrete timer circuits.

Some of the experiments, especially those in Chapter 4 (digital electronics) are designed to train you in basic techniques. These experiments are designed to be run on the AP Products *Powerace 102* digital breadboard with power supply. Any of the popular digital breadboards could be used instead, but the designations given in the experiment instructions are for the *Powerace 102*.

Joseph J. Carr

Chapter 1
Bipolar IC Devices and Operational Amplifiers

What is a linear integrated circuit? Is a device that contains a couple of transistors an IC, or is it a dual transistor? An old edition of the *RCA Linear IC Book* defined linear ICs as a "selection and interconnection of an optimum combination of active and passive components to accomplish a signal processing function with maximum efficiency and minimum cost . . ." with the components fabricated simultaneously from common materials. A definition from the *Illustrated Dictionary of Electronics*, by Dr. Rufus Turner (TAB Book No. 1066) tells us that an IC is "a circuit whose components and connecting wires are made by processing distinct areas of a "chip" of semiconductor material, such as silicone. Integrated circuits are classified according to construction, a few being monolithic, thin-film and hybrid."

The integrated circuit was first invented in the early 60s, and has taken off like a rocket in the intervening years. From the earliest ICs (was it the uA702 or the uA703?), circuits have grown steadily in complexity. The early uA703, for example, contained only a few transistors and a couple of resistors; it was used extensively as an FM IF amplifier—and not much else. The uA703 was a real hit, and we thought that it was the cat's . . . errrr . . . meow. But that device is considered low technology by engineers today. We now have MSI and LSI (medium scale integration and large scale integration) devices that may contain hundreds of transistors and other components. They are even talking about VLSI (very large scale integration) in the computer market!

The biggest IC market is the digital IC, but don't count analog devices down and out quite yet. They are alive and kicking in many areas of technology and can outperform devices made only a few years ago. This book is about IC timers; and timers can be analog, digital, or a synergism of both analog and digital technology.

Several technologies, are used for creating transistors and other components on a semiconductor *substrate*. (There may be several layers of material in the integrated circuit chip, of which the substrate is the bottom-most.) The usual substrate is approximately 6 mils thick, with the typical cross-sectional area being 50 × 50 mils, with some up to 160 × 160 mils.

In Fig. 1-1 the substrate (in a typical IC) is shown as p-type semiconductor material. The second layer is made of n-type material, and is grown as an extension of the p-type substrate crystal. This *pn junction* must be maintained at a reverse bias, or the IC will be destroyed. The junction region is approximately 5 to 30 micrometers (μM) thick. The next region is of p-type material, while the uppermost is again of n-type material.

An equivalent circuit for the IC is shown in Fig. 1-1; in this case it is a transistor emitter in series with a resistor and a capacitor. The transistor is formed from elements of the second, third, and fourth layers. Since the transistor is NPN, we must connect the collector and emitter terminals to n-type (second and fourth) regions of the device. The series resistor is formed by the ohmic resistance of an element of the n-type second layer. It is strapped to the emitter region by metallization over the top layer. A similar connection is made to the sections of material that form the capacitor. The capacitor itself is formed between the metal elements and the n-type second layer. The dielectric of the capacitor is an oxide layer that covers the top (fourth) region of the device; and Metallized contacts are formed through holes in this oxide layer.

Integrated circuits do not often contain capacitors because of size and cost factors. In general, only small values (picofarad range) can be accommodated in the integrated circuit. Hybrid circuits, a combination of integrated circuit and discrete component techniques, often use small chip capacitors on the ceramic substrate containing the several IC chips that most such circuits require.

Many IC timer devices (such as the 555) use a combination of analog and digital circuitry to perform the timer function. We can,

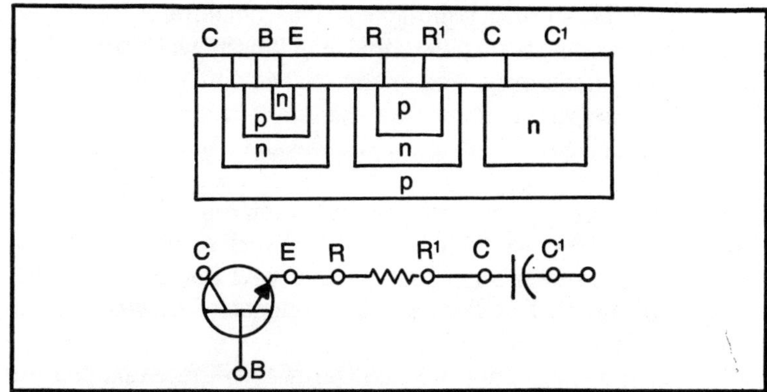

Fig. 1-1. Structure of the integrated circuit.

however, make timers from nothing but analog (linear) IC devices; or we can make other timer circuits from digital integrated circuits. Some of the best circuits, incidentally, are combinations that utilize both analog and digital circuits.

THE OPERATIONAL AMPLIFIER

The operational amplifier (op-amp) is a linear integrated circuit with an immense variety of applications. The op-amp can be used in relatively complex circuits, yet the design rules are simple enough for even beginners to apply them successfully. Many science students with no other introduction to electronics regularly use hybrid or discrete operational amplifiers and op-amp manifolds in their experiments. It seems that an "instrumental analysis" course in the chemistry department of many universities is typically a course in applying operational amplifiers. The operational amplifier has no direct bearing on the subject of timers, except that operational amplifiers are used as timers and as secondary elements in timer circuits. In this chapter, therefore, we will introduce you to the integrated circuit operational amplifier; some of the circuits using the operational amplifier will be covered in a later chapter.

OP-AMP INPUT CIRCUIT

The typical operational amplifier has two inputs, called *inverting* and *noninverting*. This is not a requirement in the definition of the operational amplifier but almost all op-amps have two inputs! Only few devices on the fringes of the linear IC market function like op-amps, yet have only one input (e.g., the LM302 device).

Figure 1-2 shows a typical operational amplifier input circuit. This circuit comprises a differential amplifier in which transistors are the differential pair. The operational amplifier was originally designed to perform mathematical *operations* in analog computers. They operated on an assortment of data values and had to produce an output of either polarity depending upon the required answer. As a result, the operational amplifier requires a *bipolar* power supply. The V+ power supply is positive with respect to ground, while the V– power supply is negative with respect to ground. Notice, however, that there is no actual *ground terminal* on the operational amplifier.

The two transistors (Q1 and Q2) of the differential pair are connected such that their emitters are fed from a single *constant current source* (I_3). The constant current source (CCS) will produce a constant current despite changes in the load resistance. Most IC constant current sources are bipolar transistors biased in a CCS manner; however, some discrete operational amplifiers use a junction field effect transistor (JFET) operated with its source and gate terminals connected together. We can then take advantage of the knee in the voltage-vs-current curve for the JFET.

In Fig. 1-2, the two emitter currents (I_1 and I_2 are derived from a single constant current (I_3). We know, therefore, that the following relationship holds true:

$$I_1 + I_2 - I_3 = 0$$

$$I_1 + I_2 = I_3$$

These relationships are known as Kirchoff's current law (KCL), which you will see again in a few moments. For the purposes of this discussion, we are going to assume that the emitter and collector currents of the transistors are equal (i.e., $I_{E_1} = I_{C_1}$ and $I_{E_2} = I_{C_2}$). In actual fact, they are different by 2 to 5 percent, but this is not a problem here (sigh).

Let's assume that a voltage (V1) is applied to the base of transistor Q1. What happens? How does the circuit work? When V1 is made positive, the collector and emitter current of Q1 (I_1) will *increase*. This increase in current I_1 must (by Kirchoff's law) cause a decrease in current I_2. Current I_2 is the collector-emitter current to transistor Q2. Since I_2 is now decreased, we know that the voltage drop across resistor R3 is also decreased (i.e., I_2R3 goes down). The output voltage V_o is the difference between the collector potential and the drop across resistor R3, so a reduction in I_2R3 causes an *increase* in V_o. Here we have a *positive increase* in V1

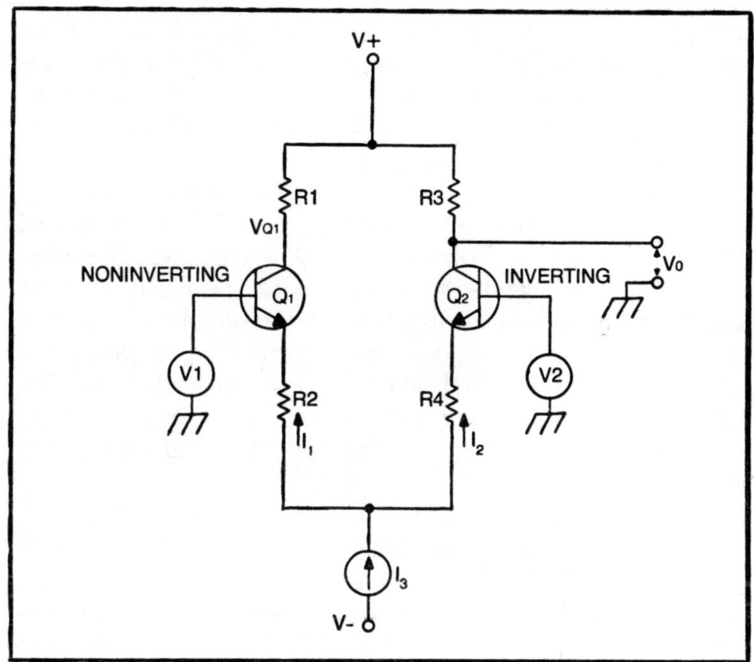

Fig. 1-2. Bipolar differential input amplifier.

causing an *increase* in output potential V_o. We may conclude, therefore, that the base of transistor Q1 is the noninverting input.

Now let's examine what happens when the voltage (V2) is applied to the base of transistor Q2. When a positive increase in potential is applied to the base of Q2, current I_2 is increased (causing a concommitant decrease in I_1. Increasing current I_2 will increase the voltage drop across collector load resistor R3. The output voltage will then be reduced by the amount of the increase in I_2R3. We may conclude, then, that the base of transistor Q2 is the inverting input: a positive increase in the input voltage causes a decrease in the output potential.

The output potential V_o is normally zero under two conditions: when both input potentials are zero, and when V1 = V2. The output voltage at any given time is a summation of the contributions of V+ and V− power supplies. This bipolar operation allows the four-quadrant behavior required of the op-amp.

OPERATIONAL AMPLIFIER SYMBOL

The operational amplifier symbol (Fig. 1-3A) used in circuit diagrams is a triangle, usually oriented horizontally, with the output at one apex. In some texts, they follow the IEEE convention of using the straight-back symbol (as shown here) for linear amplifier IC devices, and a version with a curved back for the operational amplifier. In this book, however, we will use the more common version shown in Fig. 1-3A.

The two inputs of the operational amplifier are labeled (–) for the inverting input and (+) for the noninverting input. The V+ and V– power supply terminals may not be shown in some schematics. Many draftsmen delete the power supply terminals on their drawings for the sake of simplicity. Make no mistake about it, however, the power supply terminals are still to be connected. The pinout numbers shown in the illustration are originally for the 741 device, but this particular arrangement is now considered the "industry standard." Most commercial integrated circuits use these pinouts.

The V+ and V– power supply terminals are *not* V+ and ground, but two separate power supplies of opposite polarity with respect to ground. The V+ power supply is positive with respect to ground, while the V– power supply is negative with respect to ground. A battery version of the typical operational amplifier power supply is shown in Fig. 1-3B. Note that this power supply has a ground connection, but there is no ground terminal on the operational amplifier symbol! The input and output potentials are measured with respect to ground, so the ground must come from the power supply.

PROPERTIES OF THE OPERATIONAL AMPLIFIER

We can analyze operational amplifier circuit design by using Kirchhoff's current law (KCL) and the *basic properties of the operational amplifier.* There are six basic properties of the differential input operational amplifier:

- ☐ Infinite input impedance
- ☐ Infinite open-loop (no feedback) gain
- ☐ Zero output impedance
- ☐ Infinite bandwidth
- ☐ Zero noise contribution
- ☐ Both inputs stick together

An infinite input impedance means that the input neither *sinks* nor *sources* current. The input impedance of any amplifier is given by V_{in}/I_{in}. If the input will neither accept nor generate a current, then I_{in} is zero and the input impedance is infinite. In real operational amplifiers, which are not ideal, the input impedance is

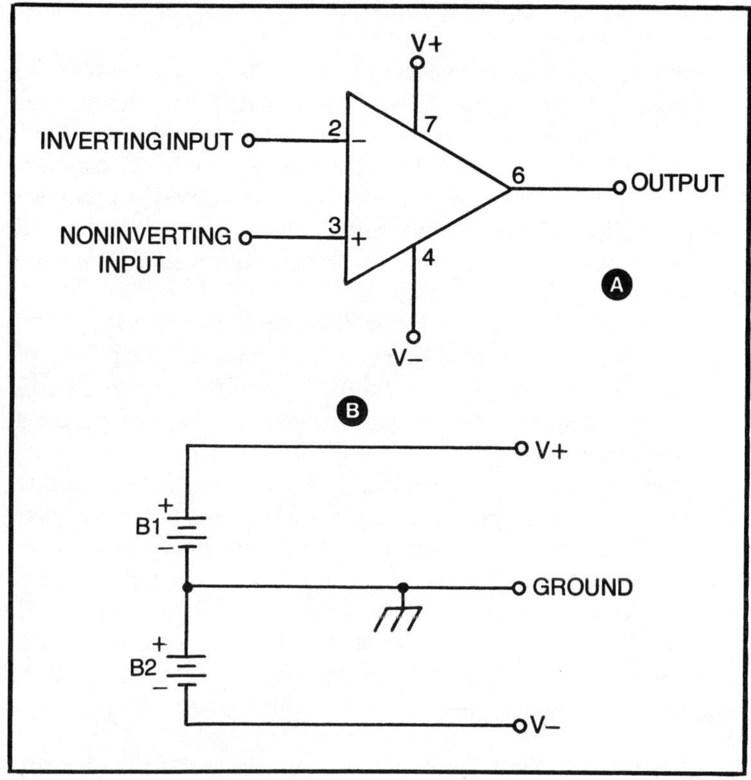

Fig. 1-3. The operational amplifier: A) Operational amplifier symbol, B) Typical op-amp power supply.

not quite infinite, but is high enough to be very exciting. Some low-cost operational amplifiers are able to boast 100 kohms, while more expensive devices can usually go 1 megohm or more. A few devices, such as the RCA BiMOS devices (which have MOSFET input transistors!) can boast input impedances of 1.5 terraohms (1.5×10^{12} ohms)! This means input currents of picoamperes or nanoamperes.

Infinite open-loop gain means that the gain is infinite when there is no negative feedback present. In real operational amplifiers the open loop gain will be about 20,000 for cheapies and over 1,000,000 for premium devices.

The output impedance of a perfect voltage amplifier is zero, thus the ideal operational amplifier is said to have a zero output impedance. In real operational amplifiers we typically find output impedances of less than 200 ohms, with many under 75 ohms. In contrast, many amplifiers have output impedances of 1000 ohms—hardly zero.

Infinite bandwidth means that the device will amplify all signals applied to the input; but this is hardly the case in real operational amplifiers. Some devices operate into the low VHF region, and many operate into the HF region. Most common operational amplifiers, however, are severely limited in frequency response. The 741 device, for example, has a frequency response of only a few kilohertz! These devices are called *frequency compensated* operational amplifiers.

Zero noise contribution means that the IC adds no noise to the signal. Unfortunately, most operational amplifiers fall far short of this ideal! Some premium (high cost) devices offer very good noise performance, but most common operational amplifiers are not good as low noise amplifiers.

Both inputs stick together? What does that mean? It means that applying a voltage to one input requires us to treat the other input mathematically as if it too were connected to the same voltage. This is not merely some theoretical device used in mathematical formulas; it is real. If you apply +4 volts to the noninverting input, then you will measure +4 volts at the inverting input also! This phenomenon is one of the most important of the ideal properties!

INVERTING FOLLOWER CIRCUITS

Figure 1-4 shows the circuit for the *inverting follower* configuration. An inverting follower uses the inverting input of the operational amplifier. Note that the noninverting input is *grounded* in this circuit. This is the same as saying that the noninverting input is at zero potential. A result of ideal property number 6 is that the inverting input must now be treated as if it were also at zero potential! The inverting input is essentially grounded, even though not physically. This confusing state of affairs is not made too much clearer by the name usually given this condition: *virtual* ground. The virtual ground is treated as a ground even though it is not officially and obviously grounded through a piece of wire.

What do we know about the circuit in Fig. 1-4? We know the currents are as follows:

$$I_1 = V_{in}/R1$$

and,

$$I_2 = V_o/R2$$

We also know from KCL that

$$I_1 = - I_2$$

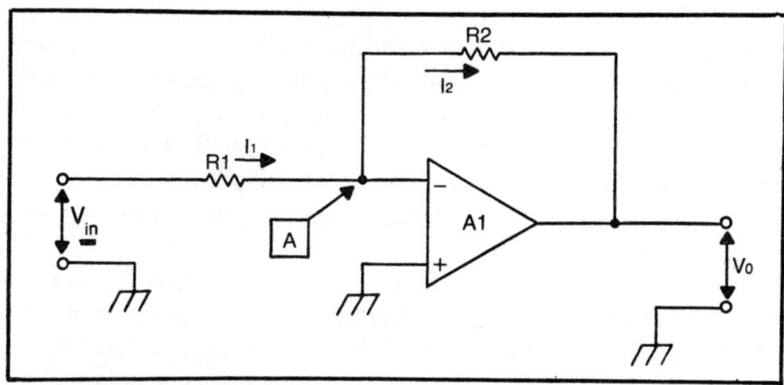

Fig. 1-4. Inverting follower.

So by substituting $I_1 = V_{in}/R1$ and $I_2 = V_o/R2$ into $I_1 = -I_2$, we obtain:

$$\frac{V_{in}}{R1} = \frac{-V_o}{R2}$$

We can solve $\frac{V_{in}}{R1} = \frac{-V_o}{R2}$ for V_o, thereby obtaining the transfer equation for the inverting follower:

$$V_o = \frac{-V_{in} R2}{R1}$$

The gain of any amplifier is the quotient V_o/V_{in}, which in this case is R2/R1. This relationship means that we can set the gain of the inverting follower operational amplifier circuit by setting the ratio of two resistors! The absolute values of the resistors are not important, only their *ratio*. For example, if the feedback resistor (R2) is 100 kohms, and the input resistor is 1000 ohms, then the gain R2/R1 is 100,000/1000, or 100.

The input impedance of the inverting follower is usually low, being limited to the value of the input resistor R1. This is because one end of R1 is grounded to the virtual ground.

NONINVERTING FOLLOWERS WITH GAIN

There are several problems associated with the inverting follower. For example, we find that the input impedance is limited to the value of the input resistor (R1). The apparent solution to this problem is to use an input resistor that has a sufficiently high value. But this is not always possible due to gain problems (remember, the open loop gain is not infinite!) and because certain problems with real operational amplifiers are made worse by using high-value resistors in the feedback and input circuits. Unless phase inversion is needed in the circuit, there may be good reasons to opt for the noninverting follower shown in Fig. 1-5.

In this circuit, the signal is applied directly to the noninverting input of the operational amplifier. The feedback network is the same as in the inverting follower, except that the "input" end of resistor R1 is grounded.

Recall property number 6: applying signal voltage V_{in} to the noninverting input has the effect of placing the other input (inverting) at the same potential V_{in}. Since there is no phase inversion in this circuit the output signal has the same polarity as the input signal, so

$$I_1 = I_2$$

By the same sort of reasoning as in the previous case:

$$I_1 = \frac{V_{in}}{R1}$$

and,

$$I_2 = \frac{V_o - V_{in}}{R2}$$

We can obtain a relationship that ends in the transfer equation for this circuit by the following substitution:

$$I_1 = I_2$$

substituting $I_1 = \frac{V_{in}}{R1}$ and $I_1 = I_2$ into $I_2 = \frac{V_o - V_{in}}{R2}$, yields

$$\frac{V_{in}}{R1} = \frac{V_o - V_{in}}{R2}$$

Algebraically rearranging $\frac{V_{in}}{R1} = \frac{V_o - V_{in}}{R2}$

$$\frac{V_{in} R2}{R1} + V_{in} = V_o$$

Factoring out V_{in}

$$V_{in}\left[\frac{R2}{R1} + 1\right] = V_o$$

This is the transfer equation for the noninverting follower with gain circuit of Fig. 1-5. Note that the circuit will always have a gain of at least unity (1). The gain factor is ((R2/R1) + 1).

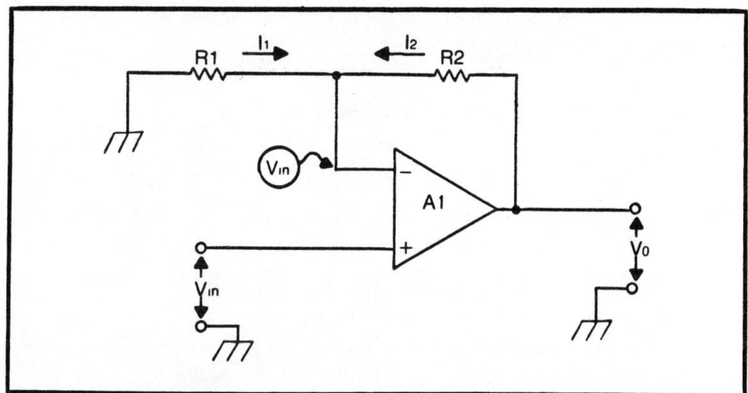

Fig. 1-5. Noninverting follower with gain.

The input impedance of the noninverting follower is very high. Recall that an ideal operational amplifier has an input impedance of infinity. Practical operational amplifiers usually have an input impedance greater than 1 megohm, with some models offering impedances as high as 10^{12} ohms! This condition is better suited to dealing with high input impedance situations. In the case of timer circuits, it also relieves us of the problem of considering the effects of the input impedance of the amplifier on RC time constants.

RECAPITULATION

There are two basic operational amplifier configurations: inverting and noninverting. These are used in all of the practical circuits that we will consider in the chapter on op-amp timers. The transfer function of the inverting follower is

$$V_o = \left[\frac{-R2}{R1}\right] V_{in}$$

The gain factor R2/R1 is sometimes represented by A_v, so the above expression would be written

$$V_o = -A_v V_{in}$$

The transfer function for the noninverting amplifier is

$$V_o = \left[\frac{R2}{R1}\right] + 1 \ V_{in}$$

The input impedance of the inverting follower is limited to the value of R1 if the noninverting input is grounded. The input impedance of the noninverting follower is essentially the input impedance of the operational amplifier device.

UNITY GAIN NONINVERTING FOLLOWERS

The ultimate in feedback is 100 percent feedback. In both the previous circuits the feedback was a fraction of the output signal and was set by a voltage divider, R1 and R2. The portion of the output signal fed back was R1/(R1 + R2). We obtain 100 percent feedback in Fig. 1-6 by connecting the output of the operational amplifier directly to the inverting input.

From the last equation we can see that the effect of removing the feedback network resistors is to reduce the gain to:

$$\begin{aligned} A_v &= \frac{R2}{R1} + 1 \\ &= 0 + 1 \\ &= 1 \end{aligned}$$

We can see this is conceptually true from consideration of ideal property number six. We know that if V_{in} is applied to the noninverting input, then this same potential also exists on the inverting input. But the inverting input is also connected to the output, so the potential V_{in} must also exist on the output terminal of the operational amplifier.

OK, what's the use of a unity gain amplifier? Aren't amplifiers supposed to amplify? The voltage amplification is nearly unity in a real operational amplifier (0.9999999), yet the output impedance is very low (implying a power gain). Since the input impedance of the operational amplifier is very high (as high as 10^{12} ohms), then it becomes obvious that the unity gain follower can be used for *impedance transformation* without loss of amplitude between a high impedance source and a low impedance load. Also, the unity gain noninverting follower can be used as a buffer stage to provide some isolation without loss of amplitude (or addition of amplitude) or phase inversion.

DIFFERENTIAL AMPLIFIERS

The two inputs of the typical IC operational amplifier are complementary: i.e., they produce equal but opposite effects on the output signal. If a voltage is applied to the noninverting input, then the output will have the same polarity as the input signal. But, if that same signal voltage is applied to the inverting input, then the

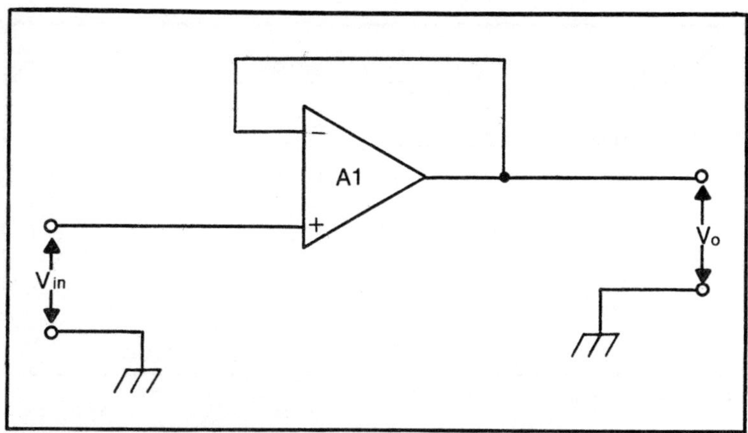

Fig. 1-6. Noninverting follower (unity gain).

output signal will have the opposite polarity. The gain seen from each input is the same, but the polarity is not the same. The gain for the inverting input would be $-A_{vol}$, while from the noninverting input it is $+A_{vol}$ where A_{vol} is the open-loop (no feedback) gain of the operational amplifier. The inputs of the operational amplifier are, therefore, *differential* inputs. The actual input voltage seen by the op-amp is the difference between the potentials applied to the inverting and noninverting inputs. We can, therefore, use the operational amplifier IC to make a differential amplifier circuit.

An example of a simple DC operational differential amplifier is shown in Fig. 1-7. This circuit is the simplest form of diff-amp. Other circuits, that are beyond the scope of this book, will be found in the texts listed at the end of this chapter. Those who want to know more about operational amplifier circuits than is necessary to understand the timer circuits in this text are referred to those works.

The gain of this circuit is

$$A_{vd} = \frac{R3}{R1}$$

Provided that

$$R1 = R2$$
$$R3 = R4$$

The ideal differential amplifier will not respond to common-mode signals (i.e., signals applied equally and simultaneously to both inputs). The real differential amplifier, however, will respond

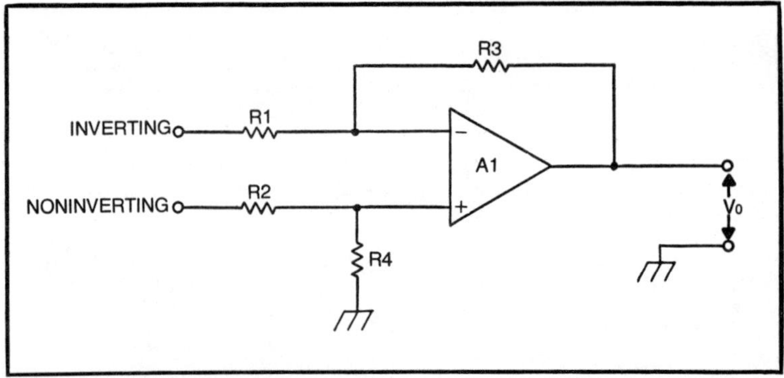

Fig. 1-7. D.C. Differential amplifier.

somewhat to common-mode signals. But the ratio of the differential gain to the common-mode gain (called the *common-mode rejection ratio*—CMRR) is quite high. Some operational amplifiers have a CMRR of 120 dB (i.e., 1,000,000:1!), while even garbage-grade blister-pack op-amps have CMRR ratings in the neighborhood of 60 dB!

The differential amplifier is used extensively in electronic instrumentation circuits where noise pickup is a factor. The human electrocardiogram (ECG) signal, for example, is very weak (1 mV), yet interference from 60 Hz power lines can be several volts on an 8-foot patient cable. The result is to have the ECG signal completely swamped by a noise signal that is 1000 times greater! However, if we acquire the ECG biopotential from the surface of a patient's body as a differential signal, we find that the 60 Hz signal affects both leads equally. This means that the 60 Hz signal will be seen by the differential amplifier as a common-mode signal, whereas the ECG signal is differential. The gain of the amplifier may be 1000 for the differential ECG signal and only 1 to 10 for the 60 Hz component. The result is an ECG trace that is free of 60 Hz artifact.

OPERATIONAL AMPLIFIER PROBLEMS

The ideal properties of the operational amplifier are never realized in practical devices. They are closely approximated in some premium devices but there are always some discrepancies. These problems fall into several categories. We have already discussed frequency response and common-mode rejection ratios. We also find that there are several *offset* problems which cause the output voltage to be nonzero at times when it should be zero. One source of offset voltage is the bias current of the transistors used in the input circuit. Remember that the ideal operational amplifier has an infinite input impedance, which means that there will be no current generated in the input circuit. But real operational amplifiers are built from real transistors that require bias currents. The bias on some premium-grade devices may be nanoamperes, or even picoamperes, but in some *cheapies* it may be a hearty fraction of 1 mA. The offset potential caused by this current is $I_{off} \times R_f$, where R_f is the resistance of the feedback resistor R2. Also, the bias current will cause a voltage drop at the input that is equal to the product of the current and the parallel combination of the feedback and input resistances (R1R2/(R1 + R2)).

One method for reducing the offset caused by the input bias current is shown in Fig. 1-8. This circuit works because the operational amplifier input circuit is symmetrical, meaning that identical currents will flow in both input terminals. If we place a resistor (R3) , with a value equal to the parallel combination of R1 and R2, between the inverting input and ground, then the noninverting input sees an equal offset input voltage. The gain of the circuit for both potentials is equal, but opposite, so common-mode potentials are cancelled. The net effect of resistor R3 is to cancel out the output offset that is due to the input bias currents.

But input bias currents are not the only source of offset voltages on the output line. There are actually several causes, some of them from inside the operational amplifier and others from outside sources (i.e., other parts of the circuit). The circuits of

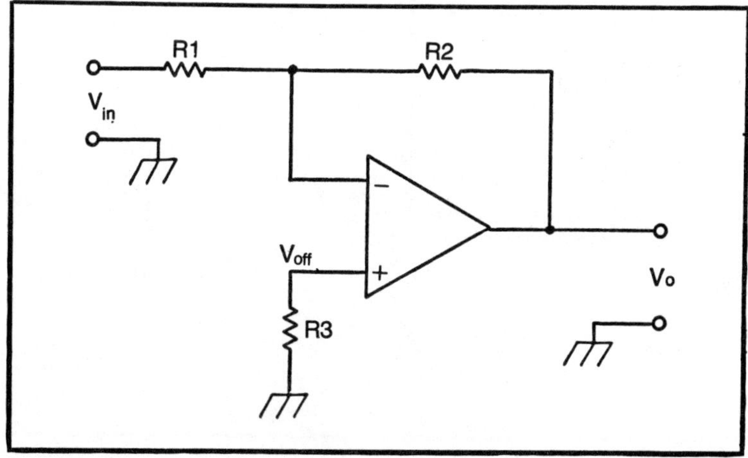

Fig. 1-8. Use of a compensation resistor.

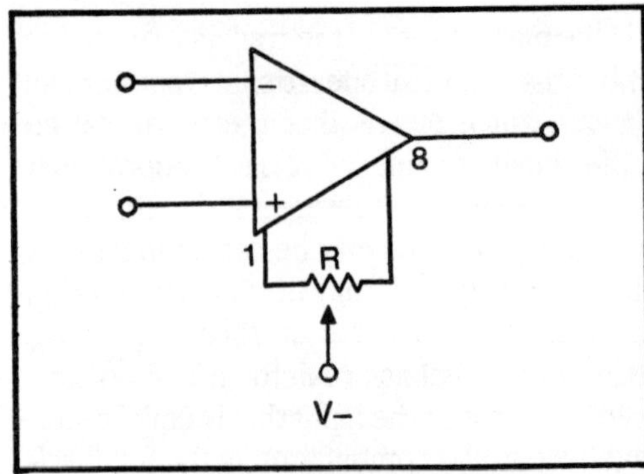

Fig. 1-9. Offset terminals.

Figs. 1-9 and 1-10 will allow us to cancel any type of offset potential, including the input bias current offset.

The circuit of Fig. 1-9 is used on operational amplifiers that have offset compensation terminals. These inputs allow us to balance the emitter currents of the input amplifier transistors for offset variations. The wiper of the potentiometer is connected to the V− power supply, while the ends of the pot are connected to the pins for the offset compensation inputs.

Another, more common, method for cancelling the offset is shown in Fig. 1-10. In this circuit, which can be used whether or not the operational amplifier has offset compensation terminals,

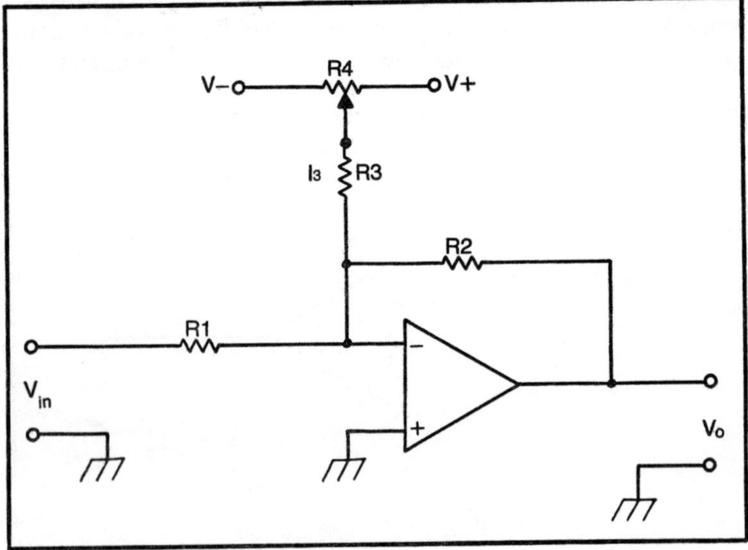

Fig. 1-10. Universal offset compensation.

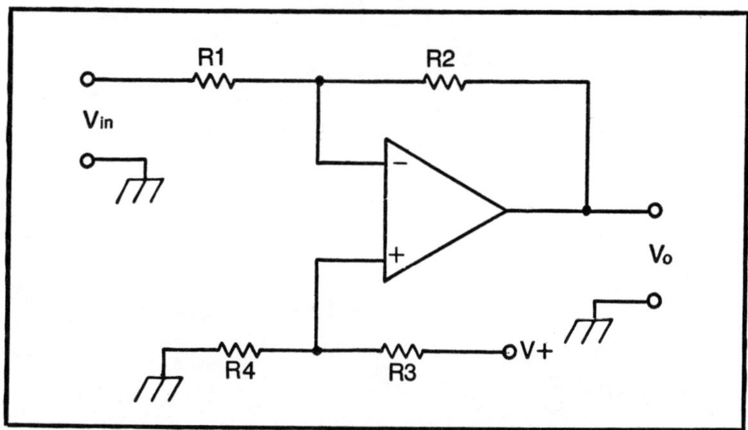

Fig. 1-11. Operating the op-amp from a monopolar power supply.

we add a counter current to the input summation junction of the inverting follower (or the noninverting follower if desired). Resistors R1 and R2 are the normal resistors used in the amplifier, while R3 and R4 are part of the offset cancellation circuit. We select a potential from the V– and V+ power supply lines that will introduce a current into R3 that exactly cancels the output offset. The output potential due to I_3 is I_3R2, and is adjusted to exactly cancel the normal offset potential, and no more (otherwise, the offset cancellation potential becomes an offset in its own right).

The operational amplifier is designed to operate from bipolar power supplies. In normal operation, the V– and V+ power supply contributions to the output voltage depend upon the polarity of the input signal. When the input signal is zero the respective contributions are zero. But what happens when the operational amplifier must be operated from a monopolar power supply? In that case we must bias one input (usually the noninverting input) to some potential between zero and the power supply potential. The output signal will be equal to this potential when the input signal is zero, and will swing about the bias potential when the input signal is nonzero. In most cases (see Fig. 1-11) the noninverting input is placed at a potential of ½V+ by making resistors R2 and R3 equal to each other.

Chapter 2
Discrete-Device Timers

We lied. This is not just a book of IC timers; we are also going to cover some discrete-device timer circuits as well. This decision was made because: 1) most readers would like to know; and 2) the basic principles are a lesson in timer operation, whether IC or discrete. We therefore, have included a few brief notes on the operation of discrete-device timers using bipolar and unijunction transistors.

First of all, let's discuss just what a timer is. This may seem like an elementary notion, but it bears some definition. A timer is any circuit that allows us to mark time. A timer might be an astable multivibrator; or it might be a monostable multivibrator. The astable circuit has no stable output states, so it will oscillate back and forth between the two unstables states (i.e., output HIGH or output LOW) at a fixed rate. The astable multivibrator, then, is a squarewave oscillator. Since the frequency of the astable multivibrator is fixed, we can use the output to mark time—especially when it is being used to drive a circuit as a digital clock source.

The monostable multivibrator, also called the one-shot circuit, has only one stable state. It will normally maintain the stable state (usually output LOW, but not always). When a trigger pulse is received at an appropriate input terminal, the output goes to the unstable state (i.e., output HIGH). It will remain in this state for a predetermined period of time, and will then revert back to the stable state. The one-shot will, therefore, serve as a timer circuit to mark the passage of a predetermined period of time.

There have been several different types of circuits designed over the years that allow us to mark time. Most circuits are dependent on the series RC network shown in Fig. 2-1A.

The maximum voltage that the capacitor will achieve is E_m, which is the full battery voltage. Assuming that the capacitor is totally discharged, we know that the voltage across the capacitor will increase slowly at a rate that is determined by the value of the resistor and the capacitor. The charging curve of the capacitor is shown in Fig. 2-1B. The voltage across the capacitor at any given instant is given by the expression:

$$E_c = E_m (1 - e^{-T/RC})$$

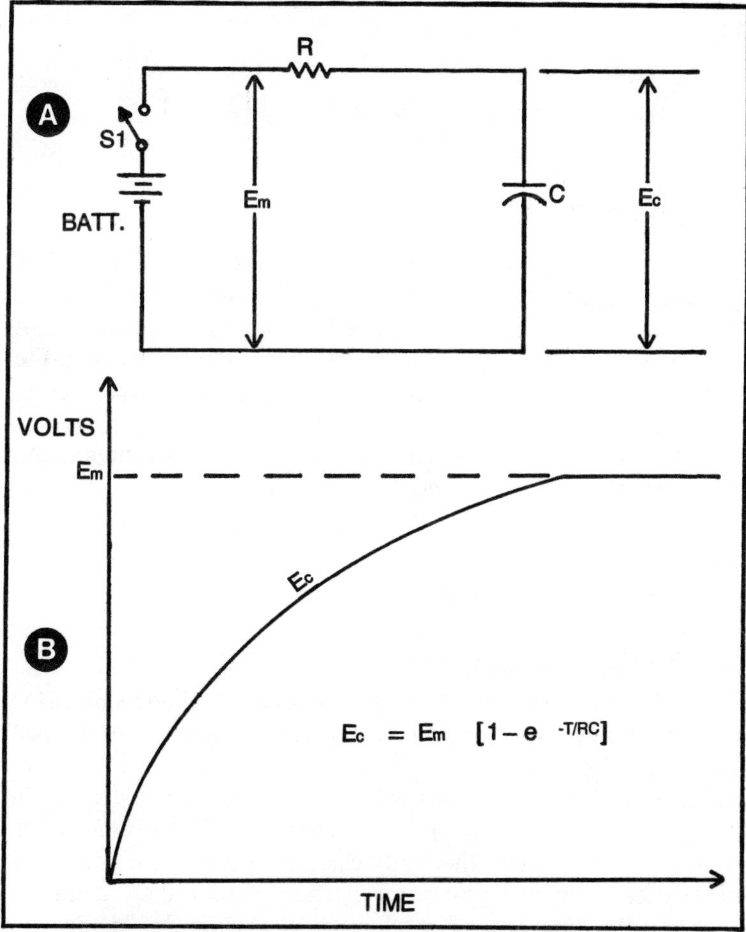

Fig. 2-1. Series RC timing network: A) RC network, B) Charging waveform.

Where:

E_c is the voltage across the capacitor
E_m is the battery voltage
e is the base of the natural logarithms
R is the resistance in ohms
C is the capacitance in farads
T is the time since the switch was closed, in seconds.

☐ **Example:**

Find the voltage in a circuit 25 microseconds after the switch is closed if C = 0.005 μF and R = 12 kohms and E_m is 15 volts. Assume that the capacitor was totally discharged when the switch was closed.

Solution:

$$E_c = E_m (1 - e^{-T/RC})$$
$$= (15V)(1 - e^{-(2.5 \times 10^{-5})/(1.2 \times 10^4)(5 \times 10^{-9})})$$
$$= (15 V)(1 - e^{-.416})$$
$$= (15 V)(1 - 0.66)$$
$$= (15 V)(0.34) = 5.11 \text{ volts}$$

The above relationship for capacitor charging tells us that it is possible to make a timer based on the RC network. We could, for example, design a circuit that caused the capacitor voltage to rise to a trip value in the required length of time, and then discharge the capacitor. We will do just that in a circuit to follow and in other circuits later in this book.

The *time constant* of a series RC network is defined as the product of the capacitance and the resistance:

$$T = RC$$

Where:

T is the time in seconds
C is the capacitance in farads
R is the resistance in ohms

We can further define the RC time constant as the time required for a totally discharged capacitor to charge to 63 percent of its final voltage, or for a fully charged capacitor to discharge to 37 percent of its initial value (either one will do). The capacitor is deemed to be fully charged after five RC time constants. Theoretically, the voltage never reaches the "fully charged" point but it comes so close after a sufficiently long time that no electronic instrument can measure the difference. For all practical circuits the "sufficient" time is 5RC.

SIMPLE TIMERS—RELAXATION OSCILLATORS

Probably the simplest form of timers is a species of astable multivibrators called the *relaxation oscillator*. Two examples of relaxation circuits are shown in Fig. 2-2. The version shown in Fig. 2-2A is based on a discrete device called a *neon glow lamp* (such as the NE-2).

The neon glow lamp consists of two electrodes inside a glass envelope that has been first evacuated and then filled with the inert gas neon. The vapor pressure of the neon in the lamp is less than atmospheric and is controlled to set the point at which a voltage will ionize the gas.

As long as the potential across the electrodes is below the voltage that will ionize the gas inside the envelope there will be no current flow. The resistance of the neon lamp at this time is very high, being limited to a high leakage value. But once the potential crosses the point required for ionization of the gas inside the lamp, the electrical current can flow between the electrodes because of the gas ions in the path. The resistance of the lamp at this time is very low. Most neon lamps have a *firing voltage* in the 40 to 80 volt range, and will maintain the ionized state of the internal gas at potentials down to 20 volts or so.

The relaxation oscillator is formed by placing the neon lamp across the capacitor in a series RC network. The voltage across the lamp is the capacitor voltage E_c. When this voltage reaches the firing potential of lamp I1, the gas ionizes causing the lamp resistance to drop to a very low value. In this condition, the current stored in the capacitor will discharge through the lamp until the capacitor voltage drops below the sustaining potential of the lamp.

Figure 2-2B shows the timing diagram for the neon lamp relaxation oscillator. Before we analyze this particular circuit, let's cover a few more basics. We already know the expression to calculate the voltage at any given time after a switch is closed allowing an uncharged capacitor to begin charging. The expression for a circuit is which the capacitor already has some charge is:

$$E_c = E_m - (E_m - E_0)e^{-T/RC}$$

Where:

E_c is the voltage across the capacitor at time T
E_m is the supply voltage
E_0 is the initial charge voltage across the capacitor
T is the time required to reach E_c

R is the resistance in ohms

C is the capacitance in farads

To calculate the timing of this circuit, we will need to know the value of the supply voltage (E_m), the firing voltage of the neon lamp (E_c), and the sustaining voltage (E_0) of the neon lamp. At time T_0 the switch is closed and the capacitor begins to charge from zero at a rate determined by the RC time constant. When the voltage reaches 60 V, the lamp will turn on and cause the capacitor to discharge down to 23 volts. At that point, the lamp deionizes and the capacitor can begin once again to charge. We will ignore the initial startup period and concentrate on the period T_1 to T_2. During this period the capacitor will charge from 23 volts (sustaining) to 60 volts (firing), and then rapidly discharge back down to 23 volts. This relaxation oscillation will continue until power is turned off. We can rearrange the capacitor charging equation in order to gain some idea of how this circuit is timed:

$$-\frac{E_c - E_m}{E_m - E_0} = e^{-T/RC}$$

(all terms previously defined)

$$-\frac{60 - 80}{80 - 23} = e^{-T/RC}$$

$$-\frac{-20}{57} = e^{-T/RC}$$

$$0.351 = e^{-T/RC}$$

$$\text{Ln}\,(0.351 = \text{Ln}\,(e^{-T/RC})$$

$$-1.05 = -T/RC$$

We may conclude, therefore, that the period will be approximately

$$T = 1.05RC$$

The usefulness of this circuit in timer applications is limited to the cases where the accuracy and stability of the timer is not too critical. As you can see from the above expressions, this timer is

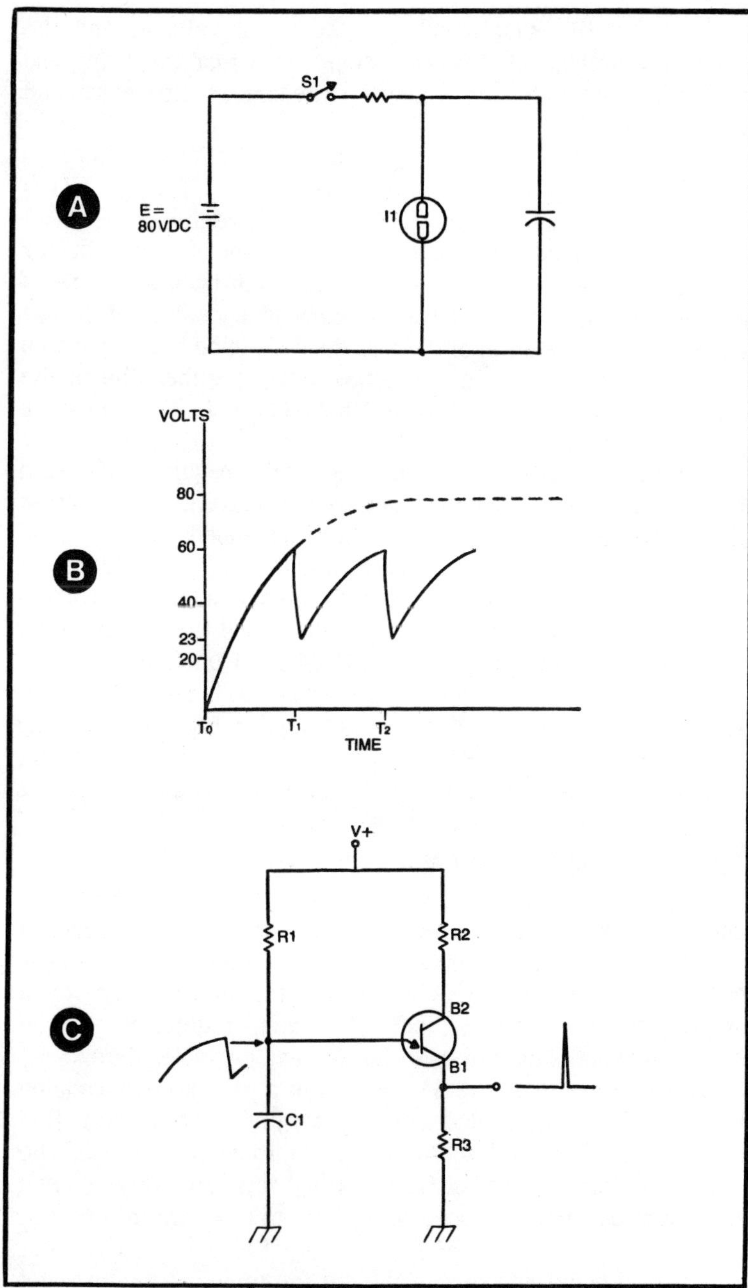

Fig. 2-2. Relaxation oscillators: A) Neon glow-lamp; B) Waveform; C) Unijunction transistor.

sensitive to the supply voltage, the firing voltage, and the sustaining voltage of the neon lamp. Note that the firing and sustaining potentials vary with age of the lamp, temperature, and *applied light*!

Figure 2-2C shows another form of relaxation oscillator, in this case it is based on the *unijunction transistor* (UJT). The UJT is constructed similarly to the JFET in that a continuous channel contains an embedded region of the opposite type of semiconductor material. For example, an n-channel UJT will have a channel of n-type silicon and will have an embedded *emitter* of p-type material. The operation of the UJT depends upon the pn junction and its normal properties. The terminals to either end of the channel are called *bases*, and are labelled base-1 (B1) and base-2 (B2).

If the emitter-base-1 voltage is below that required to forward bias the pn junction, then there will be little current flowing across the pn junction. The capacitor will see a high impedance and will be charged from the V+ power supply through resistor R1. When the voltage across the capacitor reaches the point that it forward biases the pn junction, then the impedance drops and the charge in the capacitor is dumped through the emitter-base-1 junction.

The output is a slim pulse of current across the base-1 resistor (R3). The capacitor will have the regular capacitor-charging waveform. If resistor R1 is replaced with a constant current source, then the capacitor will charge at a constant rate so the capacitor waveform will be a linear ramp sawtooth.

TRANSISTOR MONOSTABLE MULTIVIBRATOR

Figure 2-3A shows the circuit for a transistor monostable multivibrator. Recall that a monostable, or one-shot, will remain in one state indefinitely unless a trigger pulse is received. When the pulse is received, the one-shot goes to the unstable state for a predetermined period of time. When this predetermined time expires, the one-shot returns to the dormant, or stable, state.

The circuit in Fig. 2-3A consists of a pair of cross-coupled NPN transistors. In the dormant state, we find that transistor Q1 is reverse biased by the V− power supply through resistor R5. The voltage on the collector of Q1 is V+. At this same time we find transistor Q2 forward biased sufficiently to place Q2 into saturation.

Capacitor C1 and resistor R1 set the timing of the one-shot circuit. In the dormant state, the collector end of C1 is at a potential of V+ and the base end of C1 is at the normal base-emitter voltage

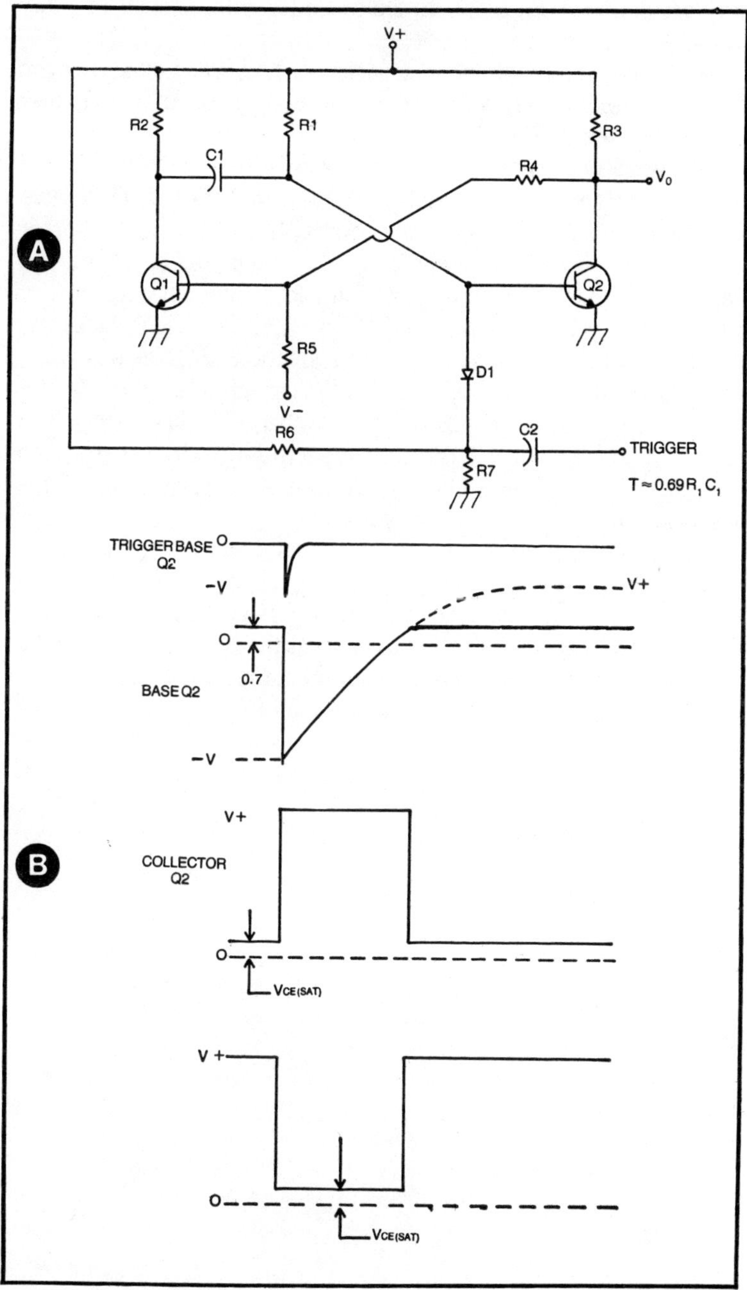

Fig. 2-3. Monostable multivibrator: A) Bipolar transistor circuit; B) Timing waveforms.

(0.7 volts, approximately), so the capacitor voltage is (V+) – 0.7 volts.

The collector voltage of transistor Q2 is very close to zero (i.e., the saturated collector voltage V_{cesat} of the particular transistor used for Q2).

A negative-going trigger pulse applied to capacitor C2 will pass a negative spike pulse to the base of transistor Q2. This pulse will momentarily turn off transistor Q2, causing its collector voltage to rise to V+. The collector voltage applies a current to the base of transistor Q1, through resistor R4, causing Q1 to turn on and grounding the collector end of capacitor C1. This places the charge of C1 across the base-emitter junction of Q2 as a reverse bias. (Grounding the opposite end of the capacitor has the effect of reversing the apparent polarity of the charge!) This change will cause Q2 to remain cut off until capacitor C1 charges through resistor R1 to a point where it can once again turn on Q2. This action is shown in the second waveform in Fig. 2-3B.

During the time that Q1 is on, its collector voltage is nearly zero (less only the V_{cesat} potential of the transistor used). At this same time the collector potential at Q2 is HIGH (i.e., V+). We can, therefore, designate the collector of Q1 as the NOT-Q output of the one-shot, and the collector of Q2 at the Q output of the one-shot.

Chapter 3
Some TTL and CMOS Devices

You will be given a detailed view of the TTL and CMOS devices in another chapter. In this chapter, we will provide you with a view of some of the timer integrated circuits from each line, mostly monostable multivibrators. These devices are in more or less standard use in the electronics industry. We will also give you the information needed to interface the TTL and CMOS lines with each other.

INTERFACING CMOS AND TTL

The TTL logic chip requires a +5 volt power supply. In addition, the power supply must be regulated because the voltage range over which the TTL device operates is limited. The CMOS line, on the other hand, will operate from either single or dual polarity power supplies, and can accept a range from ± 4.5 to ±15 volts DC. The TTL input terminal acts as a current source and is equivalent to the open emitter of a bipolar NPN transistor. The CMOS input, on the other hand, is the gate of a MOSFET transistor, so it will have a very high impedance. That CMOS input, then neither sinks nor sources current.

Obviously, with such differences it is not likely that we could interface the CMOS and TTL lines without some special tactics. Figure 3-1 shows the two basic interface methods. In Fig. 3-1A we see the situation for +5 volt TTL-compatible power supplies. In this example, both the TTL and the CMOS devices are operated from a common +5 volt power supply.

To interface the output of a TTL device to the input of the CMOS device, provide a pull-up resistor to +5 volts DC. This will simulate the current source from the input that the TTL output is looking for, while not affecting the CMOS input. The CMOS input will then be free to respond to the voltage levels of the TTL output.

We also see the method for interfacing the CMOS output to the TTL input. Recall that the TTL input is a current source and wants to see a low impedance current sink for proper interfacing. The normal CMOS output provides a 180 ohm resistance to V+ when HIGH, and a 180 ohm resistance to ground when LOW. This scheme is not exactly what the TTL input wants to see. We can, however, take advantage of two special chips in the CMOS line that offer TTL outputs (when the power supply is +5 volts). The 4049 device is a TTL-output hex inverter; while the 4050 is a noninverting hex buffer. Both devices contain six independent sections. We can apply the output of the CMOS gate to the input of either device (4050 is shown in Fig. 3-1A). If the power supply operating the 4050 is +5 volts, then the output will be TTL-compatible, so it can interface directly with the TTL input terminal. The use of the 4050 preserves the sense of the logic used in the CMOS device; i.e. HIGH is still HIGH and LOW is still LOW because the 4050 is noninverting. If we want, or will tolerate, inversion of the logic level then the 4049 device could be used instead.

The interfacing of the two logic lines in situations where the CMOS devices are operated at any potential other than +5 volts DC is shown in Fig. 3-1B. To interface a TTL output to a CMOS input requires that we use either an inverter or noninverting buffer TTL device that has a high voltage open-collector output. The 7416 is a hex inverter with open collector output; while the 7417 is a noninverting hex buffer with open-collector output. The collector of each device can operate up to +15 volts, and must be provided with a pull-up resistor to the V+ power supply that operates the CMOS devices.

Interfacing the output of the CMOS device to the TTL input is the same as in the previous case. In this situation, however, it is imperative to remember that the 4049 or 4050 device, although CMOS, must be operated from the +5 volt DC TTL power supply in order to make the interfacing complete.

All digital logic circuits require substantial amounts of bypassing in order to rid the power supply lines of noise and other glitches. The TTL devices, however, are somewhat more prone to

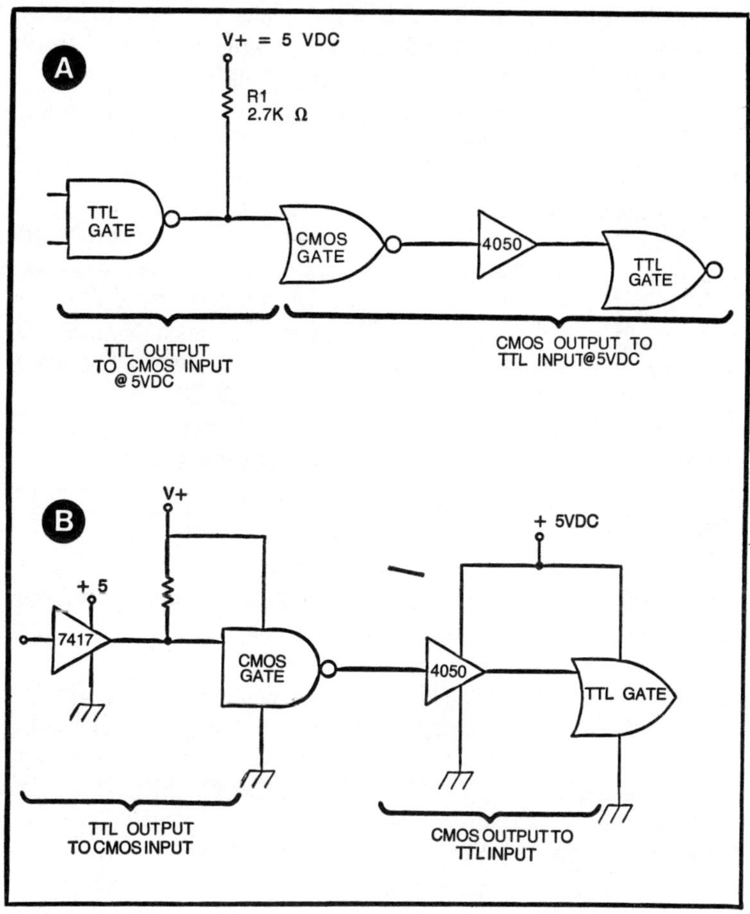

Fig. 3-1. CMOS-TTL interfacing: A) +5 volt power supply; B) other power supply voltages.

producing glitches and are, in some cases, more sensitive to them. Glitches can cause counters to increment erroneously or reset, can lock gates, and can cause other generalized havoc in electronic circuits. Where substantial numbers of TTL devices are used, make it the practice to place a 0.001 to 0.01 μF bypass capacitor between every two chips—or, if you can afford it, between every chip. There are now suitable bypass capacitors on the market that are very small. These "blue caps" have a 25 volt rating and are specially designed for use in TTL bypassing applications.

It is also wise to provide somewhat greater bypassing every so often in the power distribution sytem of a digital project—

especially if more than 20 or 30 TTL devices are used. In hobby computers, for example, it is the normal practice to place a 1 to 10 μF tantalum capacitor at each power supply entrance point on each printed circuit board. In addition, there might be a 100 to 250 μF capacitor at critical points on the PC board. These capacitors improve the immunity of the circuit from noise pulses that are inevitably present in all digital projects.

Another solution to the noise problem, at least in multiboard projects, is to use distributed regulation. An LM-309, 7805, or LM-340-5 will regulate to +5 volts DC (needed by TTL) and supplies up to 1 ampere of current. In projects such as an S-100 computer the regulation is assigned to each PC board. The main high-current power supply provides +8 volts DC unregulated. Each PC card in the computer has its own +5 volt regulator (and some devices have multiple regulators on the same board).

CMOS TIMERS

Only a few devices in the CMOS line qualify as timers. There are several special devices available under proprietary numbers from specific manufacturers, but for the most part we have very little to choose from.

We can, of course, build timer circuits from ordinary CMOS gates (and this is frequently done). Figure 3-2 shows the half-monostable multivibrator made from CMOS inverters (or inverter-connected NAND or NOR gate sections). In Fig. 3-2A we see two versions of the positive-edge triggered monostable. The 4049 device is used when we want an active-LOW output (i.e., the pulse drops LOW from its dormant HIGH position). If we need an active-HIGH condition (i.e., the pulse goes HIGH from the dormant LOW condition) then the 4050 device is used.

The timing network for the positive-edge triggered monostable circuit consists of capacitor C1 and resistor R1. The resistor is connected from the CMOS device input to ground. The capacitor is in series with the trigger line. When a positive step is received at the input of the capacitor, the output will go active for a period of time approximately equal to:

$$T = 0.8 R_1 C_1$$

Where:

T is the output pulse duration in seconds

R_1 is the resistance of R1 in ohms

C_1 is the capacitance of C1 in farads

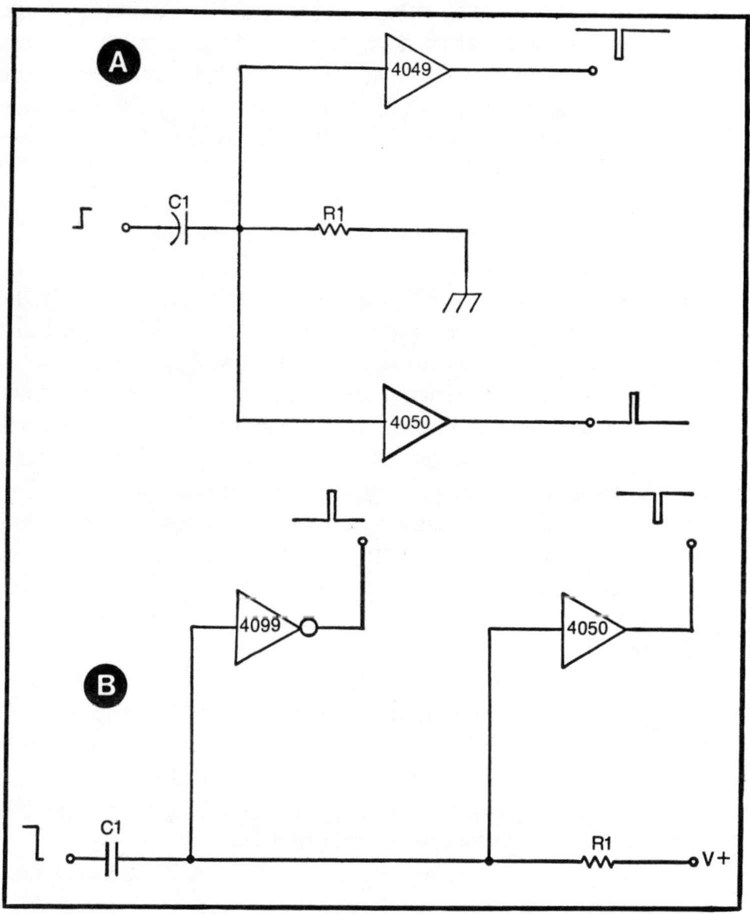

Fig. 3-2. Half monostables using CMOS devices: A) Positive-edge triggered; B) Negative-edge triggered.

The equivalent circuit for negative-edge triggered applications is shown in Fig. 3-2B. In this case, the resistor in the timing network is returned to V+ instead of ground. This accounts for the reversal in the triggering edge, but it also causes an inversion of the output sense. In the case of the inverter (4049 version) the input is held HIGH when the circuit is dormant because of the resistor to V+. This situation means that the normal output level is LOW and that it snaps HIGH during the on-period. Similarly, the 4050 (which is noninverting) output will remain HIGH until it is triggered. At that time, the output drops LOW for the duration specified in the last equation.

There are several rules that must be followed in the application of the quasi-monostables: the trigger pulse duration must be longer than the output pulse duration indicated by the timing equation; the trigger pulse must have a very fast rise time; i.e., on the order of ten times faster than T; and the monostable has a long refractory period in which it cannot be retriggered. The device must be left in the dormant state long enough between trigger pulses to allow the refractory period to expire (this is the time required for the RC network charge to change and for the CMOS device to recover from its last triggering). Also, only the B-series CMOS devices will normally work in this application without a lot of fretting and moaning on the part of the builder.

The CMOS 4528 device is shown in Fig. 3-3. This device is a dual retriggerable monostable multivibrator; both sections are independent of each other. In Fig. 3-3A the pinouts for both sections are shown, even though only one circuit is present. The pinouts for section A are shown without parenthesis, while those for section B are shown within the parenthesis: A(B). For example, the Q output for section A is on pin number 6, while the Q output for section B is on pin number 10. The pinouts, therefore, are written in the form 6(10).

The 4528 device will tolerate either positive- or negative-edge triggering, depending upon the configuration of the A and B inputs. If the positive edge triggering is desired, then tie the B input to V+ and use the A input. If, on the other hand, negative-edge triggering is desired, then do the opposite: use the B input and tie the A input to ground (i.e. LOW).

The duration of the output pulse is set by an RC network (see Fig. 3-3B), and will obey the formula

$$T = 0.2\,RC\,\mathrm{Ln}\,(V_{dd} - V_{ss})$$

Where:

T is the time in seconds

R is in ohms

C is in farads

Ln indicates that the natural logarithms are used

V_{dd} is the drain supply voltage (+) in volts

V_{ss} is the source supply voltage (−) in volts

This equation is valid for C greater than 0.01 μF. For values less than 0.01 μF, use the manufacturer's chart in the specifications data sheet for the 4528. Values of resistance from 10 kohms to 10 megohms can be used and capacitors from 20pF (and up up up)

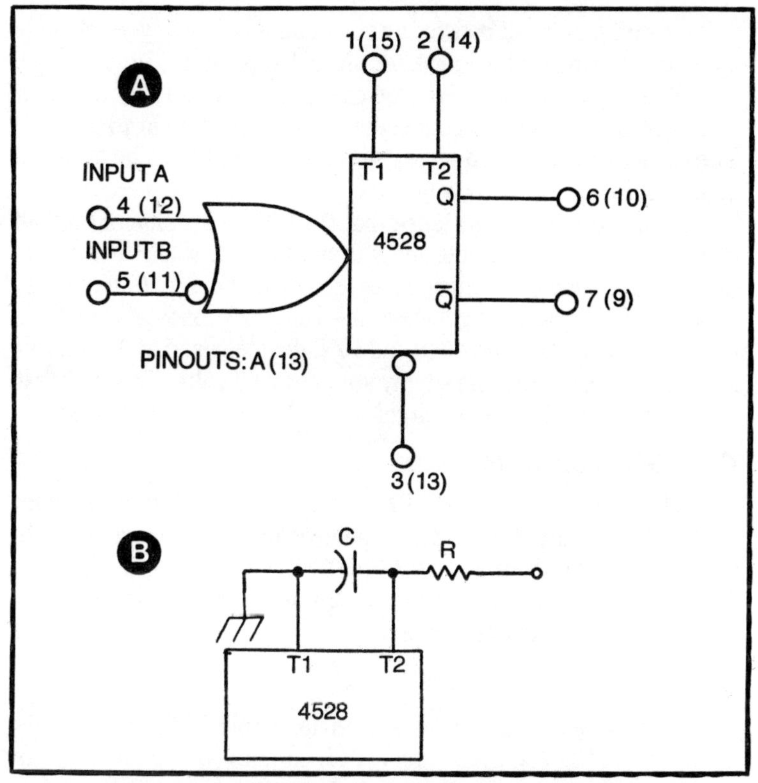

Fig. 3-3. 4528 CMOS Monostable multivibrator: A) Pinouts; B) RC network.

can be used. The triggering pulse should be greater than 75 nanoseconds and the minimum output pulse width is 0.5 microseconds.

TTL TIMERS

There are several devices in the TTL line that can be used as timers. Most of these are monostable multivibrators or astable multivibrators. Again, there are a number of special-function devices available strictly under proprietary house numbers. We may also use the half-monostable idea, but in the TTL line the circuit is a little flakey and doesn't always work properly. Figure 3-4 shows the two versions of the TTL half-monostable multivibrator. The active element is an inverter or (as shown) an inverter-connected NAND, or inverter-connected NOR gate.

In Fig. 3-4A we see the negative-edge triggered version of the circuit. The input of the TTL inverter is biased with a resistor voltage divider circuit consisting of R1 and R2. The trigger pulse is applied to the input of the inverter through capacitor C1.

The positive-edge triggered circuit is shown in Fig. 3-4B. In this case, the resistor is connected from the gate input to ground. The capacitor is, again in series with the trigger line. The duration of the output is approximately 330 C_1.

It has been my experience that the CMOS version of the half monostable is more reliable and is easier to make work (in fact, it is disgustingly easy if the three rules are remembered). In that case, it is my preference to forget about the TTL half monostable and use the CMOS version. The 4049 and 4050 devices used in this circuit are TTL-compatible if the supply voltage is +5 volts. This makes it easier to use the device in a circuit with TTL devices. Good luck.

TTL MONOSTABLE CHIPS

There are three popular TTL monostable multivibrator chips: 74121, 74122, and 74123. These devices provide a programmable output pulse width of 35 nanoseconds (typical) with the internal resistance and 40 nanoseconds to 28 seconds with an external resistor-capacitor combination.

74121

This device is a single non-retriggerable monostable multivibrator. It has complementary TTL outputs (Q and NOT-Q). If we use the internal resistor for timing, then the nominal output pulse will be 30 nanoseconds. If we want to use the internal timing resistor, we connect the R_{int} terminal (pin number 9 in Fig. 3-5A) to

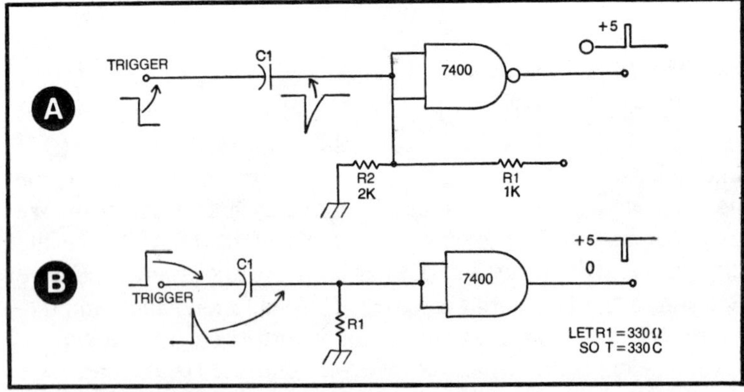

Fig. 3-4. TTL Half monostable multivibrators: A) Negative-edge triggered; B) Positive-edge triggered.

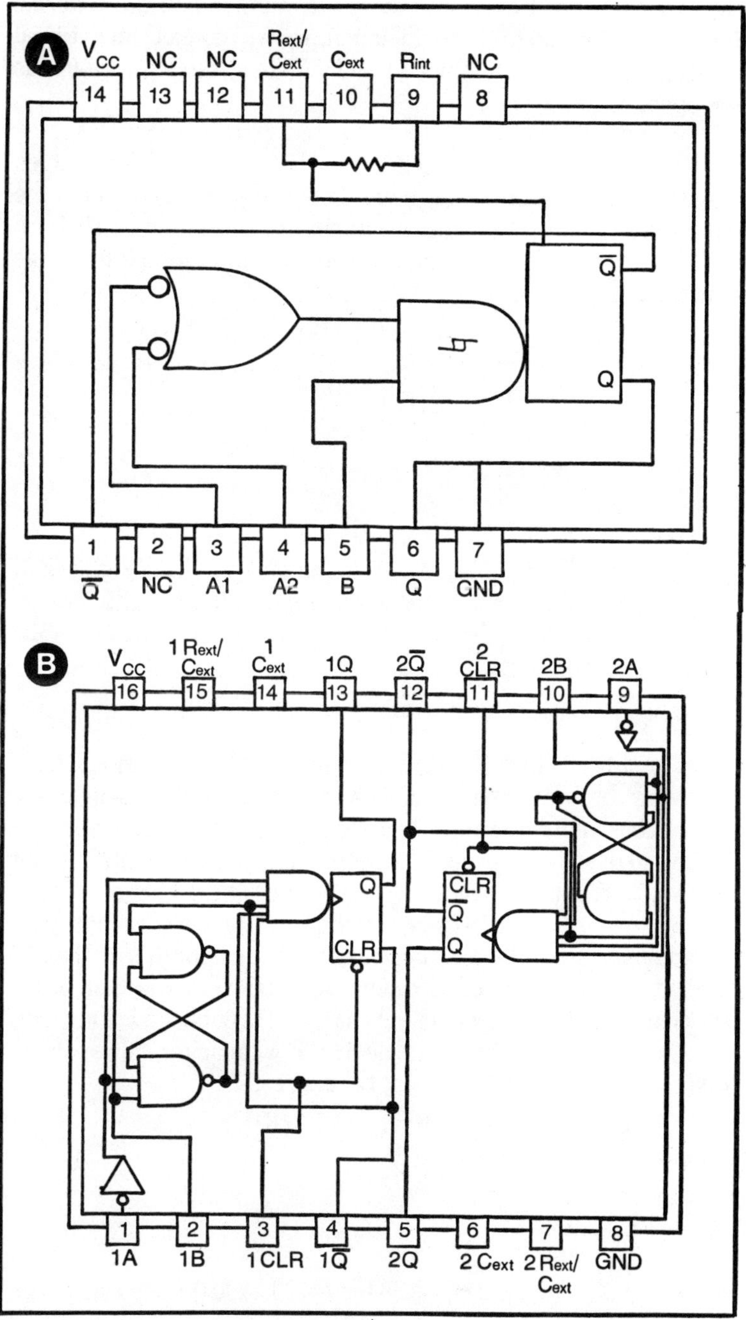

Fig. 3-6. TTL monostables: A) 74122; B) 74123.

the +5 volt line. For cases where we want to use the external timing resistor-capacitor combination, we will connect an external resistor between pin number 11 and +5 volts; the R_{int} terminal is left open.

Permissable values of the timing resistor range from 2 kohms to 40 kohms, and for the capacitor from 10 pF to 10 μF. The values may be extended somewhat if we do not require precision in the output duration. In that case, we might use capacitors to 1000 μF and resistors down to 1.4 kohms. The output duration follows the expressions:

$$T = RC\,Ln\,(2)$$

and,

$$T = 0.69\,RC$$

Where:

T is in nanoseconds
R is in kohms
C is in picofarads

The 74221 device is a dual version of the 74121 (Fig. 3-5B). The 74221 is a dual non-retriggerable monostable multivibrator. The pinouts for this device are identical to those of the 74123 device discussed below.

74122/74123

The 74122 (Fig. 3-6A) is a single retriggerable monostable multivibrator, while the 74123 is a dual (Fig. 3-6B) retriggerable monostable multivibrator. A principle difference between the 74123 and the 74221 (which shares the same pinout scheme) is that the 74123 can be retriggered, while the 74221 cannot. These devices can have duty-cycles close to 100 percent but it is recommended that the duty cycle normally be kept to less than 75 percent. When the external timing capacitance is less than 0.001 μF we use the timing component chart that is published along with the specifications sheet for the device. For capacitor values above 0.001 μF, however, we must use the expression

$$T = k\,R\,C\,[1 + (0.7/R)]$$

Where:

T is in nanoseconds
R is in kohms
C is in picofarads
k is 0.32 for 74122 and 0.28 for 74123. Add 0.05 to each if the LS versions (e.g., 74LS123) are used.

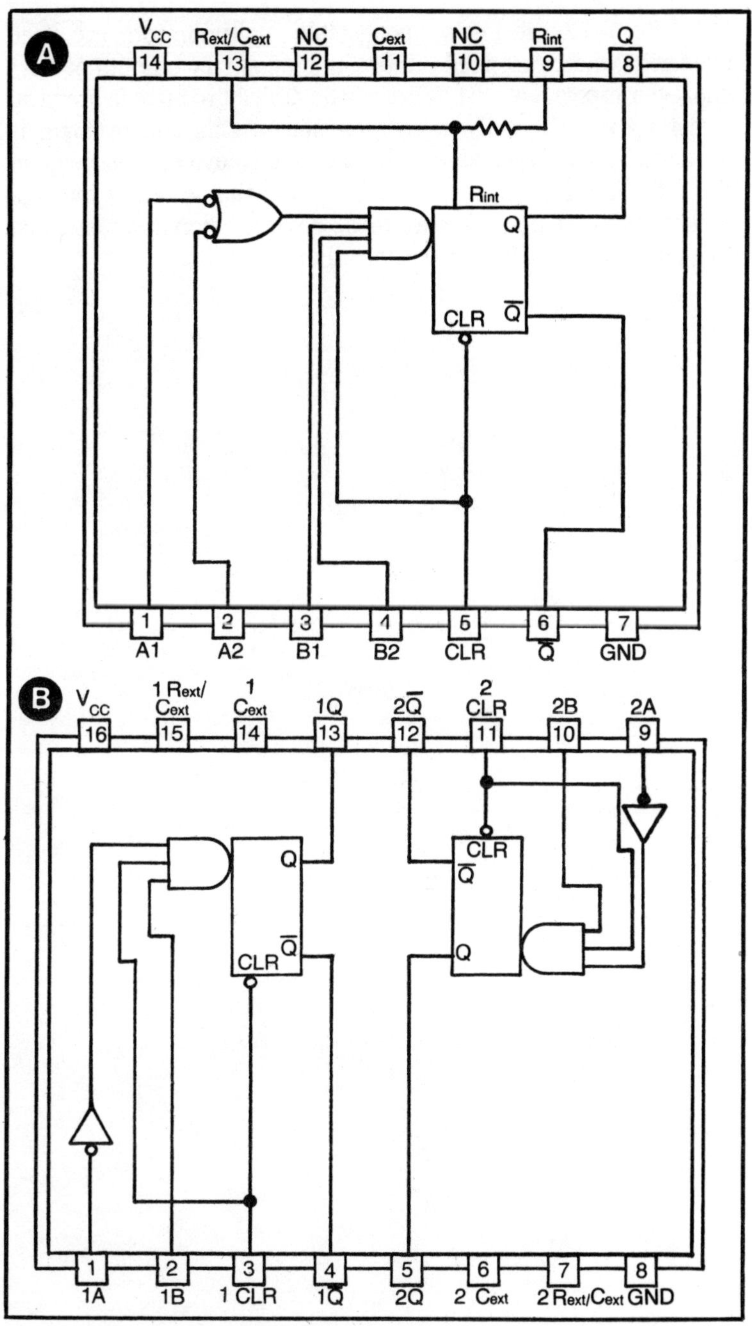

Fig. 3-6. TTL monostables: A) 74122. B) 74123.

The 7412x series of monostable multivibrators are used primarily where we need either complementary outputs or very short-duration pulses. They are a little difficult to tame in practical circuits and will easily trigger on noise. This last problem is partially because these are high speed logic devices, i.e. being of the TTL line. If the pulse width of the one-shot is nearer the high end of the range permitted for these devices, then use something like the 555, XR-2240 or one of the other timer ICs.

Chapter 4
Basics of Digital Electronics

Digital electronics was once an arcane area inhabited solely by computer people and some military electronics technicians. But today no one can safely ignore the world of digital electronics. Even radio and television technicians find digital circuits in their products. Computerized color TV tuning and stereo FM tuning are almost "the rule" rather than the high priced exception. Digital electronics is, in a way, simpler than analog electronics because the devices recognize only two states, i.e., "on" and "off." This makes digital circuits akin to relays and switches! In fact, some digital circuits are little more than high-frequency *electronic* versions of simple switches. It is this author's opinion that anyone who can understand simple relay and switch logic circuits can understand digital electronics. Certainly anyone who can understand color television, SSB FM two-way radios, and other complex products can understand digital electronics. Besides, electronic timers of the sort which we are studying are primarily *digital* devices.

LOGIC STATES

We have mentioned that digital circuits respond only to two different input states. These states are called "1" and "0" (after the two permissable digits of the binary, i.e., base 2, number system), HIGH and LOW, or (in older texts) "true" and "false." These designations are merely two different voltage states. In this chapter we will stick to the HIGH/LOW designation because it graphically describes what is going on in the circuit.

Transistor-transistor-logic (TTL) responds only to 0 and +5 volts for logic levels. If any other voltage levels are used the TTL device will: fail to work; work unpredictably; or burn out. Figure 4-1A shows the TTL logic levels.

Complementary metal-oxide silicon (CMOS) IC logic devices can use the same 0 and +5 volt logic levels as the TTL devices, but they may also be used at any combination of voltage levels between ± 4 volts and ± 15 volts. In Figure 4-1B, we see ± 7 volts used. The voltage levels that represent HIGH and LOW conditions need not be equal. In fact, many circuits exist which use a positive voltage (e.g., +12 volts) and zero voltage for HIGH and LOW, respectively. Some CMOS digital devices, notably the complex function devices, will fail to operate properly and predictably if the applied voltage is less than 7 volts. These devices will not always operate properly on TTL levels of 0 and +5 volts.

POSITIVE AND NEGATIVE LOGIC

You may sometimes hear the terms "positive" and "negative" logic. These terms tend to confuse the newcomer, but mean nothing more than how the HIGH and LOW logic states are related to voltage levels. In *positive logic* the HIGH is logical-1 and will be a positive voltage (e.g., +5 volts in the case of TTL). The LOW, logical-0, is the 0-volts condition in TTL but may be a negative voltage in some CMOS circuits. In *negative logic* these designations are reversed (i.e., HIGH = logical-0 and LOW = logical-1). In the vast majority of uses positive logic is specified. In fact, the descriptive names given to digital IC devices reflect a bias toward positive logic. This potential confusion is why I prefer HIGH/LOW designations. The "1/0" designation will be retained for illustrations and truth tables—but recall that positive logic is implied.

LOGIC FAMILIES

A logic family is a series of IC devices that may be interconnected without regard for interfacing, and which use similar technology in their construction. All of the devices within a given family will have the same input and output circuits so that direct interconnection is possible.

The major consideration is whether an output can supply sufficient current to drive all of the inputs that are connected to it. But, in any given logic family, output voltage and current levels as

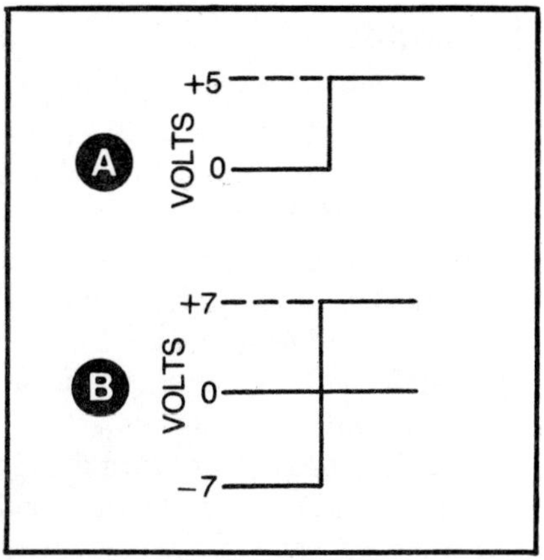

Fig. 4-1. Digital logic levels: A) TTL voltages; B) Other voltage levels.

well as input voltage and current requirements are fixed by agreement, and are defined in terms of *fan-in* and *fan-out*. The *unit* used to describe these terms in most cases is the current requirement of a single standard input at the fixed voltage level. Such an input has a "fan-in" of one unit. If an IC is said to have a fan-out of five, for example, this means that the device will drive five standard inputs. The device, therefore, can supply sufficient current to drive all five inputs satisfactorily. The total fan-in of all devices connected to any output must be equal to, or less than, the rated fan-out of the output.

The logic families which we will consider are: RTL, DTL, TTL, HTL, ECL, and CMOS. Of these families, CMOS and TTL are the most popular today; RTL and DTL are obsolete and no longer used in new designs (although plenty of older equipment, still in use, contains RTL and DTL devices).

SPEED VS POWER

The principal factors governing the speed (i.e., maximum operating frequency) of a digital IC are the internal resistances and capacitances. If resistances are increased, so that power consumption drops, then the RC time constant of the device is longer. Long RC time constants mean slower operating speeds. As a general rule, higher speed logic families require greater power consumption. CMOS devices, which require very little current (hence are low power), operate well only to 4 or 5 megahertz (MHz) with some devices tooting along to 10 MHz. TTL devices, on the other hand, usually work up to 18 or 20 MHz with some devices going to well over 80 MHz.

RTL DEVICES

Resistor-transistor-logic (RTL) is an obsolete logic family that was popular in the mid-60s. Figure 4-2 shows a typical RTL inverter circuit; i.e., a circuit that produces a LOW output when the input is HIGH, and a HIGH output when the input is LOW.

RTL logic IC devices used 0 volts for logical-0 and +3.6 volts for logical-1. If the input of the RTL inverter is grounded (i.e., placed LOW) then the output voltage will be HIGH, which in this case means +3.6 volts. But if the input voltage is +3.6 volts, then the output will be 0 volts.

RTL devices usually carry type numbers in the uL900 range (mostly 8- and 10-pin metal cans) and MC700 series (mostly 14-pin DIPS).

DTL DEVICES

The next popular IC logic family was the *diode-transistor-logic* (DTL) family. These devices operated at speeds greater than most RTL devices. Figure 4-3 shows a typical DTL inverter.

When the DTL input is HIGH, diode D1 is reverse biased. In that condition R1 will forward bias transistor Q1, which in turn

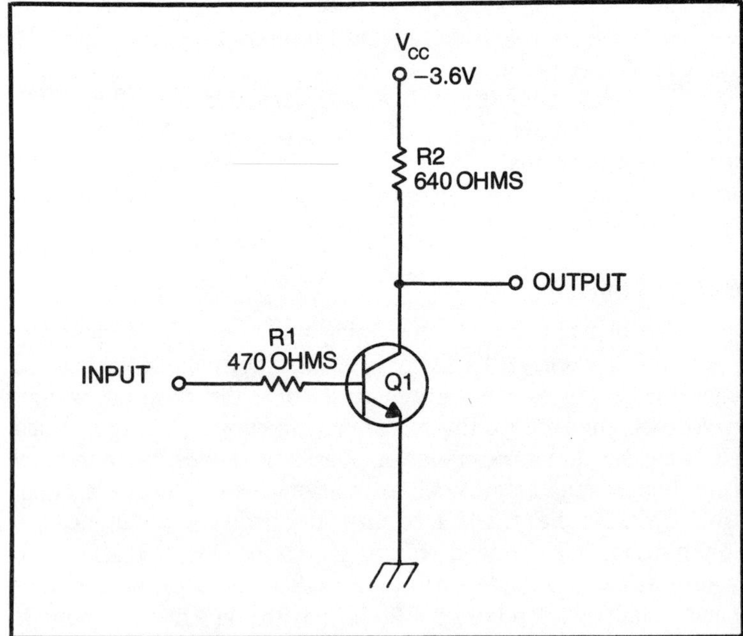

Fig. 4-2. RTL Inverter.

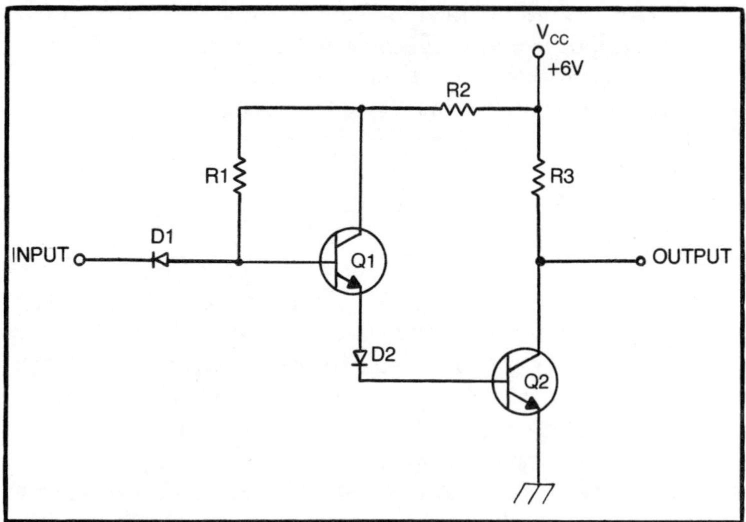

Fig. 4-3. DTL Inverter.

will forward bias transistor Q1, which in turn forward biases D2 and Q2. Voltage levels in most digital circuits are selected to *saturate* the transistors, so when Q2 is turned on it is turned on to saturation. This condition means that the output of the inverter, which is the collector of Q2, goes nearly to ground, but is actually V_{cesat} of the transistor (i.e., on the order of a few tenths of a volt, at most).

When the input is LOW, the cathode of D1 is grounded. Since D1 is now forward biased, the base of Q1 is essentially grounded. Under this condition Q1, D1 and Q2 are reverse biased. With Q2 cut off, then, the output voltage rises to that of $V_{cc(+)}$. Most DTL devices carried part numbers in the MC800 and MC900 ranges (Motorola designation).

TTL DEVICES

Probably the most widely used digital IC logic family is the *transistor-transistor-logic* (TTL) family. When most people speak of "digital ICs", they are referring to the TTL family of devices. Most TTL devices carry type numbers in the 7400 range. Those devices in the 5400 range are *military* equivalents to the 7400-series devices (e.g., a 5447 is a 7447 in uniform). The principal difference between the 5400 and 7400 devices is the operating temperature range (–55 to +125 degrees centigrade for mil-spec devices).

Figure 4-4A shows the circuit for a typical TTL inverter IC. Like the DTL device, the TTL input acts as a *current source*, while the output acts as a *current sink*. The typical TTL input will source 1.8 mA. It will be LOW if the voltage is 0 to 0.8 volts and HIGH if 2.4 to 5.0 volts are applied. Performance at values of input potentials between 0.8 and 2.4 volts are not defined, so operation of the devices in that range is unpredictable.

When the TTL input is HIGH, Q1 is cut off so point A goes HIGH. This condition turns on Q2, forcing point B HIGH and C LOW. We find, then, Q3 is turned on and Q4 is off. This forces the output LOW. Again, the transistors are operated either totally cut-off or totally saturated on. If the input is LOW, then exactly the opposite situation occurs: Q1 is turned on (forcing point A LOW); Q3 is off; and Q4 is turned on, i.e., it is connected to $V_{cc(+)}$.

TTL devices must have a regulated DC power supply of +4.75 to +5.25 volts. In fact, there are some circuit combinations that require a more limited range of voltages nearer to +5 volts DC. Voltages greater than +5.25 volts often result in a high failure rate of TTL devices.

Some TTL devices are described as being *open-collector* devices. These devices are essentially the same as regular TTL devices, except that the output circuit is modified; i.e., Q4 and D2 are missing. An example of an open-collector circuit is shown in Fig. 4-4B. These devices require an external 1 to 2 kohm resistor between the output terminal and the +5 volt DC power supply line.

CMOS DEVICES

Complementary metal-oxide silicon (CMOS) IC devices use MOSFET transistors instead of the PNP or NPN bipolar transistors that are used in other logic families. CMOS inputs, therefore, are very high impedance. Figure 4-5A shows a typical CMOS inverter circuit. Note that this family is called *complementary* because the output circuit consists of a complementary pair of MOSFET transistors; i.e., an n-channel and a p-channel in series.

CMOS devices can use a monopolar power supply, like TTL or DTL, or may use a bipolar power supply after the fashion of operational amplifiers. In bipolar supplies the V+ can be any potential between +4 volts and +15 volts, while the V− may be −4 volts to −15 volts. In monopolar cases the V+ can also be +4 to +15 volts, while the V− is actually zero volts.

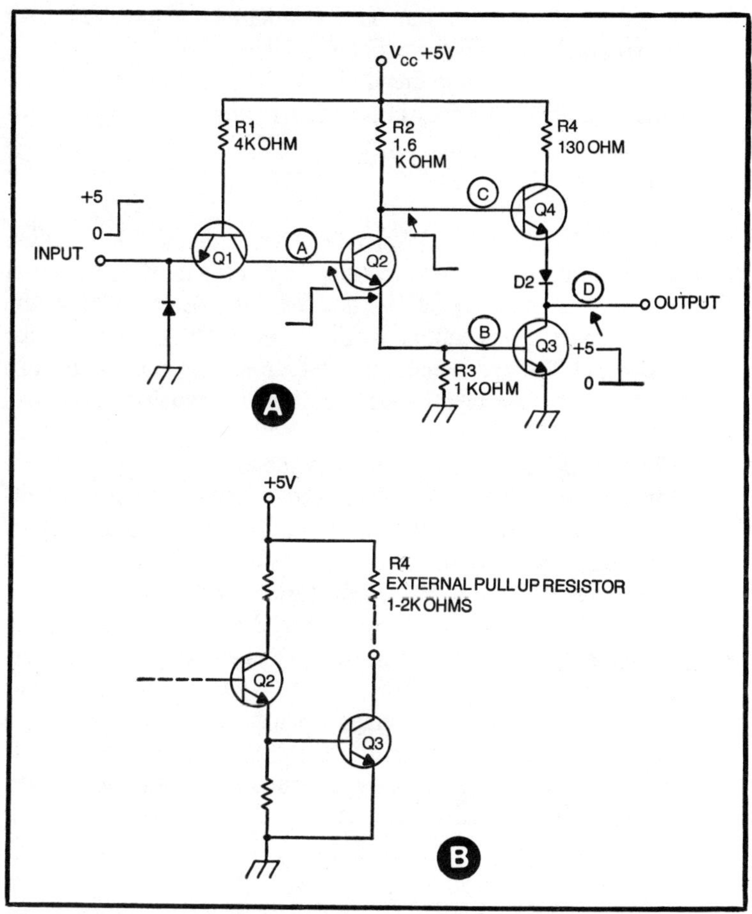

Fig. 4-4. Typical TTL circuits: A) Inverter; B) Open-collector output stage.

CMOS outputs are not directly TTL compatible, although some specific ICs in the CMOS line are designed to have a TTL output stage (e.g., the 4049 and 4050 devices). These TTL-compatible devices are often used to directly interface CMOS and TTL devices.

Figures 4-5B and 4-5C show the equivalent circuits for a CMOS inverter in both possible input conditions (i.e., input HIGH and input LOW). Recall that a p-channel MOSFET turns on when the gate is LOW, while the n-channel device turns on when the gate is HIGH.

Figure 4-5B shows the situation where the input is LOW. Transistor Q1 will have a very low (e.g., 200 ohms) channel resistance. In this case, the output would be equivalent to a 200-ohm resistor to the V+ power supply line.

In Figure 4-5C we see the situation where the input is HIGH. Transistor Q2 now has a very high channel resistance, and Q1 has a very low channel resistance (again, about 200 ohms). In this case, the output looks like a 200 ohm resistance to ground so the output is LOW.

The HIGH/LOW or LOW/HIGH output transition in a CMOS device occurs at a point where the input voltage is midway between the V+ and V− voltages. If V− and V+ are not equal, then the transition occurs at a potential of ((V+)−(V−))/2. If, on the other hand, V− and V+ are equal, then the transition occurs at zero volts. If the V− potential is zero volts, then the transition occurs at (V+)/2.

The CMOS output stage always looks like a high and low resistor in series across the power supply (reexamine Figs. 4-5B and 4-5C), so negligible current is drawn. The power supply sees a low resistance load only when the input voltage is at the transition point. The overall current drain, therefore, is very small.

But CMOS devices do have a problem: they contain MOSFETs and therefore are sensitive to static electricity (see Chapter 3). All A-series CMOS devices (e.g., 4001A) have this problem, but it is less severe in B-series devices (e.g., 4001B). The B-series have built-in diode gate protection to bypass high static potentials around the sensitive gate structure; nonetheless, they should be handled with care.

HTL DEVICES

Noise pulses are often seen by logic circuits as valid input pulses. This problem is especially bothersome in high speed TTL devices that are normally able to pass high frequency short-duration pulses. The solution in noisy environments is to use a digital IC logic family that requires a high input voltage to trigger. CMOS operated at high V− and V+ values meet this requirement, but an older bipolar *high threshold logic* (HTL) family may also be used.

HTL (sometimes called *high noise immunity logic,* or HNIL) uses V+ values of 12 or 15 volts, depending upon the series (see Fig. 4-6). As a result, the logic levels are also high so it requires a bigger noise pulse to cause trouble.

Up until now we have been talking about *saturated* logic families; i.e, the transistors in the IC are either all the way on or all

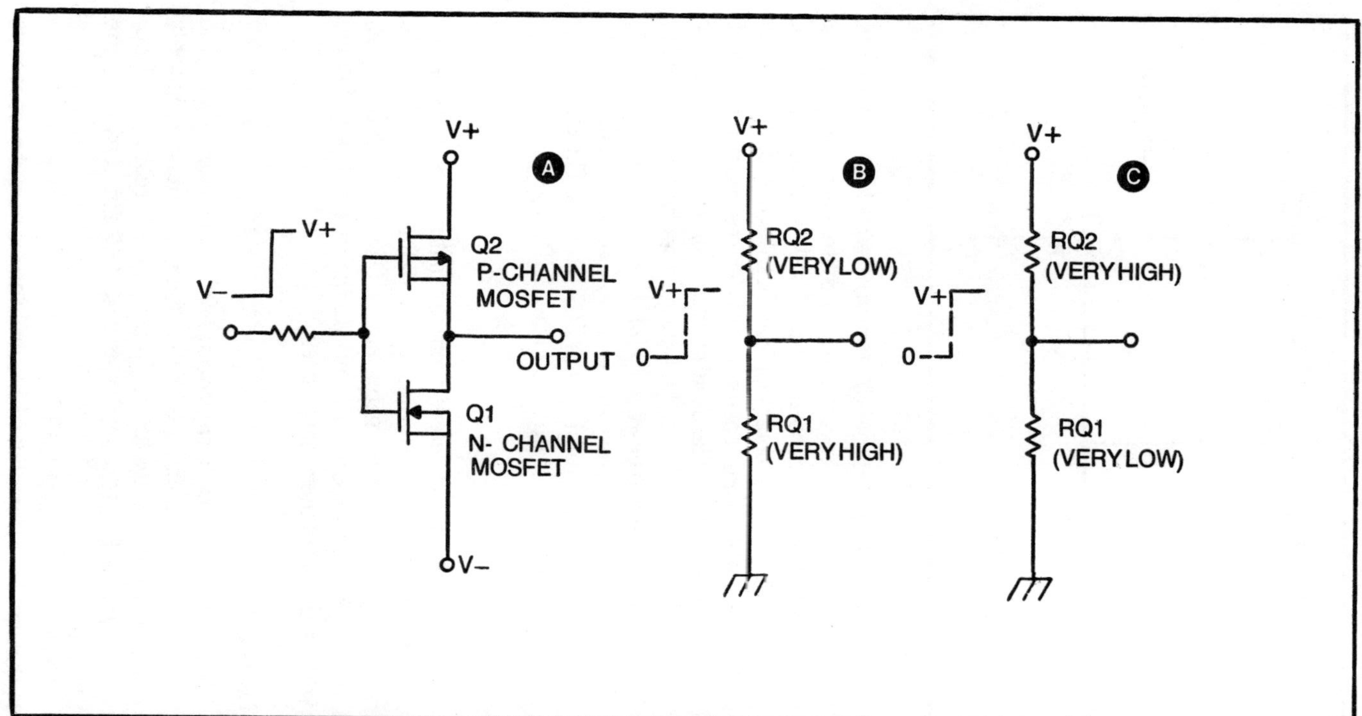

Fig. 4-5. Typical CMOS circuits: A) Inverter circuit; B) Equivalent circuit; C) Equivalent circuit.

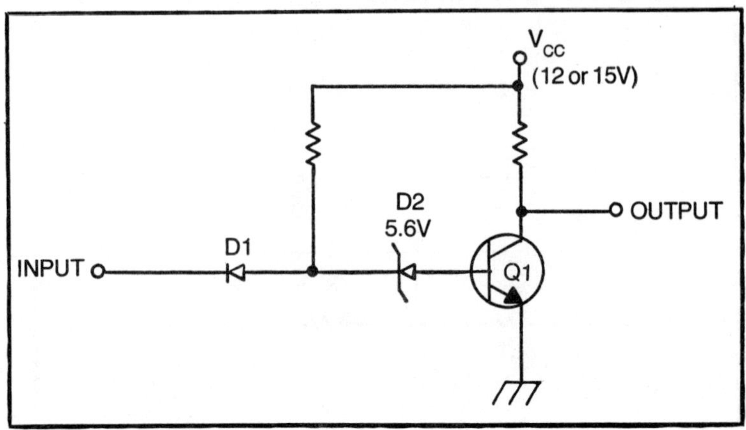

Fig. 4-6. High threshold logic.

the way off (cutoff or saturated). *Emitter coupled logic* (ECL) is called an *AC* logic family because the transistors are operated in a nonsaturated mode. As a consequence, ECL devices are capable of very fast operation. Most commonplace ECL devices operate up to 80 or 120 MHz, while some costly special devices operate to over 1 GHz (that's 1000 MHz!). The usual *prescaler* for a digital frequency counter is nothing more than an ECL frequency divider that divides the 500-MHz input signal down to 50 MHz.

Note that it is necessary to use VHF/UHF circuit design and layout techniques when working with ECL devices. The extremely high frequencies used are, after all, in the UHF range.

GATES

A digital electronic *gate* is a circuit that passes a signal, or refuses to pass a signal, according to well-defined rules. The basic forms of digital electronic gates are: NOT, OR, AND, NOR, NAND, and Exclusive-OR (or XOR). In the paragraphs to follow, we will discuss all of these basic gates.

NOT Gates

NOT gates, also called *inverters,* produce an output that is the opposite of the input signal. Recall that digital circuits respond only to HIGH and LOW voltage levels. In an inverter circuit, therefore, the output will be HIGH when the input is LOW and LOW when the input is HIGH.

The circuit symbol for the inverter is shown in Fig. 4-7A. Note that any digital symbol with a circle on the output produces an inverted output. Similarly, if an input has a circle on it, then that input is inverted. The rules for the operation of the inverter are the following:

- ☐ A HIGH on the input produces a LOW output
- ☐ A LOW on the input produces a HIGH output.

OR Gates

An OR gate will pass a signal to the output if *either* input is HIGH. The symbol for an OR gate is shown in Fig. 4-7B, while the truth table is given in Fig. 4-8B. The truth table shows the rules of operation for the gate, and these are the following:

- ☐ If both inputs A and B are LOW, then the output is LOW
- ☐ If either input A or B is HIGH, then the output is HIGH
- ☐ If both inputs A and B are HIGH, then the output is HIGH.

AND Gates

The AND gate is the opposite of the OR gate. The AND gate produces a HIGH output only when *both* inputs are also HIGH. The

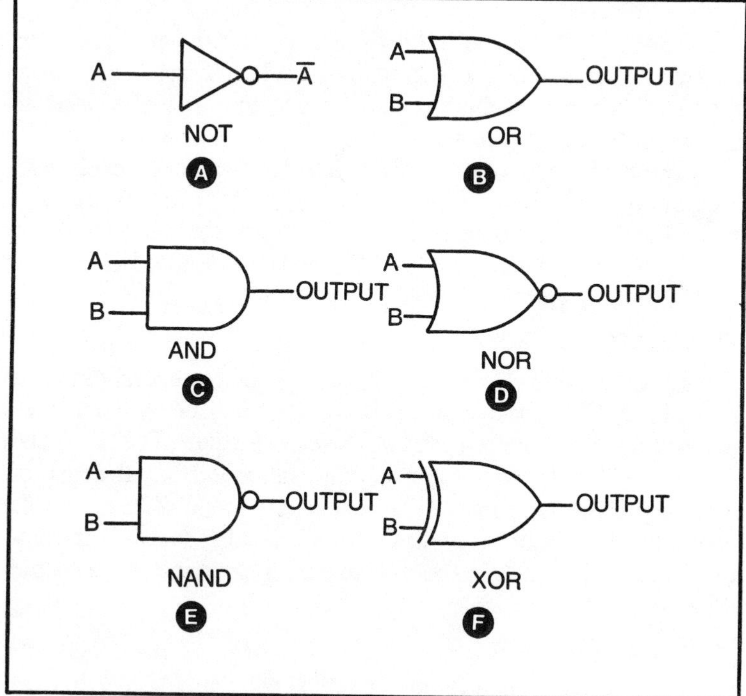

Fig. 4-7. Logic element circuit symbols.

circuit symbol for the AND gate is given in Fig. 4-7C, and the truth table is shown in Fig. 4-8C. The rules for the operation of the AND gate are as follows:

- ☐ If both inputs A and B are LOW, then the output is LOW
- ☐ If either input A or B is LOW, then the output is LOW
- ☐ If both inputs A and B are HIGH, then the output is HIGH.

NOR Gates

The NOR gate is a combination of a NOT gate (inverter) and an OR gate, hence the designation NOR means "NOT-OR." It is, therefore, an OR gate with an inverted output. The NOR gate is, in fact, sometimes represented in textbooks as an OR gate with an inverter following. The NOR gate symbol, shown in Fig. 4-7D, is an OR gate symbol with the circle denoting inversion at the output. The truth table for the NOR gate is shown in Fig. 4-8D, and the rules of operation are the following:

- ☐ If both A and B inputs are LOW, then the output is HIGH
- ☐ If either input A or B is HIGH, then the output is LOW
- ☐ If both inputs A and B are HIGH, then the output is LOW

NAND Gates

The NAND gate is a NOT-AND gate; i.e., an AND gate followed by an inverter. The symbol for the NAND gate is shown in Fig. 4-7E. This symbol is the AND gate symbol with the circle at the output to denote inversion.

The truth table for the NAND gate is shown in Fig. 4-8E, and the rules are:

- ☐ If both A and B inputs are LOW, then the output is HIGH
- ☐ If either input A or B is LOW, then the output is HIGH
- ☐ If both A and B are HIGH, then the output is LOW.

TTL AND CMOS EXAMPLES

Earlier in this chapter we introduced you to several different families of IC digital logic devices. Of these, several are considered obsolete, so will not be covered further. The TTL and CMOS families, however, are very much alive and form the basis of most digital design today. It is beyond the scope of this book to present all the members of these IC logic families (we will attempt that trick in a forthcoming book!), so we will describe only a couple of representative devices.

TTL/CMOS NAND Gates

In the TTL IC logic family, the most popular NAND gate (and probably the most popular IC!) is the 7400. This device contains

INPUT	OUTPUT
1	0
0	1

A NOT

INPUT		OUTPUT
A	B	
0	0	0
0	1	1
1	0	1
1	1	1

B OR

INPUT		OUTPUT
A	B	
0	0	0
0	1	0
1	0	0
1	1	1

C AND

INPUT		OUTPUT
A	B	
0	0	1
0	1	0
1	0	0
1	1	0

D NOR

INPUT		OUTPUT
A	B	
0	0	1
0	1	1
1	0	1
1	1	0

E NAND

INPUT		OUTPUT
A	B	
0	0	0
0	1	1
1	0	1
1	1	0

F EXCLUSIVE OR (XOR)

Fig. 4-8. Truth tables.

four two-input NAND gates, and usually sells for less than 25 cents, or about 6 cents per gate. Each of the four NAND gates in the 7400 package is an independent entity, but they share a common power supply and ground connection (pins 14 and 7 respectively).

The 7401 device is similar to the 7400, except that the pinouts are somewhat different, and it's an open-collector device. This means that pull-up resistors are needed; i.e., one 2 to 4 kohm resistor from each output to the +5-volt line.

The 7403 device is also an open-collector device but uses exactly the same pinouts as the 7400. Of these three devices, the 7400 is by far the most popular.

The 7430 is an eight-input NAND gate (one per 14-pin DIP package). The 7430 device, therefore, has eight distinct inputs—and all eight must be HIGH before the output drops LOW. If any one of the eight inputs remains LOW, then the output stays HIGH. Since most microcomputers today are eight-bit machines, the 7430 is often used as an address or I/O port decoder.

The 7410 and 7420 are three- and four-input TTL NAND gates, respectively.

In the CMOS line, we also have several different types of NAND gates. The 4011 is a quad two-input NAND reminiscent of the 7400. The current requirements of the CMOS device, however, are a lot lower than the current requirements of the TTL equivalent.

The 4012 device is a dual four-input NAND gate. All four inputs of either gate must be HIGH for the respective output to be LOW. The 4023 device is a triple three-input NAND gate, while the 4068 is an eight-input NAND gate that is reminiscent of the TTL 7430 device.

TTL/CMOS NOR Gates

The 7402 TTL NOR gate is, by far, the most common example from the TTL line, and is almost as popular as the 7400 device. The pinouts for the 7402 are shown in Fig. 4-9B (see Fig. 4-9A for the pinouts to the 7400, for comparison). Note that the pinouts of the 7400 and 7402 are different, as are their logic responses.

SOME EXPERIMENTS IN LOGIC GATES

This book promises hands-on experiments and some projects. The following experiments are designed to teach you the ordinary rules of logic for the various types of digital gate.

The experiments are designed to be performed on the AP Products Powerace 102 Digital Circuits Breadboard. Although

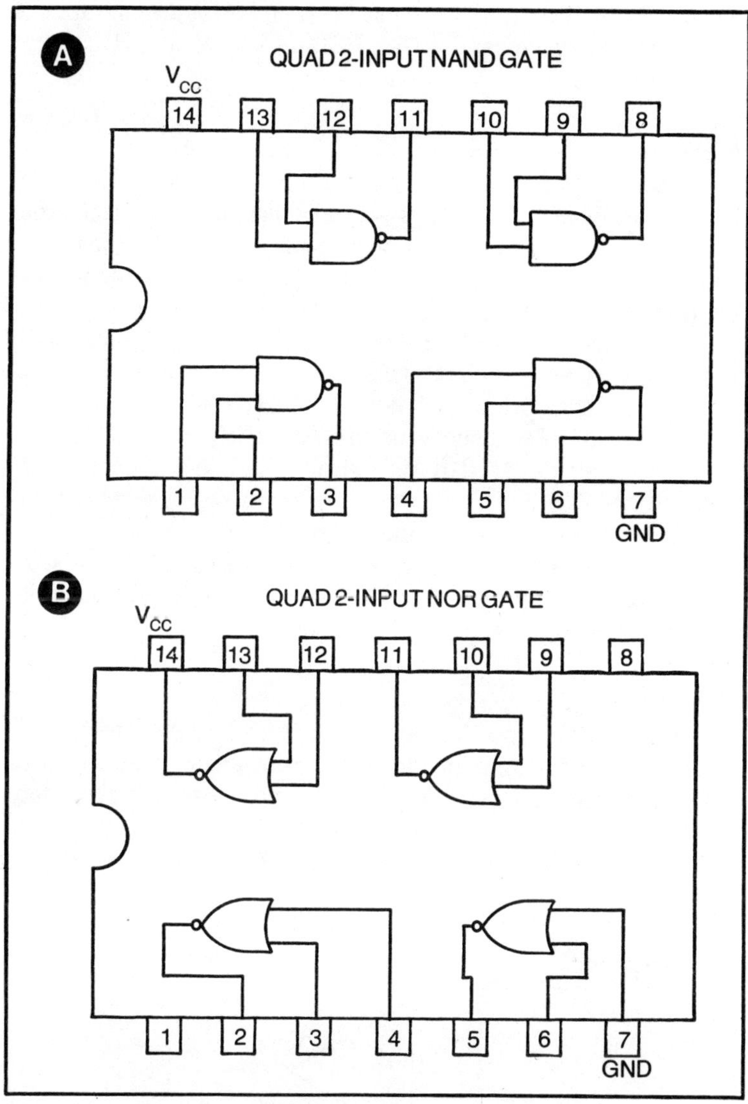

Fig. 4-9. TTL gates: A) 7400 NAND gates; B) 7402 NOR gates.

they may also be performed on digital breadboards of other manufacturers, the terminal and pinout names used in the experiment instructions are for the Powerace 102.

When an instruction tells you to connect an input terminal of the gate to "L2 and S2," this means that a piece of #22 or #24 solid hook-up wire is to be connected from the IC pin indicated to switch

S2 and LED indicator L2. Note that the protocol for the operation of the LED indicators is that the LED is lighted for HIGH and extinquished for LOW conditions.

Experiment No. 1: Inverter Circuits

Use a 7404 TTL hex inverter for this experiment. The 7404, being a *hex* inverter, contains six individual inverter stages that can be operated independently of each other, except for the power supply connection. Only one section of the 7404 is needed for this experiment:

☐ Connect the 7404 as in Fig. 4-10. The output goes to L4, and the input goes to both L2 and S2.

☐ Set S2 to "0" (i.e., LOW). L2 should be off.

☐ Observe L4, it should be on (i.e., HIGH, or logical-1).

☐ Set S2 to "1" (HIGH, logical-1), L2 should come on to indicate the HIGH.

☐ Observe L4 and note the difference.

You may wish to replace S2 with the output of the on-board clock of the Powerace 102. Set the clock frequency to 1 Hz and note the action of the LEDs.

Experiment No. 2: NAND Gates

Use a TTL type 7400 quad two-input NAND gate for this experiment. Only one of the four independent gates is used. Connect the circuit as in Fig. 4-11A, and follow the procedure below:

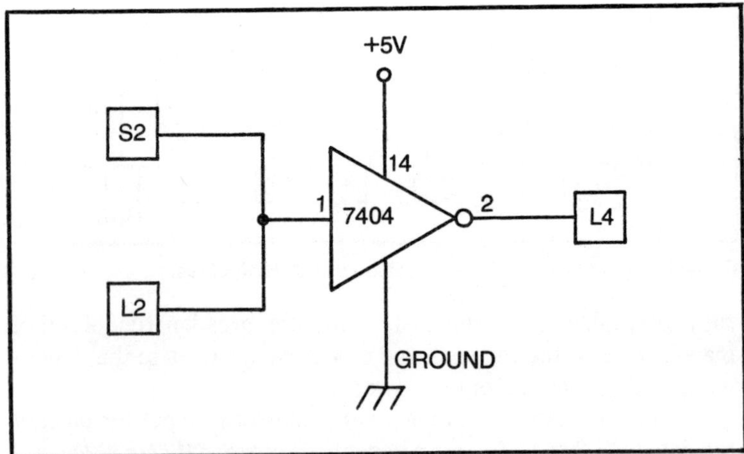

Fig. 4-10. Experiment No. 1: Inverter circuits.

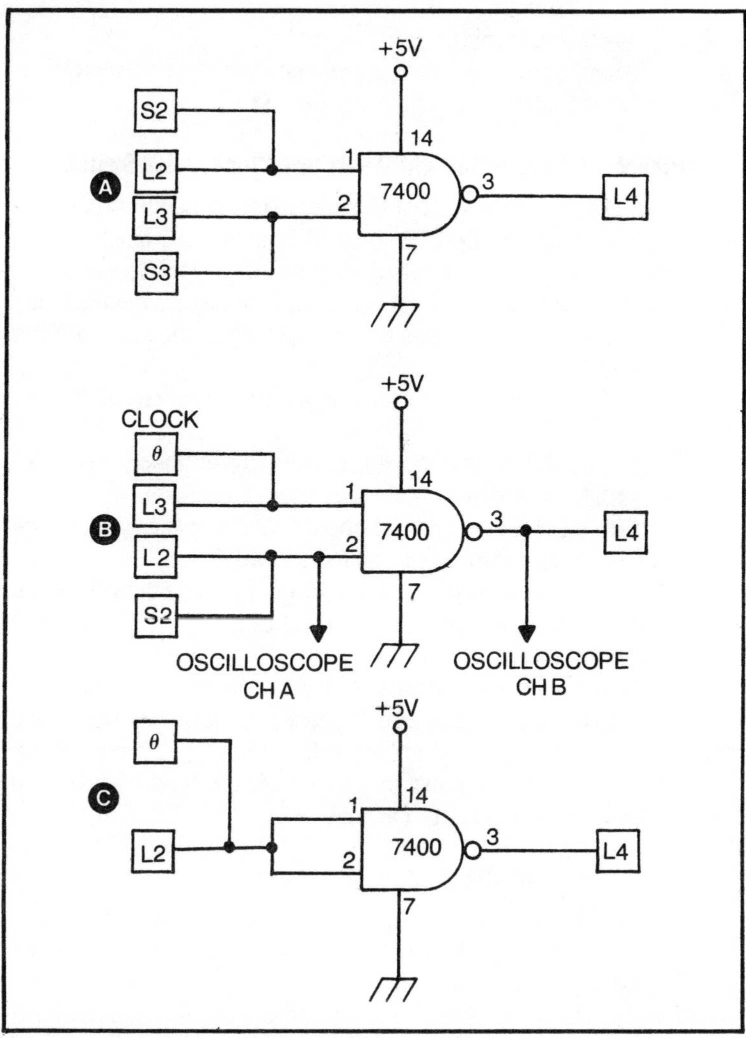

Fig. 4-11. Experiment No. 3: NAND gates.

- ☐ Set both S2 and S3 to "0"; L2 and L3 should be off.
- ☐ Note L4, which should be on.
- ☐ Set S2 to "1" and note the change in L2.
- ☐ Observe L4 (on).
- ☐ Reset S2 to "0" and set S3 to "1"; L2 should be off and L3 on.
- ☐ Observe L4 (off).
- ☐ Set both S2 and S3 to "1"; L2 and L3 should both be on.

☐ Observe L4 (off).

☐ Compare the results of these steps with the truth table for the NAND gate and the rules given in the text.

Experiment No. 3: Using the NAND Gate as a Clock-Pulse Switch.

This experiment is best performed using an oscilloscope, but the indicator lamps on the Powerace 102 can be used if no scope is available. If the oscilloscope is used, then use a clock frequency of 1000 Hz; but if the LED (L4) is used, then the clock frequency of 1 Hz is used (these frequencies are switch-selectable on the Powerace 102).

Connect the 7400 NAND gate as shown in the circuit of Fig. 4-11B:

☐ Set the clock frequency to 1000 Hz (if oscilloscope is used) or 1 Hz (if LED L4 is used).

☐ Set switch S2 to "0" (L2 should be off or the scope trace channel A should be at, or near, ground potential).

☐ Note channel B on the oscilloscope (or L2 on the Powerace 102). It should indicate a zero (LOW) condition.

☐ Set S2 to "1".

☐ Now observe Channel B (or L4) and note the action. It should be observed that the output passes clock pulses whenever the "control" input (i.e., pin number 2 in this experiment) of the NAND is HIGH. This allows us to shut off the flow of pulses to circuits such as counters, etc., at will.

Experiment No. 4: NAND Gate Used As An Inverter

Connect the circuit as in Fig. 4-11C. If an oscilloscope is used, set the clock frequency to 1000 Hz; and connect channel A of the oscilloscope to the input of the gate and channel B of the oscilloscope to the output. If the LED indicators are used, then set the clock frequency to 1 Hz.

Observe that when both inputs of the NAND gate are connected together, the circuit operates as an inverter. This fact allows designers to conserve cost when an inverter is needed, and there is already one unused section of a NAND gate available.

Experiment No. 5: NAND Gate

This one is on you! Examine the truth table and rules given for the operation of the 7400 NAND gate, and try to come up with at least one other method for using the NAND gate as an inverter.

You may want to add a little twist and select a method for turning the inverter on and off using one of the inputs.

Experiment No. 6: NOR Gates

This experiment uses the TTL type 7402 NOR gate IC and is similar in operation to the NAND gate experiment given earlier. The *results*, of course, will differ from the earlier experiment because we are using a different form of logic. Connect the circuit of Fig. 4-12A:

- ☐ Set S2 and S3 to "0"; L2 and L3 should be off.
- ☐ Note L4 (on).
- ☐ Set S2 to "1"; L2 should be on and L3 is off.
- ☐ Note L4 (off).
- ☐ Set S2 to "0" and S3 to "1"; L2 should be off, while L3 is on.
- ☐ Note L4 (off).
- ☐ Set both S2 and S3 to "1"; Both L2 and L3 should now be on.

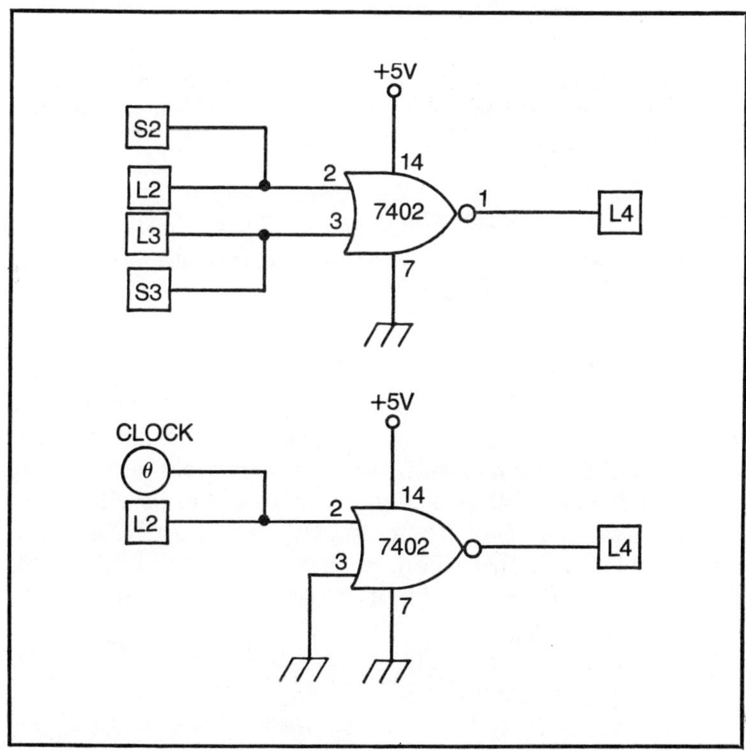

Fig. 4-12. Experiment No. 7: NOR gates.

☐ Note L4 (off).

☐ Compare the results of this experiment with the truth table and text rules.

Experiment No. 7: NOR Gate Used As An Inverter

Use the same 7402 NOR gate as in experiment number 6. Set the Powerace 102 clock frequency to 1000 Hz (if an oscilloscope is used) or 1 Hz (if the LED indicators are used). Connect the circuit as shown in Fig. 4-12B. If you use an oscilloscope, then connect Channel A to the same point as L2, and connect channel B to the 7402 output (i.e., pin number 1):

☐ Connect the circuit; applying the clock will cause the output light to blink out of phase (indicating inversion is taking place) with the input light.

☐ Compare this behavior with the rules given in the truth table and text.

Experiment No. 8: NOR Gate Inverter

This one is on you! Try to think of at least one other way to use the 7402 NOR gate as an inverter. Examine the rules of operation, both in the text and in the truth tables, and see if you can find the clue.

Experiment No. 9: Gated Inverter

Expand experiment number 8 to include finding a way to make a *gated* inverter from the 7402 device.

Experiment No. 10: Exclusive-OR Gates

The exclusive-OR (also called XOR) gate is a strange bird: it will produce a HIGH output if *either* input is HIGH but not if *both* are HIGH simultaneously. In that case, the output will be LOW. In other words, the XOR gate output is LOW if both (or all) inputs have the same logic level. If both are LOW, or both are HIGH, then the output is LOW. This experiment uses the 7486 TTL XOR gate to demonstrate the rules of XOR behavior. Connect the circuit shown in Fig. 4-13:

☐ Set S2 and S3 to "0". L2 and L3 should both be off.

☐ Note the status of L4 (off).

☐ Set S2 to "1" and S3 to "0". L2 should be on and L3 off.

☐ Note L4 (on).

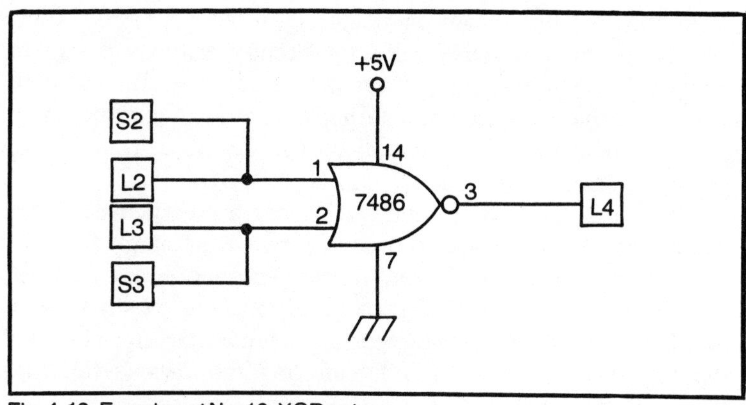

Fig. 4-13. Experiment No. 10: XOR gates.

□ Set S2 to "0" and S3 to "1". L2 should be off (indicating a LOW condition), and L3 should be on (indicating a HIGH condition).

□ Note L4 (on).

□ Turn both S2 and S3 to "1". Both L2 and L3 should be on.

□ Note L4 (off).

□ Compare these results with the truth table and text for the rules of operation for the XOR gate.

Experiment No. 11: XOR as a Non-Inverting Buffer

Examine the truth table and text for the XOR rules of operation. Perform experiment number 10 to verify these rules for yourself. Then try to design a circuit that will make the XOR gate operate as a noninverting buffer stage.

Experiment No. 12: XOR as an Inverter

Perform experiments numbers 10 and 11, and then think of a way to make the XOR gate think that it is an inverter stage.

FLIP-FLOPS

All of the digital circuits discussed so far in this chapter have operated in a "transient" manner. Once a pulse (or any other input signal) has passed, the output state reverts back to its previous condition. Gates and inverters do not have any *memory*, so once the input condition has passed then the output that resulted from those conditions also passes.

A *flip-flop* is a circuit that is capable of storing a single bit (i.e., *bi*nary digi*t*, either 1 or 0) of digital data; it will remember an input condition and hold the same output after the data has passed. There are various different types of flip-flop circuits, and they operate on slightly different (even though similar) sets of rules. But one thing that they all have in common is the ability to store a single data bit.

All common forms of flip-flops can be made from various combinations of the basic AND, OR, NAND, NOR, NOT, and XOR gates. The NAND, NOR, and NOT gates are particularly often used to make flip-flops. Except for the two most simple flip-flops presented in this installment, most electronic circuits use IC flip-flops instead of actual IC gates. It is simply too costly to *make* flip-flops from IC gates when the same IC manufacturers do all of the interconnection for you by offering the various flip-flops pre-made in IC form.

Reset-Set (RS) Flip-Flops

One of the simplest forms of flip-flop circuit is the *reset-set*, or *RS*, flip-flop. Some textbooks, especially those over ten years old, call this type of circuit a *set-reset*, or *SR*, flip-flop. This designation is merely the same thing under a different label.

The RS flip-flop can be made from either two NAND gates or two NOR gates, although the operation of the two versions is slightly different.

Figure 4-14 shows the circuit for an RS flip-flop made from a pair of NAND gates, such as the TTL 7400 device (which contains four two-input NAND gate sections).

There are two inputs required on the RS flip-flop: *set* (S) and *reset* (R). There are also usually two output terminals, and these are complementary: Q and NOT-Q. *Complementary* means that one will be LOW if the other is HIGH. For example, when the Q output is HIGH then the NOT-Q will be LOW. When the Q output is LOW then the NOT-Q will be HIGH.

The inputs of the NAND version of the RS flip-flop are active-LOW, so are sometimes designated $\overline{S}$ (NOT-S) and R (NOT-R). Whenever you see an *input* that is designated as a NOT-input, or has a bar over its symbol (i.e., $\overline{R}$), or that has a circle in the schematic diagram, then we know that it is an active-LOW input terminal. The circuit action of an active-LOW input occurs when the terminal is brought LOW. An example of a schematic that uses the circled inputs is shown in Fig. 4-14C, while the normal symbol for the RS flip-flop is shown in Fig. 4-14B.

In most cases (there are exceptions to any rule, it seems), the absence of a bar over the designating letter, or a circle at the input terminal in the drawing, indicates that the input is active-HIGH. We sometimes see the "exception" when the RS flip-flop of Fig. 4-14, made with NAND gate sections, is shown as a square (similar to Fig. 4-14B), but without the bars over the letters.

A momentary LOW on the *set* input of the NAND gate RS flip-flop causes the outputs to go to the state where the Q is HIGH and the NOT-Q is LOW. Note that the term *set* usually means Q = HIGH and NOT-Q = LOW, while *reset* indicates just the opposite: Q = LOW and NOT-Q = HIGH. The flip-flop is said to possess *memory* (and, indeed, solidstate computer memory uses arrays of flip-flops), so the outputs will stay in the set condition unless a *reset* pulse is applied to the R input.

The reset function is obtained by momentarily bringing the reset input LOW. This forces the outputs to go to a state in which the Q is LOW and the NOT-Q is HIGH.

The rules for the operation of the NAND-logic RS flip-flop are summarized in the truth table shown in Fig. 4-14D. This truth table also lists two additional conditions besides those discussed above. One of these is the condition in which both set and reset inputs are brought LOW simultaneously. This is a *disallowed* state, and the

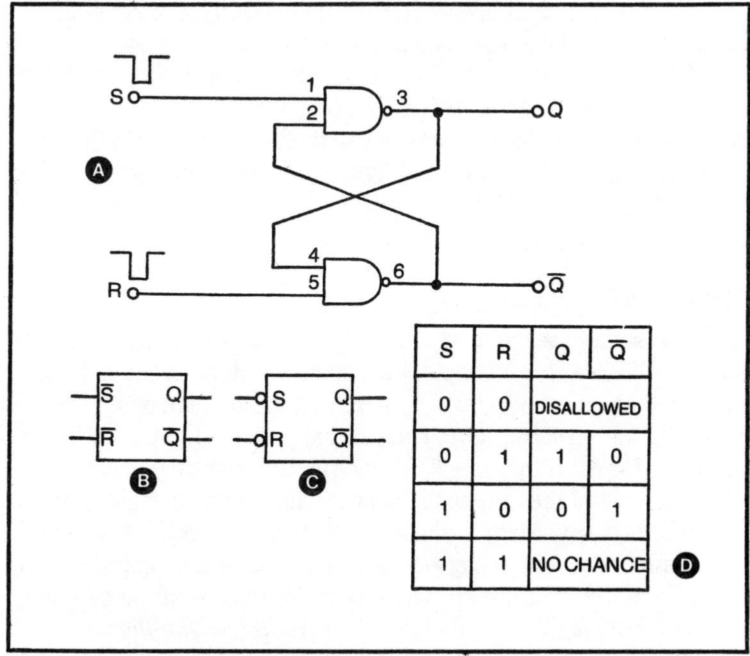

S	R	Q	$\overline{Q}$
0	0	DISALLOWED	
0	1	1	0
1	0	0	1
1	1	NO CHANCE	

Fig. 4-14. NAND logic RS flip-flop.

circuit will not know what to do; the output state will be unpredictable.

The other condition is the case where both inputs are simultaneously HIGH. In this condition we find that there is no change in the output state. The RS flip-flop simply remains in the condition present when the inputs were made HIGH.

A NOR-logic version of the RS flip-flop is shown in Fig. 4-15. This circuit may be constructed from TTL 7402 NOR gates. Like the 7400 device, the 7402 contains four independent two-input gates (in this case, the NOR variety). The circuit in Fig. 4-15 performs differently from the NAND-logic version of Fig. 4-14, but there are certain similarities (however, a slightly different set of operating rules prevails).

The rules governing the NOR-logic RS flip-flop are summarized in the truth table of Fig. 4-15C, but let us go over them briefly:

- ☐ If *both* inputs are LOW, then there is no change in the output state.
- ☐ If *both* inputs are simultaneously HIGH, then we have a disallowed state and the output condition is unpredictable.
- ☐ If the *set* input is made HIGH momentarily, then the output condition is Q = HIGH and NOT-Q = LOW.
- ☐ If the *reset* input is made HIGH momentarily, then the output condition is Q = LOW and NOT-Q = HIGH.

Note the principal difference between the two forms of RS flip-flop (examine the truth tables in Figs. 4-14 and 4-15 again). The NAND-logic RS flip-flop has *active-LOW* inputs, while the NOR-logic RS flip-flop has *active-HIGH* inputs.

Clocked RS Flip-Flops

We sometimes get into trouble with flip-flops that are too simple. We see, for example, electronic versions of the old *relay race* problem. In that problem, and its modern electronic version with digital circuits, two relays may have slightly different actuation times. If the time difference is such that they operate out of "sync" with the intended manner, then catastrophic results sometimes occur. Many of these problems are solved in the digital electronics world by using *clocked,* or *synchronous,* operation. In the case of the RS flip-flop, we obtain clocked operation by using the *master-slave flip-flop,* also called the *clocked RS flip-flop.*

The purpose of the *clock* (a train of pulses) is to synchronize the changes in the output condition by allowing them to occur only

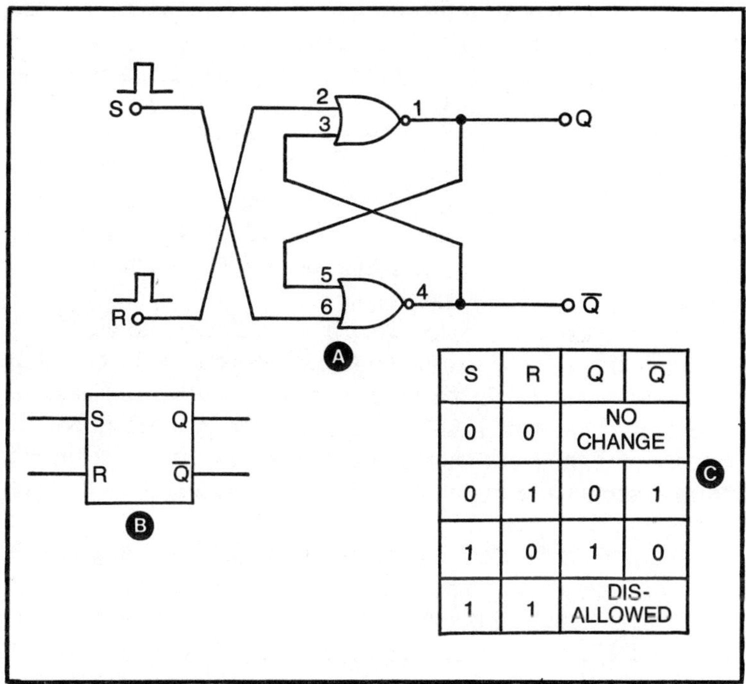

S	R	Q	$\overline{Q}$
0	0	NO CHANGE	
0	1	0	1
1	0	1	0
1	1	DIS-ALLOWED	

Fig. 4-15. NOR logic RS flip-flop.

at certain times during, or immediately following, a clock pulse. Most large-scale digital circuits will use synchronous operation in order to keep things straight.

There are two basic forms of clocking used in RS flip-flops: *level triggered* and *edge triggered.*

A level triggered flip-flop is one in which the output state changes in response to conditions on the inputs only when the clock input is HIGH (or LOW, depending upon type). Some level triggered circuits require the clock pulse to be LOW for it to be active, while others (the more usual case) require the clock pulse to be HIGH.

An edge triggered flip-flop will allow state changes only during one of the two transitions of the clock pulse. The pulse must be in the process of going from LOW to HIGH or from HIGH to LOW, (again depending upon type). A positive-edge triggered flip-flop, therefore, will allow output changes to occur only on the positive-going transition (LOW to HIGH) of the clock pulse. A negative-edge triggered flip-flop allows output transitions only on the negative-going (HIGH to LOW) transition of the clock pulse.

It is important to remember the differences between these two types of triggering, so let's reiterate:

Level triggering means that changes can take place only during the time when the clock pulse is active; i.e., either HIGH (positive-level triggered) or LOW (negative-level triggered).

Edge triggering means that output changes can take place only during the transition period of the clock pulse. A positive-edge triggered flip-flop changes only on the LOW-to-HIGH transition, while a negative-edge triggered flip-flop wants to see the negative going, or HIGH-to-LOW transition.

An example of a simple level triggered clocked RS flip-flop is shown in Fig. 4-16. The main flip-flop is the same as the circuit in Fig. 4-14, so it is shown here in block form for the sake of simplicity. The S and R inputs are controlled by a pair of NAND gates. When the clock pulse is LOW, then both inputs of the RS flip-flop section (i.e., points A and B) see a HIGH, so no changes can take place.

But, when the clock input goes HIGH the levels at points A and B (i.e., the S and R inputs of the FF section) are then controlled by the other inputs of the NAND gates. These inputs are used as the S and R inputs of the *clocked* FF. If you doubt this, then review the operation of the NAND gates discussed earlier in this chapter.

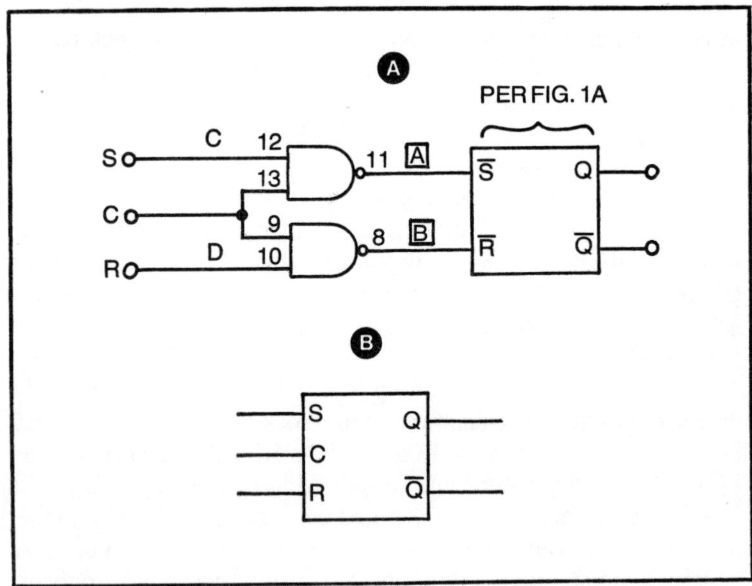

Fig. 4-16. Clocked RS flip-flop.

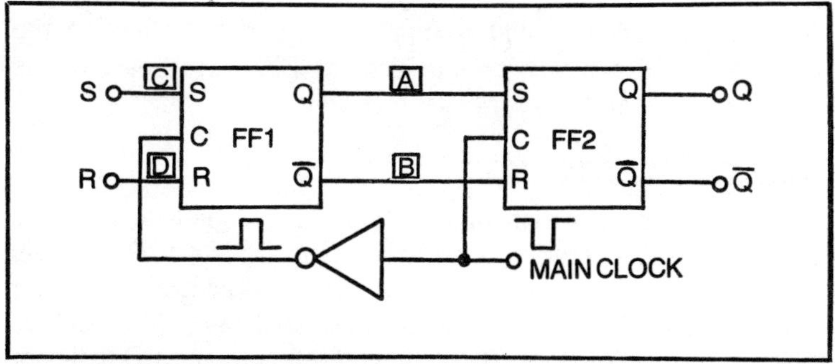

Fig. 4-17. Master-slave flip-flop.

Master-Slave Flip-Flops

The use of clocking helps a great deal in taming the RS flip-flop. But several problems, again electronic versions of the old relay race problem, can still pop up. Most of these are solved by using a slightly different approach, the so-called *master-slave flip-flop*. An example of the master-slave FF is shown in Fig. 4-17. This circuit allows only one output state change per clock pulse (the clocked RS FF allows continuous output state changes as long as the clock input is active).

The M-S flip-flop of Fig. 4-17 uses the clocked RS flip-flops of the previous example connected in cascade. The inverter shown in Fig. 4-17 allows us to drive the clock inputs of the two clocked RS FFs out of phase with each other.

Recall that the clocked RS flip-flop can change its output state only when the clock input is HIGH, and then only in response to conditions on the R and S inputs. In the M-S FF, the main clock is kept HIGH, so FF2 is active and FF1 is inactive.

When a clock pulse is applied, in this case a negative transition, FF1 will become active and FF2 will become inactive. Note that the effect of the inverter is to make the clock input of FF1 HIGH at this time. Any commands placed on the S and R inputs will cause changes in the outputs of FF1 (i.e., points A and B in Fig. 4-17).

But because FF2 is inactive at this time (its clock input is LOW), changes at A and B are not yet reflected at the Q and NOT-Q outputs of FF2. Once the clock pulse has evaporated, the clock input of FF2 goes HIGH again, so the changes that took place on A and B can be transferred to action at the Q and NOT-Q outputs.

Synchronization occurs by keeping FF2 inactive when the input stage (FF1) is being set up, and then rendering FF1 inactive (forbidding further S and R input changes from affecting the output), while transferring the data to FF2. This part of the sequence is called a *load-transfer operation.*

SOME FLIP-FLOP EXPERIMENTS

The experiments in this section are designed to help you understand the operation of certain simple flip-flops. Once again we are using the Powerace 102 digital breadboard for our experiments. Recall that designations such as "S2" refer to *switches* on the Powerace 102, while L2 refers to an LED indicator on the Powerace 102. The LED indicator will be turned off for a LOW and turned on for a HIGH.

Experiment No. 13: NAND Logic RS Flip-Flop

Make an RS flip-flop using the 7400 TTL quad two-input NAND gate. Only two sections of the 7400 are needed in this experiment:

☐ Connect the circuit of Fig. 4-14 using the pin numbers shown. The V+ of the 7400 is pin number 14, and must be connected to the +5 volt terminal on the Powerace 102. Ground on the 7400 is pin number 7, and must be connected to the ground terminal of the 5 volt supply on the Powerace 102.

☐ Connect the Q output of the RS flip-flop (IC pins 3 and 4) to indicator L3, and connect the NOT-Q output of the flip-flop (IC pins 2 and 6) to indicator L4.

☐ Connect the set input of the RS flip-flop to the NOT-Q terminal of switch S5 and the reset input of the RS flip-flop to the NOT-Q terminal of switch S6.

☐ Turn on the Powerace 102. If L4 is not lighted, then momentarily depress switch S6.

☐ Momentarily depress switch S5. This action will apply a negative-going pulse to the set input.

☐ Note L3 and L4. What has happened? Is Q (i.e., L3) HIGH or LOW, is NOT-Q (i.e., L4) HIGH or LOW?

☐ Momentarily depress switch S6 and note any changes in L3 and L4.

☐ Compare these results with the truth table for the NAND-logic RS flip-flop.

Experiment No. 14: NOR-logic RS Flip-Flop

Repeat experiment number 13 using two sections of the 7402 quad two-input NOR gate instead of the NAND gate device. The correct pinouts are shown in Fig. 4-15. Again, use pin 14 for +5 volts and pin 7 for ground.

Recall that the NOR-logic RS flip-flop uses active-HIGH inputs. This requires that we use the Q terminals of S5 and S6 instead of the NOT-Q terminals as were used in experiment 13.

Experiment No. 15: Clocked RS Flip-Flop

In this experiment we will use all four sections of a 7400 quad two-input NAND gate IC to make a clocked RS FF (see Fig. 4-16 for pinouts). Two sections of the 7400 are connected as an ordinary NAND-logic RS flip-flop, while the other two sections are used as clocking gates:

☐ Connect the clocked RS flip-flop circuit of Fig. 4-16, using a 7400 IC

☐ Pin 14 is connected to the +5 volts DC terminal on the Powerace 102, and pin 7 of the 7400 is connected to the negative side of the 5 volt power supply on the Powerace 102

☐ Connect the Q output of the RS flip-flop to L3 and the NOT-Q output of the RS flip-flop to L4.

☐ Connect the clock input of the RS FF to thc Q terminal of the pulse generator P1.

☐ Connect the set input of the RS FF to S3 and the reset input of the RS FF to S4.

☐ Turn on the Powerace 102.

☐ Note L3 and L4.

☐ Set both S3 and S4 to "0".

☐ Press P1 and release. This clocks the RS FF.

☐ Note what output changes occurred.

☐ This is either a disallowed state or a no change state; can you tell which?

☐ Set both S3 and S4 to "1".

☐ Press P1 and release. This again clocks the RS FF.

☐ Set S3 to "1" and S4 to "0".

☐ Press P1 to clock the RS FF.

☐ Does this set (L3 on) or reset (L4 on) the RS FF?

☐ Set S3 to "0" and S4 to "1".

☐ Press P1 to clock the RS FF.

☐ Compare the results of this experiment with the truth table for the clocked RS flip-flop.

Experiment No. 16: Master-Slave Flip-Flop

Connect two sections of a 4043 CMOS IC as a master-slave flip-flop, and then do experiment number 15 over again. You may use a single section of the 4043 device as the inverter (see Fig. 4-17).

Experiment No. 17

Perform experiment number 16 with a CMOS 4044. Explain the differences in operation.

Experiment No. 18: Direct Set/Direct Clear Inputs For The FF

Some commercial flip-flop circuits have *direct set* (also called *preset*) and *direct clear* (also called *preclear*) inputs. The direct set input is used to preset the master-slave flip-flop (i.e., force the outputs to a state of Q = HIGH and NOT-Q = LOW). Similarly, the direct clear input will reset the MS FF (i.e., force Q = LOW and NOT-Q = HIGH). These inputs operate independently of the clock.

In this experiment we are asking *you* to design a way to provide direct set and direct clear inputs. You will need to build an M-S flip-flop (*hint:* the output stage of FF2 in the M-S FF will need two, three-input NAND gates (i.e., 7410 instead of a 7400 device).

ADDITIONAL TYPES OF FLIP-FLOP

Thus far we have considered two versions of the RS flip-flop (NAND-logic and NOR-logic) and two flip-flops that are derivative of the RS circuits; i.e., the clocked RS flip-flop and the master-slave flip-flop. In the sections to follow we will consider some more complex types of flip-flop: Type-T FF, J-K FF, and the Type-D FF. The last circuit, the Type-D, is also sometimes called a *latch* circuit for reasons that will become apparent shortly.

Type-T Flip-Flops

The type-T flip-flop (also called the *toggle FF*) is shown in Fig. 4-18. This FF circuit can be constructed by providing feedback connections (as shown) around an ordinary master-slave flip-flop. Recall that the M-S FF was constructed from a pair of RS FFs and an inverter stage. Note that the Q output is fed back to the reset input and the NOT-Q output is fed back to the set input.

The type-T flip-flop functions as a binary divider; that is, the output signal has a frequency that is one-half (i.e., divided by 2) of the input signal. The timing diagram for this circuit is shown in Fig. 4-18B. Note that the Q output changes state only on negative-going transitions (i.e., HIGH-to-LOW) of the clock pulse. At the first

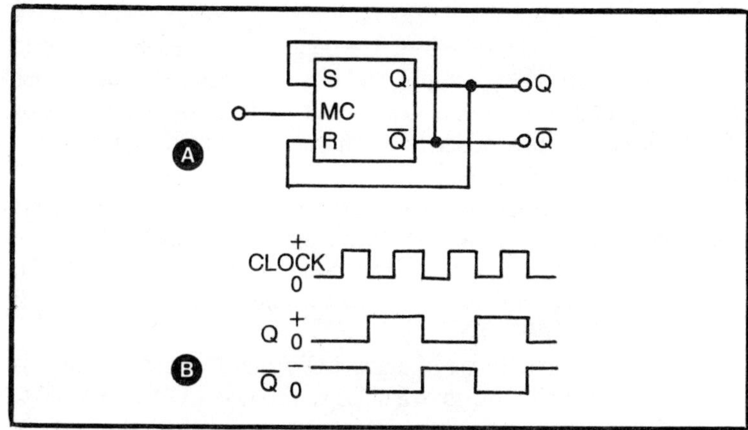

Fig. 4-18. Flip-flop wired for binary division.

negative transition, the Q output will snap HIGH and remain HIGH until the clock input sees *another* negative transition. This condition occurs at pulse number 2, at which time the Q output goes LOW again. We therefore have binary division of the input frequency; one output pulse is produced for each *two* input pulses.

There are sometimes differences in terminal designations from one text or spec sheet to another. In Fig. 4-18A, for example, we have labeled the clock input MC for "main clock." But it is also likely that you will see T for "toggle," or C_p for "clock."

J-K Flip-Flops

One of the most useful, and perhaps most common, forms of clocked FF is the J-K flip-flop. There are several advantages to the typical J-K FF: a) there are no invalid or disallowed states in the clocked mode; b) it can cause the outputs to complement; and c) it can provide non-clocked operation (in some IC versions).

Figure 4-19 shows one of several popular ways to represent the J-K FF. In this case, we see that it is a type-T FF with the feedback to the set and reset inputs controlled by a pair of two-input AND gates. One input from each gate accepts the feedback lines, while the remaining inputs of the gates are used to form the J and K inputs of the FF, respectively.

Figure 4-19B shows the circuit symbol for a J-K flip-flop. Not all versions of the J-K will have the direct mode inputs (preset and preclear). These inputs do, however, make it a more useful device. The *preset* input may also be called a *direct set* input, and the *preclear* input called a *direct clear* input.

Direct Mode Operation. The operation of the J-K flip-flop in the direct mode is very simple, and it is independent of conditions applied to the J and K inputs. The direct mode is controlled only by conditions on the preset and preclear input terminals, and the rules are summarized in Fig. 4-19D.

The direct mode inputs are active when LOW, so the only disallowed state occurs when both are simultaneously LOW.

If the preset input is LOW and the preclear input is HIGH, then the outputs immediately go to a condition where Q is HIGH and NOT-Q is LOW.

If the preclear input is made LOW and the preset input is HIGH, then the outputs go to a state where Q is LOW and NOT-Q is HIGH.

It is a general rule when dealing with flip-flops of any type that set or preset operations make the Q output HIGH and the NOT-Q output LOW, while clear or reset operations work in just the opposite manner (i.e., Q LOW and NOT-Q HIGH).

If both preset and preclear inputs are made HIGH, then the flip-flop is ready for normal clocked operation.

Clocked Operation. Whenever the preset and preclear inputs (where used) are simultaneously HIGH, the J-K FF will operate in the clocked mode. The rules for clocked operation are summarized in Fig. 4-19C.

Like the Type-T FF, the J-K FF (in the clocked mode) responds on the negative-going transition of the clock pulse. No output changes will occur, regardless of changes at the J and K inputs, until one of these negative-going clock pulse transitions is seen. The outputs will then respond according to the J-K input conditions. The rules for clocked operation are as follows:

- ☐ If both J and K are LOW, then the FF is inert and does nothing. No changes will occur in the outputs.
- ☐ If J is LOW and K is HIGH, then the clocking will make Q LOW and NOT-Q HIGH.
- ☐ If J is HIGH and K is LOW, then the clock pulse transition makes Q HIGH and NOT-Q LOW.
- ☐ If Both J and K are HIGH, then the J-K FF behaves much like a type-T FF; clocking complements the outputs. This means that negative-going clock pulse transitions force the outputs to go to the opposite state. The output waveform of the J-K flip-flop is then identical to the output waveform of the Type-T flip-flop given in Fig. 4-18.

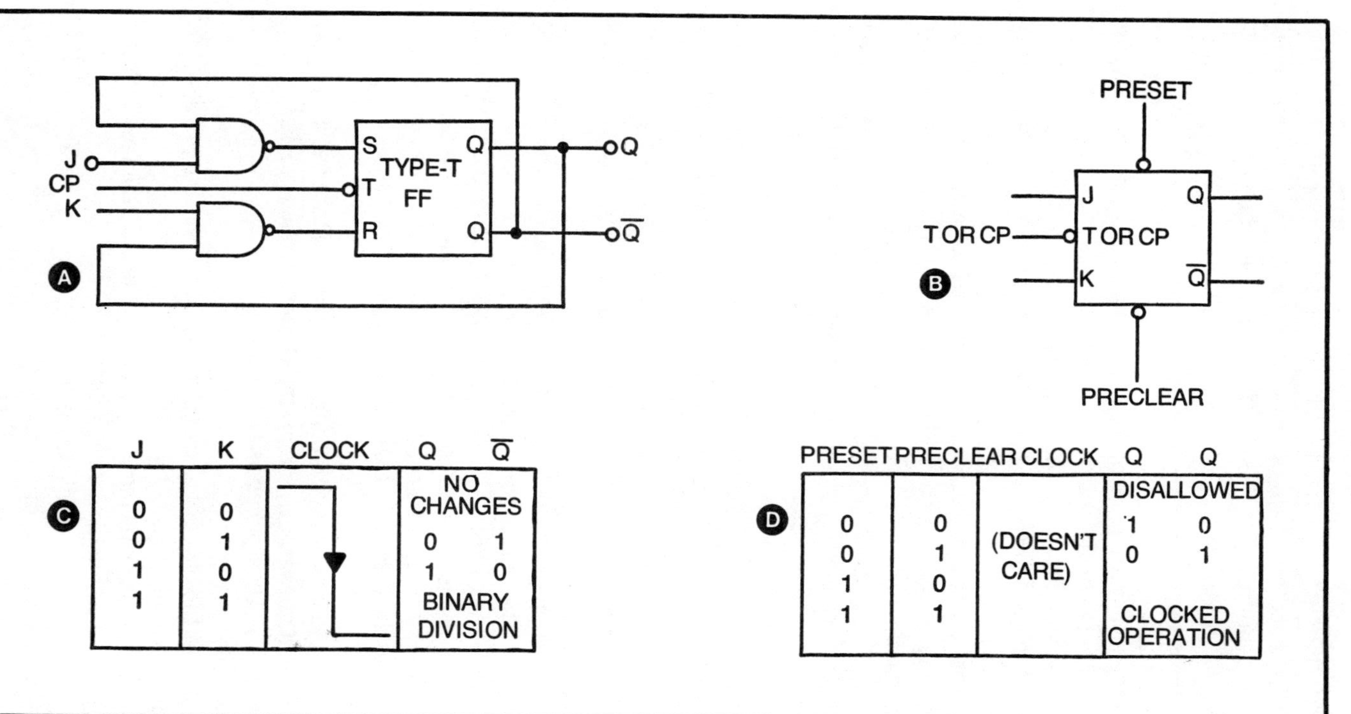

Fig. 4-19. J-K flip-flop.

Type-D Flip-Flop

The type-D flip-flop is shown in Fig. 4-20. The equivalent circuit is shown in Fig. 4-20A, while the usual schematic symbol is shown in Fig. 4-20B.

The equivalent circuit consists of a clocked RS FF in which the set and reset inputs are fed by the same signal, but are 180 degrees out of phase with each other (i.e., complementary inputs). An inverter between the S and R lines accomplishes this neat trick.

The common line to the reset-set inverter is called the *data* or D input instead of clock. This input is usually labeled D on most schematics.

The rules for operation of the type-D FF are very simple. Data appearing on the D input will be transferred to the Q output only when the *clock* line is HIGH:

☐ If the clock line is HIGH, then the output will *follow* changes in the input signal (i.e., changes on the D input). When the D line goes HIGH, then the output will go HIGH. Similarly, when the D line goes LOW the outputs follow by also going LOW.

☐ If the clock line is LOW, then the output will retain the last data that existed on the D input at the instant the clock line dropped LOW.

These rules can also be seen in the timing diagram of Fig. 4-20C. Read the description below, keeping in mind the two rules just given:

☐ When the first clock pulse arrives (T1-T2), the D input is LOW, so the Q output will be LOW.

☐ During interval T2-T3, the D input goes HIGH, but since no clock pulse is present it cannot affect the output conditions.

☐ At the beginning of interval T3-T4, clock pulse number 2 is HIGH but the D input is LOW. The output, therefore, must remain LOW.

☐ About midway through clock pulse number 2, however, the D input goes HIGH, forcing the Q output to also go HIGH.

☐ The Q output stays HIGH even after clock pulse number 2 goes LOW.

☐ At the onset of clock pulse number 3, the D input is LOW, so the Q output drops LOW also.

☐ The pulse on the D input during the interval T6-T7 cannot affect the Q output because the clock is LOW.

The so-called *data latch* is a special case of the Type-D FF. This device is used in digital readout circuits (i.e., in frequency counters) to hold current data until updated data is ready for

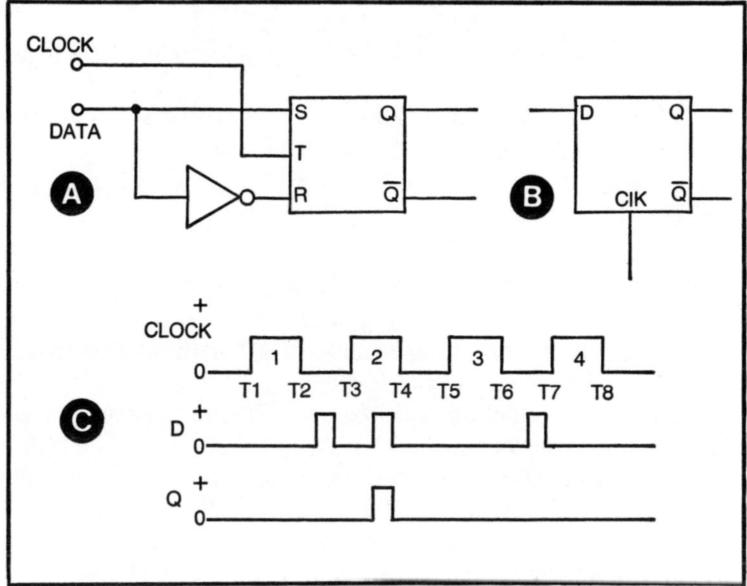

Fig. 4-20. Type-D flip-flop.

display. This gives the illusion that the data is updated instantaneously. In most cases, the clock input is called a *strobe* input. Data at the D input will be transferred to the outputs only when the strobe line is HIGH. The idea is to momentarily bring the strobe line HIGH when the data at the input is valid, and then let the strobe line go LOW again until the next instrument update is ready.

EXPERIMENTS IN CLOCKED FLIP-FLOPS

In the experiments to follow we will investigate the behavior of the various clocked flip-flops discussed in the preceding sections. A type 7476 FF is selected as "representative" because it has both preset and preclear inputs. Connect the 7476 device into the circuit of Fig. 4-21. Once again, the pin number callouts are for the AP Products Powerace 102 digital breadboard.

Experiment No. 19: Direct-Mode J-K Operation

- ☐ Connect the circuit of Fig. 4-21.
- ☐ Set the following conditions:
  - —J (S1) and K (S2): HIGH
  - —*Set* (S3) and *clear* (S4): LOW
  - —Clock: OFF.

☐ Observe Q (L3) and NOT-Q (L4). Q should be LOW (L3 off) and NOT-Q should be HIGH (L4 on).

☐ Turn on the digital clock. There should not be any changes in the LEDs.

☐ Set J to HIGH. Any change? Set J to LOW, K to HIGH. Any change? Set both J and K to HIGH. Any Change?

☐ Repeat all steps, but with *set* (S3) LOW and *clear* (S4) HIGH.

Experiment No. 20: Clocked J-K Operation.

☐ Connect the Circuit of Fig. 4-21.

☐ Set up the following conditions: a) *Set* (S3), *clear* (S4), J (S1), and K (S2): HIGH; b) Clock: off.

☐ Turn the clock on, and observe L2 (the clock signal) and L3 (Q output signal). Note the binary division action.

☐ Make J LOW and K HIGH. Observe L3/L4.

☐ Make J HIGH and K LOW. Observe L3/L4.

Experiment No. 21: Complementing the Output on Clock Pulse

Connect the circuit of Fig. 4-21, except replace the clock input of the FF with S6-Q. This change will cause the clock input to go HIGH when S6 is pressed, and then drop back LOW when S6 is released. S6 is used to allow us to manually clock the FF while observing the output.

☐ Set switches S1 through S4 HIGH.

☐ Momentarily bring S4 LOW, thereby clearing the FF. Q is now LOW and NOT-Q is HIGH (L3 off, L4 on).

☐ Set J LOW.

☐ Clock the FF by pressing S6 slowly several times.

☐ Did anything change on L3/L4? It shouldn't.

☐ Set J HIGH and K LOW.

☐ Clock the FF by pressing S6 slowly several times.

☐ Note that Q went HIGH on the first negative-going transition (release of S6), but nothing happened on subsequent clock pulses.

☐ Set S1 through S4 HIGH.

☐ Set FF (bring S3 LOW momentarily). Q should be HIGH and NOT-Q is LOW (L3 on and L4 off).

☐ Set K LOW.

☐ Clock the FF by pressing S6 slowly several times.

☐ Set K HIGH and J LOW.

☐ Clock the FF by pressing S6 slowly several times.

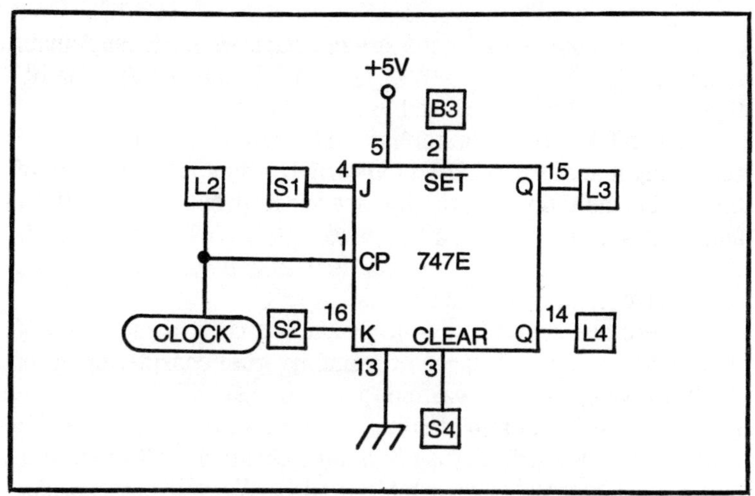

Fig. 4-21. Experiment No. 19: Direct mode J-K operation.

□ Note Q and NOT-Q (L3 and L4) changes and compare with the previous Q. Note the difference.

MULTIVIBRATORS (THE HEART OF THE TIMER CIRCUIT)

Thus far in this extensive chapter, we have discussed the various digital IC logic families, assorted types of gates, and the flip-flops. We will now turn to the topic of multivibrators. The multivibrator is the heart of the timer circuit, so some of the material may repeat (for emphasis) material found elsewhere in the book.

A multivibrator is basically a pulse-producing circuit. There are three basic forms of multivibrator: *monostable, bistable,* and *astable.* It takes little imagination to detect that these designations refer to the stable output states possible for each type of circuit.

The monostable multivibrator has but *one stable state* (usually the state in which Q = LOW—but not always). Triggering the monostable multivibrator causes Q to go HIGH for a time, but since this is not a stable state Q will drop LOW again when a predetermined time period has elapsed. Monostable multivibrators are also called *one-shot* circuits; and sometimes (erroneously, albeit graphically) *pulse stretcher* circuits. The latter label is a misnomer because the circuit does not actually stretch a pulse, but instead generates a *new* pulse that has a longer period.

The bistable multivibrator has *two stable states.* It can remain in either state (i.e., Q = LOW or Q = HIGH) indefinitely. The RS flip-flop is an example of a bistable multivibrator.

The astable multivibrator has *no stable states*. It is incapable of remaining in either Q=LOW or Q=HIGH states. The Q output of the astable multivibrator will flip back and forth between the HIGH and LOW states producing a squarewave pulse train output signal. For this reason, the astable circuit is usually used to produce the clock pulses found in digital circuits.

There are several ways to produce each of these types of multivibrator. Lack of space prevents us from considering all of them. However, we will examine a few circuits built from discrete gates and the integrated circuits. Some IC devices, like the 555 timer (which we will discuss in a later chapter), will operate in either the monostable or astable modes. We will not consider circuits in which discrete transistors and resistors form the multivibrator.

When we speak of bistable multivibrators we are actually talking about the RS flip-flop. Recall from the earlier sections of this chapter that the RS FF can remain happily in either Q = LOW or Q = HIGH states. These conditions only change when an input signal commands the circuit to change. It can remain in either state indefinitely.

The monostable multivibrator, or one-shot as it is called, has but one stable state. In most circuits, the stable state is Q = LOW. When the input of the monostable is triggered, the output will snap HIGH for a certain period of time and then drop LOW again. The monostable multivibrator, then, produces one output pulse for each pulse received at the trigger input. This is why the monostable multivibrator is sometimes called the *one-shot.*

There are a number of reasons why you might want a "one-out-for-one-in" circuit. One of the most common is to stretch pulses. A very rapid input pulse, even down to the nanosecond range, can be used to trigger monostable circuits that produce output pulse periods from nanoseconds to days. The duration of the output pulse is usually much longer than the duration of the input trigger pulse. In so-called pulse stretcher circuit applications, the output of the one-shot is used to substitute for the shorter trigger pulse.

Most monostable MVs will not respond to further input trigger pulses until the period of the output pulse has "timed out"; i.e., the output has returned to its stable state. Monostables that will not respond to further trigger commands until the output duration has expired are called *nonretriggerable* monostables.

Retriggerable One-shots

Some one-shot circuits are retriggerable, meaning that they will respond to further input trigger commands, while the one-shot is in the unstable state (i.e., before it has timed out). Consider Fig. 4-22 to see how this might work. Figure 4-22A shows the operation of the regular nonretriggerable one-shot multivibrator. The first trigger pulse causes the output to go HIGH, and it remains HIGH for period T. A second trigger pulse has no effect on the one-shot because it occurs before T expires.

Now consider Fig. 4-22B. This is a timing diagram for the retriggerable monostable multivibrator. The output goes HIGH when the first pulse arrives. But before T expires, a second trigger pulse is received. This second pulse causes the one-shot to retrigger, so the output will remain HIGH for an additional period T. Note that the total duration of the HIGH state is not 2T, but T plus the portion of the first period that expired prior to the second trigger (i.e., $T + (T_2 - T_1)$.

An example of a monostable multivibrator built from a CMOS Type-D flip-flop is shown in Fig. 4-23. Recall from earlier in this chapter the rules for the Type-D FF: 1) since D is HIGH, a HIGH will be transferred to the Q output when the *clk* line goes HIGH; and 2) when the *clear* line goes HIGH the Q output is forced LOW.

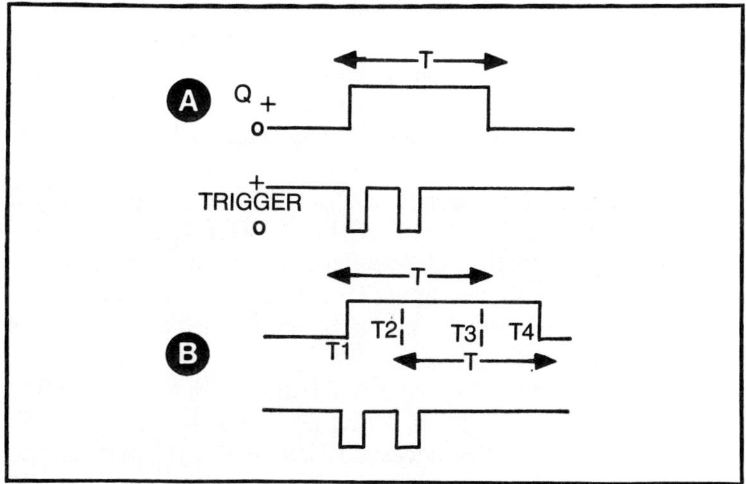

Fig. 4-22. One-shot waveforms: A) Normal; B) Retriggerable.

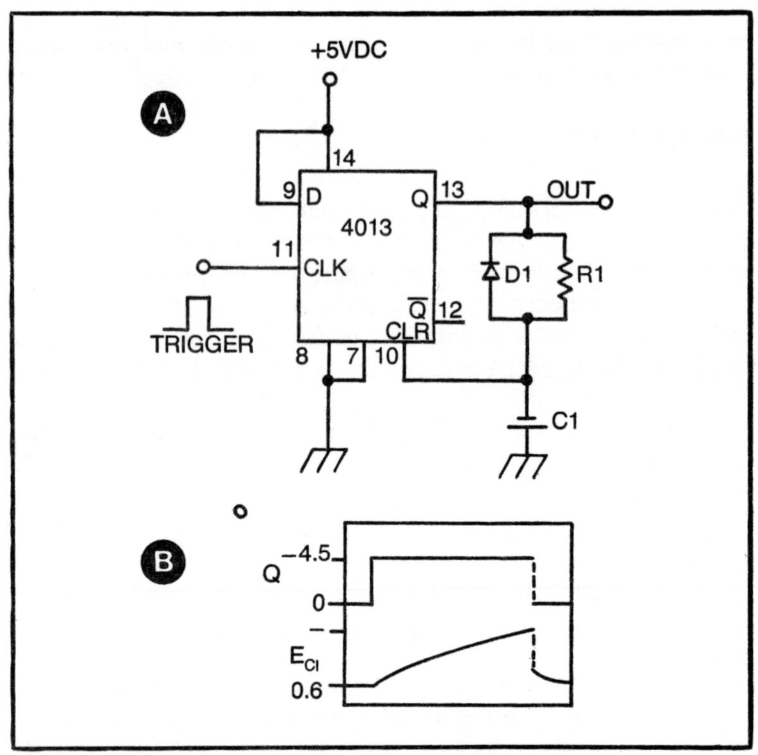

Fig. 4-23. 4013 Type-D FF used as a monostable multivibrator.

The operation of the one-shot circuit in Fig. 4-23, then, is as follows:

☐ When the circuit is at rest, Q is LOW and any charge on capacitor C1 is drained off through diode D1.

☐ When a trigger pulse is received by the *clk* input, Q goes HIGH. When Q is HIGH, capacitor C1 will charge through resistor R1.

☐ When C1 has charged to a potential of approximately 2 volts, the *clear* input thinks it is HIGH so the FF will force Q LOW.

☐ The period that Q was HIGH (i.e., the period of the one-shot) is determined by the time constant of R1C1, the potentials of the Q output, and the point at which the *clear* input thinks that it is HIGH instead of LOW.

The circuit in Fig. 4-23A uses a diode (D1) across the timing resistor (R1) to discharge C1 during the period when Q is LOW. This diode is not strictly necessary, but serves to speed up the circuit considerably. Without D1 the charge on capacitor C1 would bleed off through R1; but this would require another R1C1 time constant (or so) before the voltage across C1 would discharge enough to permit retriggering of the one-shot. The purpose of D1 is to rapidly discharge C1 so that retriggering can occur almost immediately after Q drops LOW (see the waveform in Fig. 4-23B).

But the use of D1 creates a little problem. The charge potential across C1 cannot drop lower than the function potential of the diode (200 to 300 millivolts in germanium types and 600 to 700 millivolts for silicon types). Figure 4-24 shows a modified version of the circuit that uses a switching transistor (Q1) to discharge C1. The base of transistor Q1 is driven by the NOT-Q output of the flip-flop. When the one-shot is at rest NOT-Q is HIGH, so Q1 will be forward biased. When a trigger pulse is received, however, Q goes HIGH and NOT-Q becomes LOW. This condition turns off Q1, allowing C1 to charge. When the voltage on C1 reaches the threshold point, the FF sees a *clear* command (i.e., *clear* input HIGH) and this forces Q LOW and NOT-Q HIGH. Transistor Q1 then discharges C1.

Preview of the 555 Timer IC

One of the most popular IC timers on the market is a little device called the 555, originated by Signetics, Inc. The 555 is offered in the 8-pin miniDIP package (Fig. 4-25A). A typical one-shot circuit based on the 555 is shown in Fig. 4-25B.

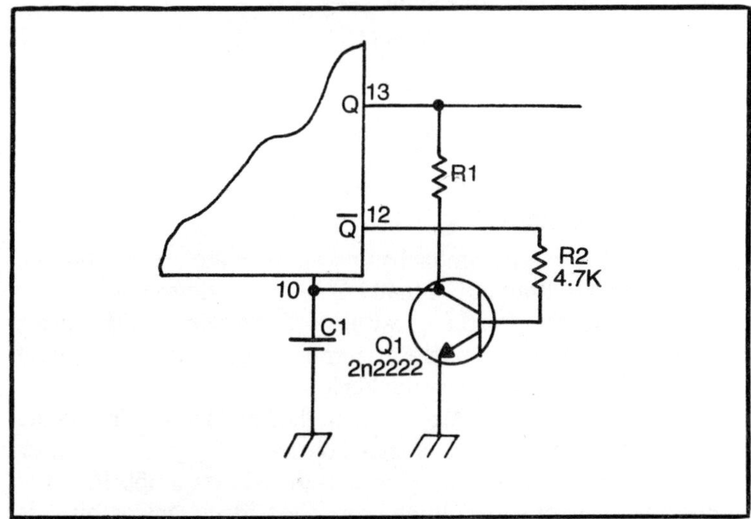

Fig. 4-24. Modified CMOS monostable multivibrator.

The 555 IC contains a control RS FF, two voltage comparators (COMP1 and COMP2), a switching transistor (Q1), and an inverter amplifier (A1).

During the rest period, when the one-shot is not active, the output is LOW. Since the output is an inverter, the NOT-Q terminal of the FF is used. During the reset period the NOT-Q will be HIGH.

The *reset* input of the 555 is active-LOW, so to disable it we must tie it to the V+ line, or a logical-HIGH point.

The trigger input (pin 2) is normally kept HIGH, which in this case is defined as a potential greater than two-thirds of V+. When triggering of the one-shot is desired, we must bring this trigger terminal down (i.e., create a negative-going transition) one-third of V+ or lower. This voltage is the threshold for COMP2.

When the trigger input is brought below the COMP2 threshold, the output of COMP2 goes HIGH forcing the FF into the *set* condition. The NOT-Q terminal, then, goes LOW forcing the output (pin 3) of A1 HIGH.

With the NOT-Q terminal of FF1 LOW, transistor Q1 is no longer turned on so capacitor C1 begins to charge from potential V+ applied to resistor R1. When the voltage across C1 reaches two-thirds of V+, then the output of COMP1 goes HIGH resetting FF1. The output of A1 then goes LOW again, and Q1 is turned back on forcing C1 to discharge rapidly to zero. The HIGH period of the output is given by the expression:

$$T_{sec} = 1.1\,(R1\,C1)$$

Where:

T is in seconds
R1 is in ohms
C1 is in farads

Astable Multivibrators

An astable multivibrator has no stable state. It will bounce back and forth (output HIGH then LOW), producing a squarewave signal. The astable MV, then, may be used to produce clock signals for digital instruments or as the master oscillator in certain types of signal generator or frequency marker.

One view of the astable, and a method used to produce some astable circuits, is that it is a one-shot that retriggers itself after the output pulse times out. Figure 4-26A shows a 555 IC timer connected as an astable multivibrator. The trigger control (pin 2) is connected across the capacitor so the 555 will retrigger: 1) when

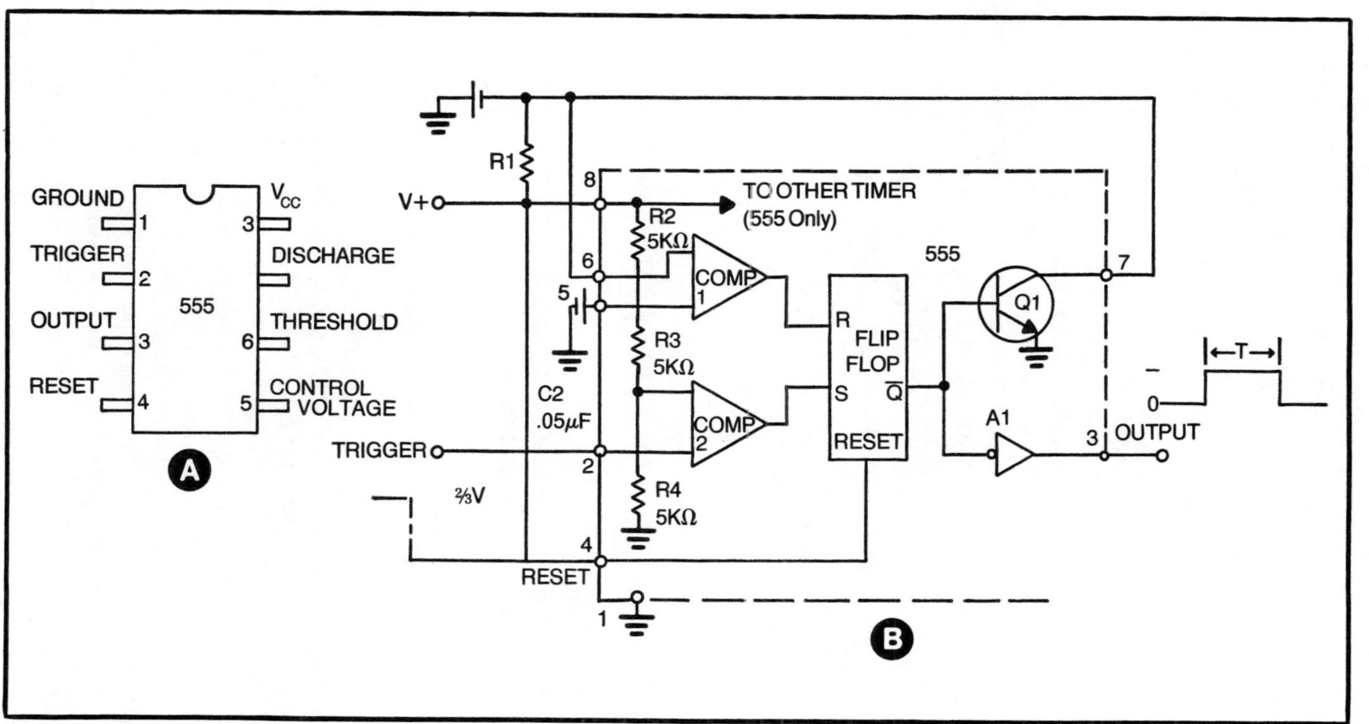

Fig. 4-25. The 555 IC timer.

power is initially applied; and 2) every time the device times out and the capacitor charge is dumped through the internal transistor.

In this circuit, capacitor C1 is charged through resistors R1 and R2 but is discharged through R2 only. This makes the LOW time of the output signal shorter than the HIGH time. These time periods, and the frequency of oscillation, are given by the expressions shown in the figure.

When capacitor C1 charges to two-thirds V+, the output goes LOW, and the discharge transistor Q1 is turned on. This will cause C1 to discharge through R2. When the voltage across C1 drops to one-third V+, the 555 will retrigger and the cycle is repeated.

But we do not necessarily need a 555 or similar device to form an astable multivibrator. We may, for example, use either TTL or CMOS gates or inverter circuits as the active circuit elements. In the former case, incidentally, it is common practice to use NAND gates (i.e., 7400) connected with both inputs shorted together, effectively forming an inverter. Figures 4-26B and 4-26C show typical TTL astable multivibrators capable of oscillating up to frequencies of 20 or 25 MHz (although 5 to 10 MHz are more reasonable). Both of these circuits are RC oscillators, so their stability at high frequencies is poor (at best). In many cases, however, we find that the capacitor is replaced with a piezoelectric crystal, so the circuit will oscillate on the crystal frequency.

Shift Registers

A flip-flop is able to store a single bit of digital data. When two or more flip-flops are organized to store multiple bits of data they constitute a *register*. Most registers are merely specially connected arrays of flip-flops.

There are several different circuit configurations that one would call a register, and we classify them according to the manner in which data is input to and output from the register. We have, for example, *serial-in-serial-out*(SISO), *serial-in-parallel-out* (SIPO), *parallel-in-parallel-out* (PIPO), and *parallel-in-serial-out* (PISO).

Figure 4-27 represents both SISO and SIPO shift registers. The only significant difference is that the parallel output lines used on the SIPO register would be absent on the SISO register.

The SIPO shift register consists of a cascade chain of Type-D flip-flops that have their clock lines connected together. Recall the rules for Type-D flip-flops: data can be transferred from the D-input to the Q output only when the clock input is HIGH. The input can change at will, and the output will remain the same as long

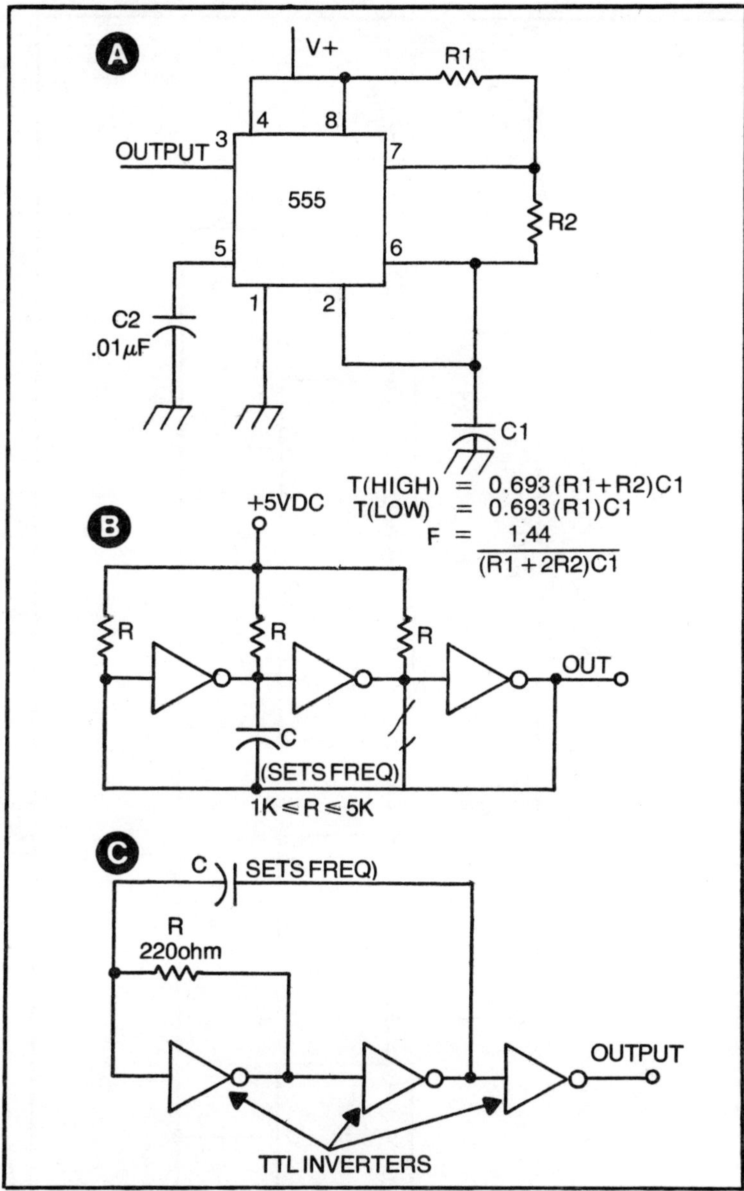

Fig. 4-26. Typical astable multivibrators.

as the clock line is LOW. But if the clock line goes HIGH, the Q output will follow the D input. The Q output will retain the last valid data present before the clock dropped LOW again.

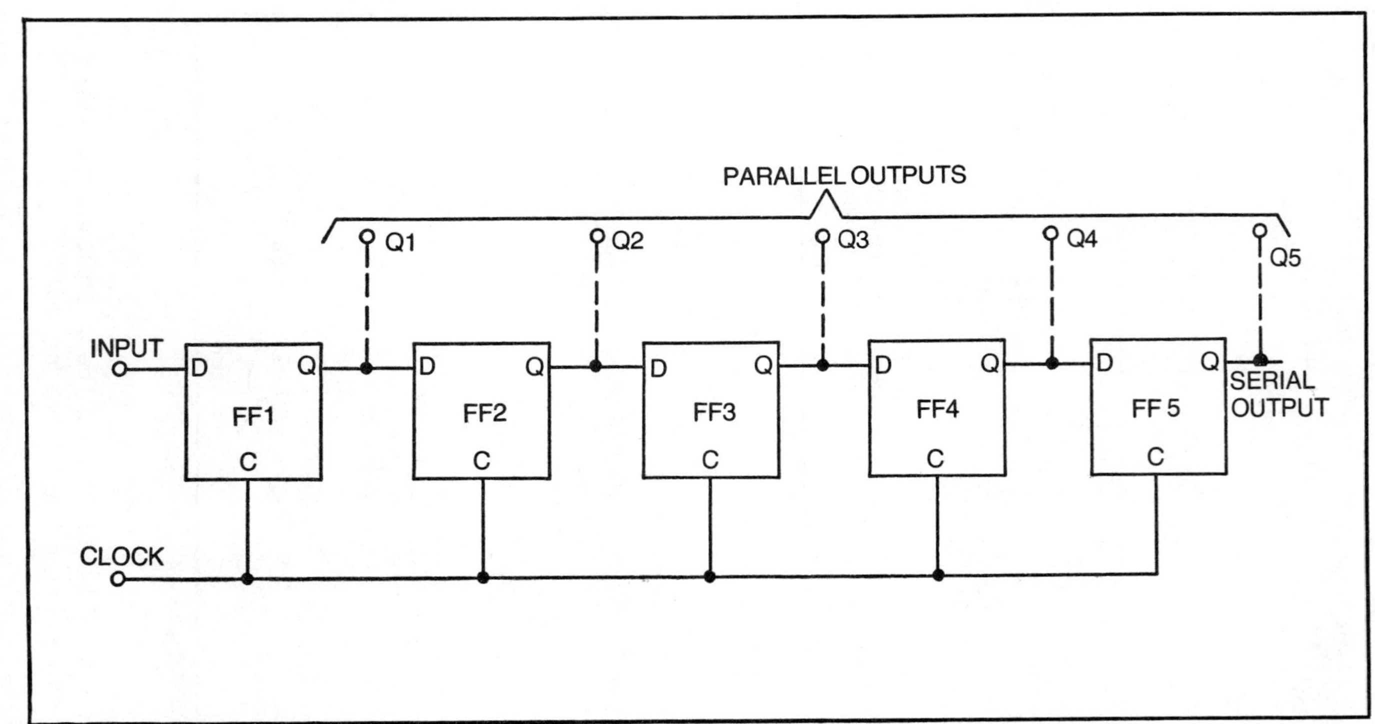

Fig. 4-27. Shift register.

This rule can be applied to the situation shown in Fig. 4-28, where we show the transmission of a single bit of data, from left to right, through a SISO shift register. At the occurence of the first clock pulse, the input line is HIGH. This point is the D-input of FF1, so a HIGH which is applied to the D input of the second flip-flop (FF2) remains after the clock pulse disappears.

When the second clock pulse arrives, FF2 sees a HIGH on its D-input, and FF1 sees a LOW on its D-input. This situation causes a LOW at Q1 and a HIGH at Q2. The third clock pulse sees a LOW condition on the D inputs of FF1 and FF2 and a HIGH at the input of FF3. The third clock pulse, then, causes Q1 and Q2 to be LOW and Q3 to be HIGH.

Note that the SISO input remains LOW after the initial HIGH during clock pulse number 1. This means that the single HIGH condition will be propagated through the entire SISO shift register, one stage at a time. The HIGH bit will shift one flip-flop to the right each time a clock pulse arrives. If the data at the input had changed, then the bit pattern at that input would have been propagated through the shift register.

The shift register in Fig. 4-27 is a five-bit, or five-stage, register (any bit length could be selected). On the sixth clock pulse, therefore the HIGH is propagated out of the register, so all flip-flops are now LOW.

The SISO shift register can be made into a SIPO device by adding parallel output lines at Q1, Q2, Q3, Q4, and Q5.

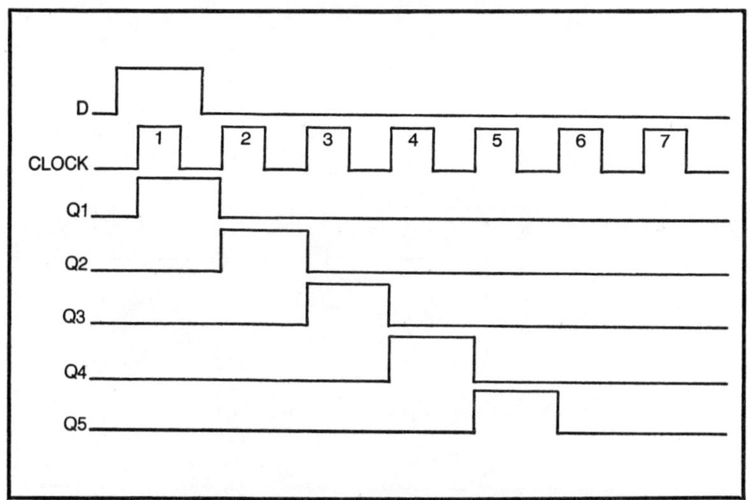

Fig. 4-28. Timing diagram for shift register operation.

One use for the SIPO register is serial-to-parallel binary code conversion. For economic reasons, digital data is usually transmitted as a serial stream of bits; i.e., the bits of the digital word are sent over a communications link. But most computers and other digital instruments use a parallel form of data entry. Parallel data transfer is more expensive but is considerably faster than serial transmission. If, for example, we have an eight-bit system, we would need an eight-stage SIPO shift register to convert the serial code to parallel form. The code is entered into the SIPO register one bit at a time, so that after eight clock pulses the first bit will appear at Q8 and the last bit at Q1.

Parallel entry shift registers are faster to load than serial input shift registers. This is because a single bit can be changed, if needed. In the serial type, to change a single bit of data requires us to ripple through the entire contents.

There are two basic forms of parallel data entry: *parallel* and *jam.* In *parallel entry,* shown in the partial schematic of Fig. 4-29, the register must first be cleared (i.e., all bits set to zero) by bringing the *reset* line momentarily LOW. The data that is applied to inputs B1 through B_n can be loaded into the register by momentarily bringing the *set* line HIGH.

The *jam entry* circuit of Fig. 4-30 is able to load data from bits B1 through B_n onto the other inputs. While this may not look superior at first glance, it is, because IC shift registers using this technique have internal inverter stages at the complement inputs.

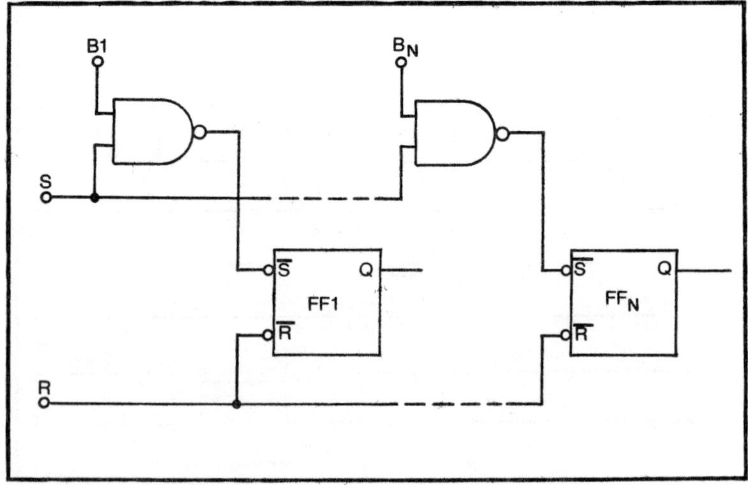

Fig. 4-29. Parallel entry.

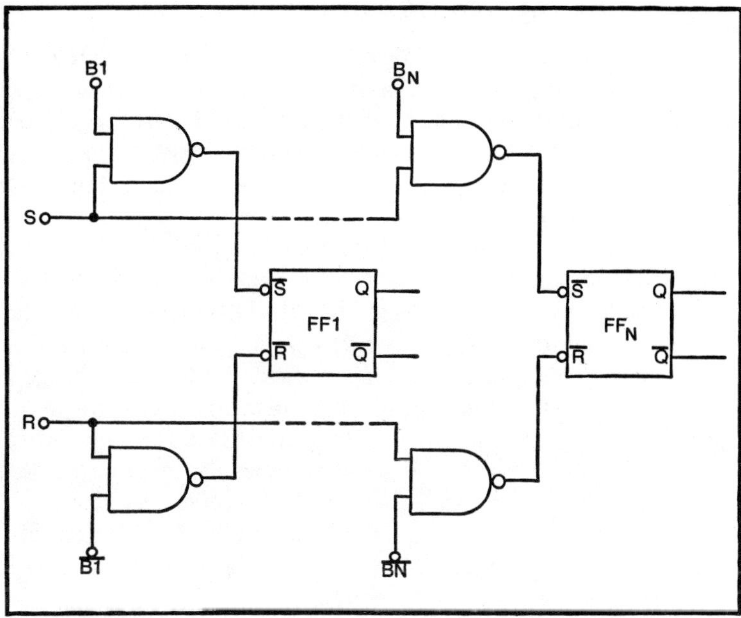

Fig. 4-30. Jam entry.

These have their inputs connected to the non-complemented inputs, so the outside user never sees the complementing process.

A recirculating shift register is shown in Fig. 4-31. Since the output of a serial shift register allows the outside world to see only one bit at a time, we must empty the entire contents of the shift register in order to read these contents. But that would ordinarily destroy the data because the input would be HIGH or LOW during the entire operation. A single read operation, then, would fill up the register with all one's or zero's. The recirculating shift register connects the output (serial) back to the input, so that a read operation would automatically rewrite the data back into the shift register.

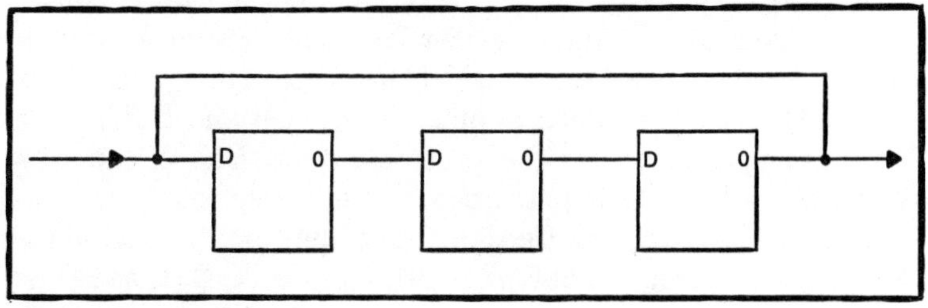

Fig. 4-31. Recirculating shift register.

DIGITAL COUNTERS

A digital counter is a device or circuit that operates as a frequency divider. The most basic form of digital counter is the J-K flip-flop connected with the J and K inputs tied HIGH (i.e., placed in the clocked mode). This makes the output produce one output pulse for every two input pulses. It is, then, a binary or divide-by-two counter.

Those fancy digital frequency/period counters are nothing more than digital divide-by-10 counters connected such that the binary coded output is converted to a decimal display. There are two basic classes of digital counter circuits, *serial* and *parallel*. The serial counters are called *ripple counters* because a change in the input must ripple through all stages of the counter to its proper point. (In a ripple counter the data is transferred serially, which means that the output of one stage becomes the input of the next stage.) Parallel counters are also called *synchronous counters*.

The basic element in most counters is the J-K flip-flop (Fig. 4-32A). Note in the figure that the J and K inputs are permanently tied HIGH, so they will remain active.

A timing diagram for this divide-by-two circuit is shown in Fig. 4-32B, and it shows the action of the circuit. J-K FF outputs change state on negative-going transitions of the clock pulse. In Fig. 4-32B the first negative-going transition causes the Q output to go HIGH. Q will remain HIGH until the input sees another negative-going clock pulse. At that time the output will drop LOW. The action required to make a complete output requires two clock pulses, so this J-K flip-flop is dividing the input frequency by two.

We can make a binary ripple counter by cascading two or more stages, as shown in Fig. 4-33A. This particular circuit uses four J-K FFs in cascade. Any number, however, could be used. The major problem with this type of counter is that only those division ratios that are powers of two can be accommodated. In the four-stage circuit shown, the possible division ratios are 2, 4, 8 and 16.

Frequency division is one major use for a counter circuit. In some electronic instruments, for example, we may want to prescale a frequency, i.e., divide it to a lower frequency that can be handled by a digital counter or other digital instrument. But this is only one application for the counter circuit. One of the most common applications is to *count;* i.e., tell us the total number of pulses that have passed. Consider again the circuit of Fig. 4-33A and the timing diagram of Fig. 4-33B. Outputs A, B, C and D are coded in binary, with A being the least significant bit and the D the

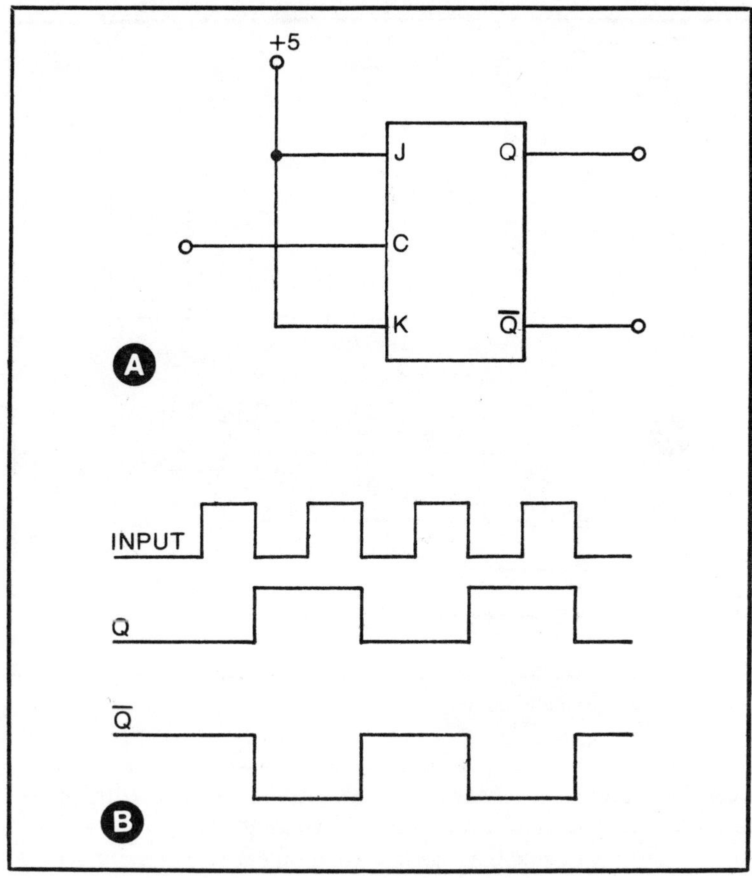

Fig. 4-32. J-K flip-flop wired for binary division.

most significant. These are weighted in a 1-2-4-8 code system to represent decimal (or hexadecimal) digits 0 to 15. These are the normal weights of the binary number system.

Consider the timing diagram of Fig. 4-33B. Note that all W output changes occur following the arrival of each pulse. After pulse number one has passed, the Q_A line is HIGH and all others are LOW. This means that the binary word on the output lines is 0001_2 (i.e., 1_{10}); one pulse has passed.

Following pulse number 2 we would expect 0010_2 (i.e., 2_{10}) because two pulses have passed. Note that Q_B is HIGH and all others are LOW. The digital word is, indeed, 0010_2. If you follow each pulse, you will find that the binary code present at the outputs of the counter is as shown in Table 4-1.

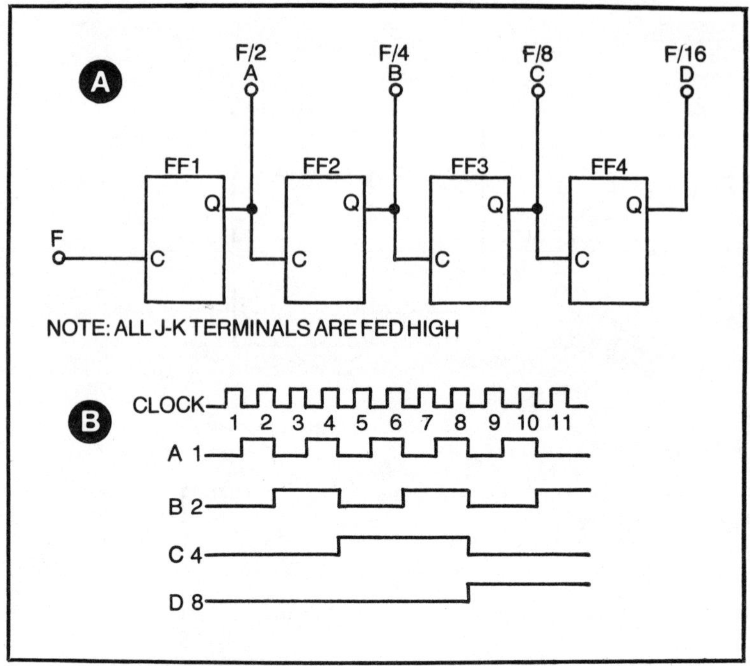

Fig. 4-33. Four-bit binary counter.

The counter in Fig. 4-33A is called a *modulo-16* counter, or a *base-16* counter, or a *hexadecimal* counter (all meaning the same thing). The output of a hexadecimal counter can be decoded to drive a display device that indicates 0 through 9 (i.e., decimal) or 0 through F (hexadecimal). In most applications where a real, live, human-type "people" is to read the display a decimal readout is provided.

Decimal Counters

A decimal counter operates in the *base-10*, or *decimal,* number system. The most significant bit of a decimal counter produces one output pulse for every ten input pulses. Decimal counters are also sometimes called *decade counters.* The decimal counter forms the basis for digital event, period, and frequency counter instruments. Thus, the hexadecimal counter in Fig. 4-33 is not suitable for decimal counting unless it is modified for base-10 operation.

Figure 4-34A shows a TTL hex counter modified by adding a single TTL NAND gate. Recall that a TTL J-K FF uses inverted inputs for the clear and set functions. As long as the clear input remains HIGH, the flip-flop will function normally. But when the clear input is momentarily brought LOW, then the Q output of the FF goes LOW.

The decade counter in Fig. 4-34A is connected so that all four clear inputs are tied together to form a common clear line. This line is connected to the output of a TTL NAND gate (i.e., one section of a 7400 device). Recall the rules of operation for the TTL NAND gate: if either input goes LOW, then the output goes HIGH. But if both inputs are HIGH, then the output goes LOW.

The idea behind the circuit of Fig. 4-34 is to clear the counter to 0000 following the tenth input pulse. Let's examine the timing diagram in Fig. 4-34B to see if the circuit does the correct thing. Up until the 10th pulse, this diagram is the same as for the base-16 counter discussed previously.

The output of the NAND gate will keep the clear line HIGH for all counts through 10. The inputs of this gate are connected to the B and D lines. The D line stays LOW, forcing clear HIGH, up until the 8th input pulse has passed. At that time (i.e., T_0 in Fig. 4-34B), D will go HIGH and bit B drops LOW, so the clear line remains HIGH for the 9th pulse.

The clear line will remain HIGH until the end of the 10th pulse. At that point (T_2) both B and D are HIGH, so the NAND gate output drops LOW clearing all four flip-flops (i.e., forcing them to the state where all four Q outputs are LOW). The counter is therefore reset to 0000.

Table 4-1. Four-bit Binary Counter Output.

After Pulse	Word
0	0000
1	0001
2	0010
3	0011
4	0100
5	0101
6	0110
7	0111
8	1000
9	1001
10	1010
11	1011
12	1100
13	1101
14	1110
15	1111

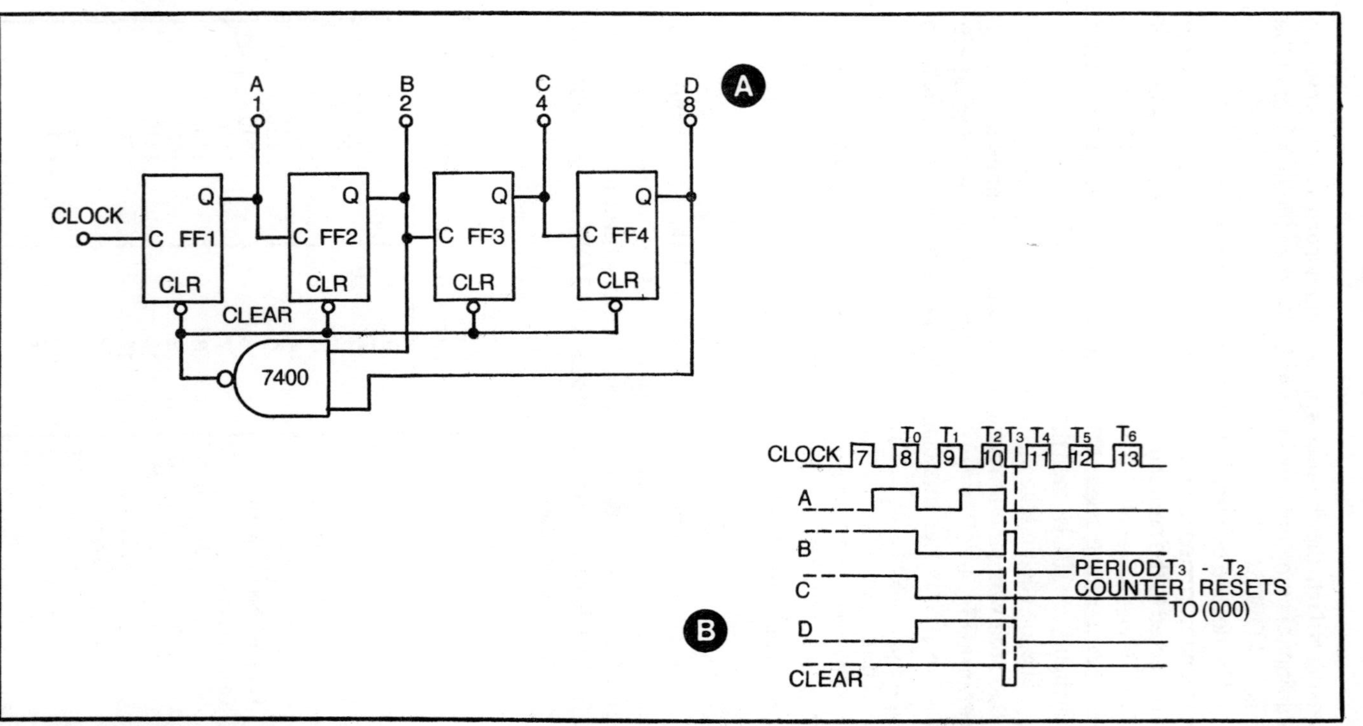

Fig. 4-34. Decade counter.

R. LANGEVIN

The reset counter produces a 0000 code, so the B and D outputs are now LOW, forcing the clear line HIGH again. The entire reset cycle occurs during period ($T_3 - T_2$). This period has been expanded greatly for graphic illustration in the figure but actually takes only nanoseconds or microseconds.

The 11th pulse will increment the counter one time, so the output will be 0001_2. The count sequence, in decimal, then is 0-1-2-3-4-5-6-7-8-9-0-1 etc. The output code is a ten-digit version of four-bit binary (hexadecimal), and is called *binary coded decimal* (BCD).

Synchronous Counters

Ripple counters suffer from one major problem—speed. The counter elements are wired in cascade, so an input pulse must ripple through the entire chain before it affects the output. A synchronous counter feeds the clock input to all flip-flops in parallel, and this results in a much faster operation.

Figure 4-35 shows a partial schematic for a synchronous binary counter. We accomplish synchronous operation by using four flip-flops with clock inputs tied together and a pair of AND gates.

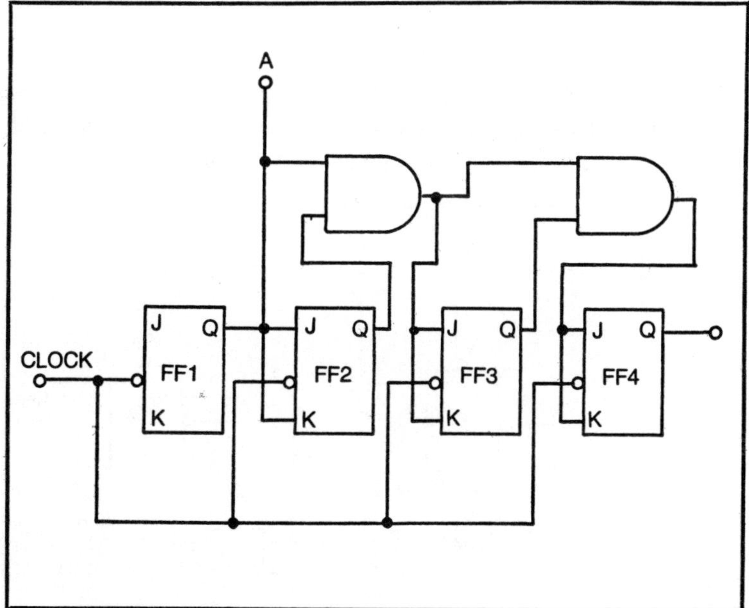

Fig. 4-35. Counter.

After Pulse	Word
0	0000
15	1111
14	1110
13	1101
12	1100
11	1011
10	1010
9	1001
8	1000
7	0111
6	0110
5	0101
4	0100
3	0011
2	0010
1	0001
0	0000

Table 4-2. Preset Counter Output.

One AND gate is connected so that both Q1 and Q2 are HIGH before FF3 is active. Similarly, Q2 and Q3 must be HIGH before FF4 is made active. On a clock pulse, any of the four flip-flops scheduled to change will do so simultaneously. Synchronous counters attain faster speeds, although ripple counters seem to predominate most applications.

Preset Counters

A preset counter increments from a preset point instead of 0000. For example, suppose we wanted to count from 5_{10} (0101_2).

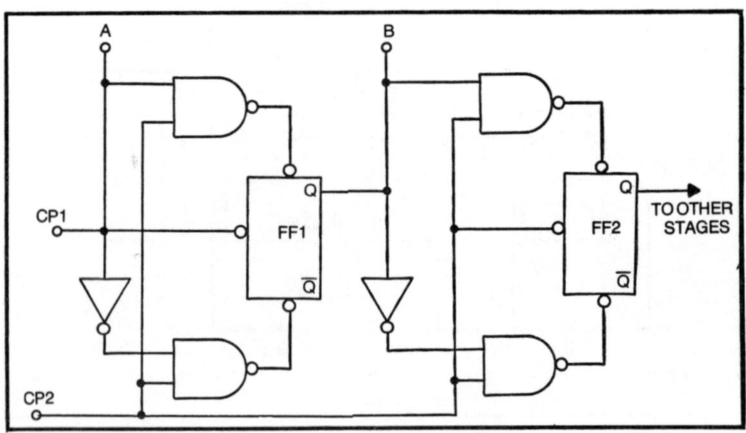

Fig. 4-36. Preset counter.

We could preset the counter to 0101 and increment from there. The count would be as in Table 4-2.

Figure 4-36 shows a common method for achieving preset conditions: the *jam* input. Only two stages are shown here, but adding two additional stages will make it a four-bit counter. Of course, any number of stages may be connected in cascade to form an *n*-bit preset counter.

In Fig. 4-36, the preset count is applied to points A and B, and both bits will be entered simultaneously when clock line CP2 is brought HIGH. Line CP2 is sometimes called the enter or jam terminal. Once the preset bit pattern is entered, the counter will increment from there with every transition of the clock line CP1.

Down Counters

A down counter decrements, instead of incrementing, the count for each excursion of the input pulse. If the reset condition is 0000, then the next count would be 0000 – 1, or 1111. It would have been 0001 in an up counter. The count sequence for a four-bit down counter is shown in Table 4-3.

We use basically the same circuit as before, but toggle each FF from the NOT-Q rather than Q, of the preceding FF. An example of a four-bit binary down counter is shown in Fig. 4-37A. Note that the outputs are taken from the Q outputs of the FFs, but that toggling is from the NOT-Q.

The preset inputs of the flip-flops are connected together to provide a means to preset the counter to its initial (i.e., 1111) state. This counter is also called a *subtraction counter,* because each input pulse causes the output to decrement by one bit.

Table 4-3. Four-bit Down Counter Output.

Counter	Count
5	0101
6	0110
7	0111
8	1000
9	1001
0	0000
1	0001
2	0010
3	0011
4	0100
5	0101

A decade version of this circuit is shown in Fig. 4-37B. As in the case of the regular decade counter, a NAND gate is added to the circuit to reset the counter following the 10th count. We detect the states where outputs C and D are HIGH, and then clear the two middle FFs. This action forces the output to 1001_2 (i.e., 9_{10}). The counter then decrements from 1001 in the decimal sequence 9-8-7-6-5-4-3-2-1-0-9, etc.

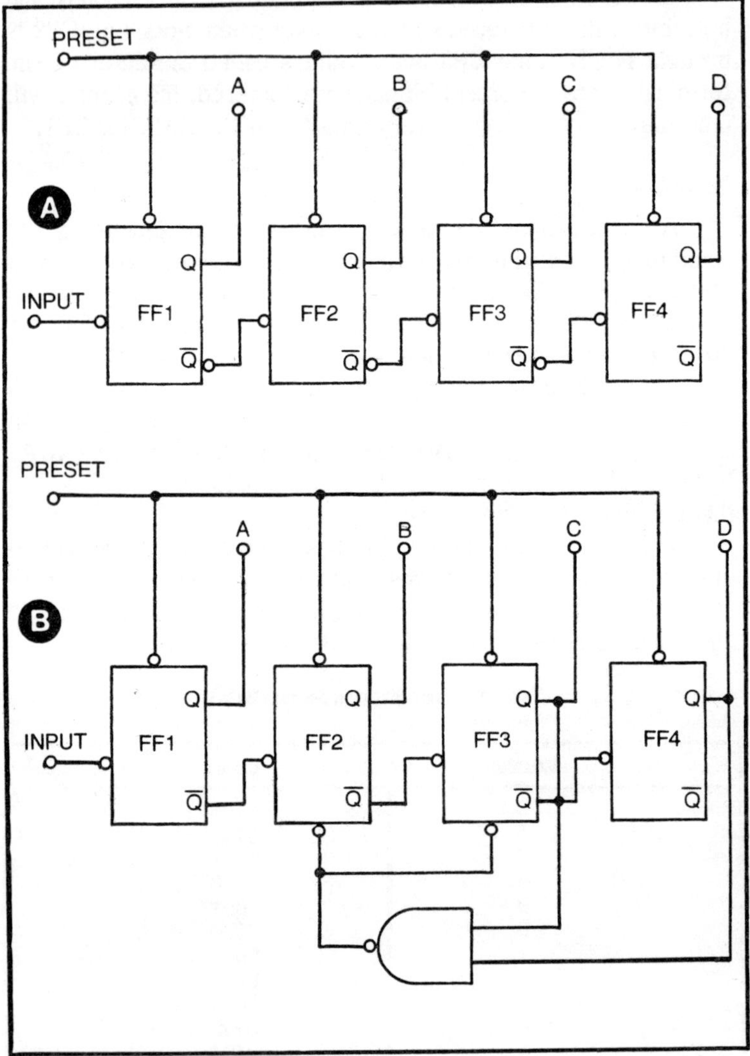

Fig. 4-37. Down counters.

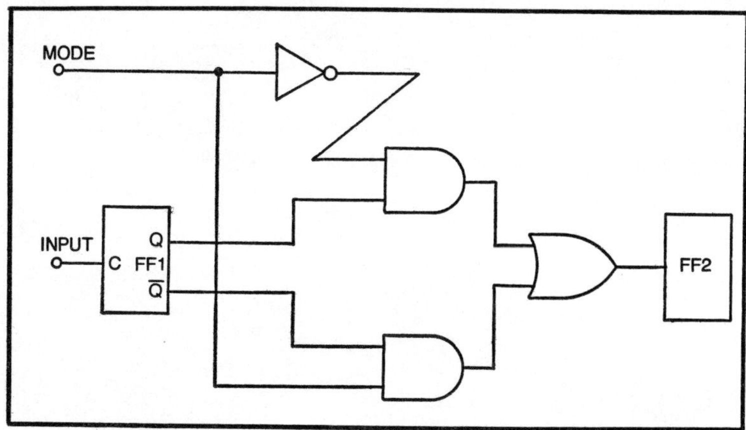

Fig. 4-38. Up/down counter.

Up/Down Counters

Some counters will operate in both up and down modes depending upon both the logic level applied to a *mode* input. Figure 4-38 shows a representative circuit in which the first two stages of a cascade counter are modified by the addition of several gates. If the mode input is HIGH, then the circuit is an up counter. But if the mode input is LOW, then the circuit operates as a down counter.

Chapter 5

The 555 Timer

Most IC timers suffer from several problems that limit the operating frequencies or period attainable and keep the stability marginal at best. Unijunction and bipolar transistors, for example, can be used in timer circuits but suffer from problems such as the frequency and period stability due to changes in power supply potential. It is sometimes simply not adequate to have a timer that changes parameters merely because the battery voltage drops! The 555 integrated circuit timer was one of the first of a new generation of timer chips that solve the problem rather nicely. The 555 operates using the *ratio* of the supply voltage and certain IC terminal voltages (i.e., the trigger potential is always a fraction of the supply potential).

The 555 is neither bipolar, TTL nor CMOS, yet can be compatible with either family of IC logic! The 555 device will operate with any supply potential from +4.5 to +15 volts DC.

The 555 is sold in an 8-pin miniDIP IC package, and a dual version (556) is sold in a 14-pin DIP package. The ubiquitous bipolar 555 design is still with us and is now quite low in cost. Some companies now offer low-current CMOS versions.

Figure 5-1 shows the block diagram of the 555's internal circuitry. The principle stages of the 555 are: RS control flip-flop, two comparators (COMP1 and COMP2), a discharge transistor (Q1) and an inverter amplifier (A1).

The principle of operation for the RS flip-flop was discussed at length in Chapter 4, so will not be repeated here. The principle

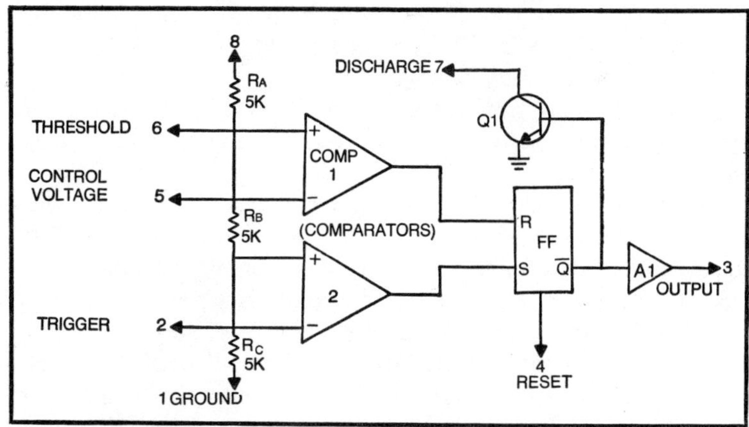

Fig. 5-1. Internal block diagram of the 555.

behind the comparator can be stated as "an amplifier with too much gain." The comparator is like an operational amplifier without a feedback resistor. The open-loop gain of the operational amplifier then becomes the stage gain for the comparator. This gain is typically from 20,000 to over 1,000,000 depending upon the quality of the operational amplifier. When the voltage applied across the inverting and noninverting inputs is zero, then the output voltage is also zero. But, when the voltage is nonzero, then the output will be either positive (or negative) high. Once the differential input voltage exceeds a few millivolts, the gain of the comparator causes the amplifier to saturate so only zero (LOW) and maximum (HIGH) outputs are allowed. In most digital ICs the comparator is designed to allow only zero and a single-polarity HIGH (usually positive) output. While this imposes some constraints on the permissible polarity of the differential input voltage, it is usually no big problem. In the 555 comparator, the inputs see a monopolar situation anyway, so the point is moot.

The discharge transistor (Q1) will be used to discharge the timing capacitor at the end of each cycle. Note that transistor Q1 is open-collector, meaning that an external circuit is needed to operate the transistor. The bias for the comparators is supplied from a series voltage divider R_a, R_b, and R_c. The voltage applied to the inverting input of COMP1 is ⅔(V+), while the bias at COMP2 is ⅓(V+). The voltage at COMP1 is called the *control voltage,* while the voltage at COMP2 sets the trigger point. The operation of the 555, then, is determined by the relationships set by the voltage divider, not by the absolute voltage level present. In the next chapter, we will demonstrate monostable multivibrator operation of the 555. That discussion will demonstrate the operation of the 555 very clearly.

The 555 is probably the timer of choice for any period from microseconds to minutes. Since the device is a simple RC design, the operating range is limited by reasonable choices of these resistors and capacitors. In the following two chapters we will discuss both monostable and astable operation of the 555. The equations that determine the values of the timing components (a capacitor and either one or two resistors) will be given.

The inverter amplifier at the output of the 555 is not open-collector, so it will produce an output pulse of zero when LOW and close to V+ when HIGH. The output will be TTL-compatible only when the 555 is operated from a +5 volt DC power supply, or when some output device is used that will convert a non-TTL pulse to the TTL level (e.g. a pull-up transistor).

Chapter 6

Monostable Operation Of The 555

A monostable multivibrator (also called a one-shot circuit) will produce a single output pulse of fixed duration when triggered by an input pulse. The output of the one-shot will snap HIGH following the trigger pulse, and will remain for a fixed, predetermined period of time. When this time expires, the output will drop LOW again. The output of the one-shot will remain LOW indefinitely unless another trigger pulse is applied to the circuit. The 555 can be operated as a monostable multivibrator.

Figure 6-1 shows the operation of the 555 as a monostable multivibrator. The version of the circuit shown in Fig. 6-1A uses the block-diagram method of the last chapter, while the version in Fig. 6-1B is more conventional (and is how you will see it in some of the projects given in this and other chapters). For purposes of discussing the operation of the circuit, however, we will use the version in Fig. 6-1A.

Recall from the last chapter that the two comparators are biased from a resistor voltage divider network R_a, R_b, and R_c. The inverting input of COMP1 is connected to the ⅔(V+) point on the voltage divider, while the noninverting input of COMP2 is connected to the ⅓(V+) point on the resistor voltage divider. The output of the 555 is taken from an inverting amplifier connected to the NOT-Q output of the RS flip-flop. When the FF is set, then, the NOT-Q output of the RS is LOW, so the output of the 555 is HIGH. Similarly, when the FF is reset the NOT-Q output of the FF is HIGH, so the inverted version at the output of the 555 is LOW.

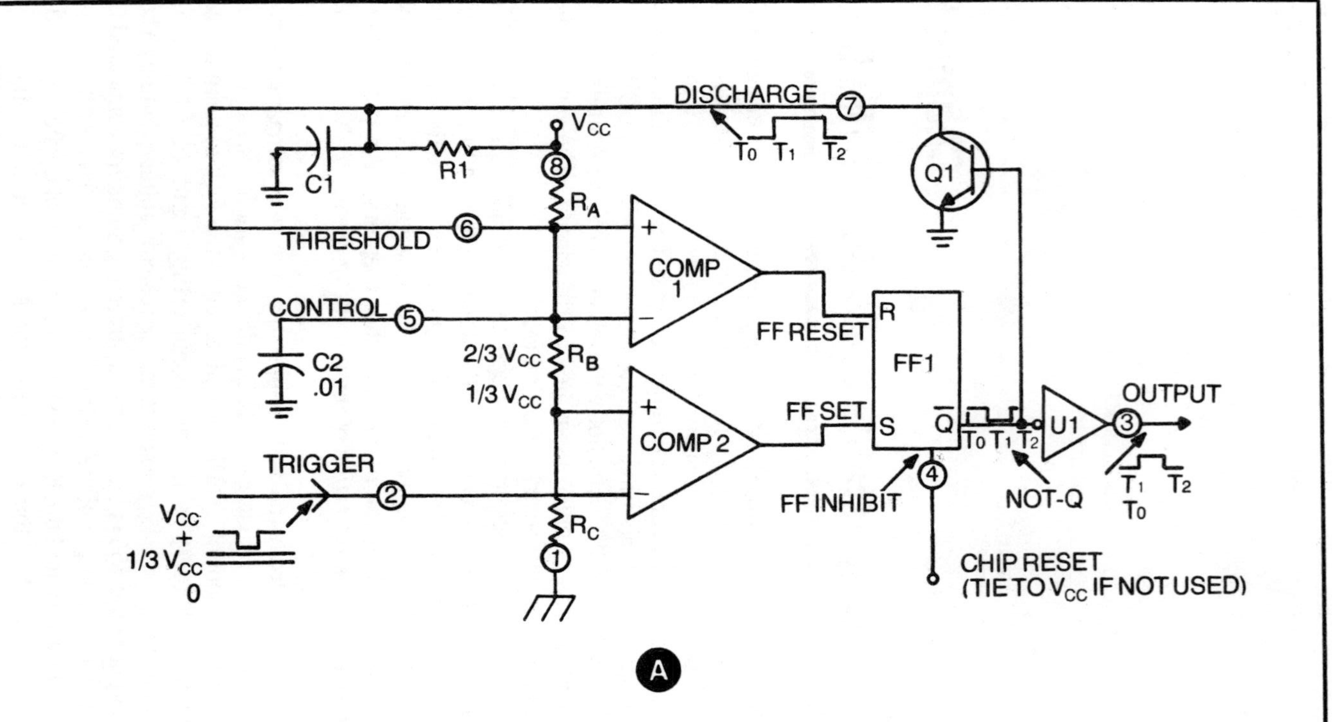

DISCHARGE
7
T0 T1 T2
VCC
C1
R1
8
RA
Q1
THRESHOLD
6
+
COMP 1
−
CONTROL
5
FF RESET
R
FF1
C2
.01
2/3 VCC
RB
1/3 VCC
+
COMP 2
−
FF SET
S
Q̄
T0 T1 T2
U1
3
OUTPUT
TRIGGER
2
FF INHIBIT
4
NOT-Q
T1 T2
T0
VCC
+
1/3 VCC
0
RC
1
CHIP RESET
(TIE TO VCC IF NOT USED)
A

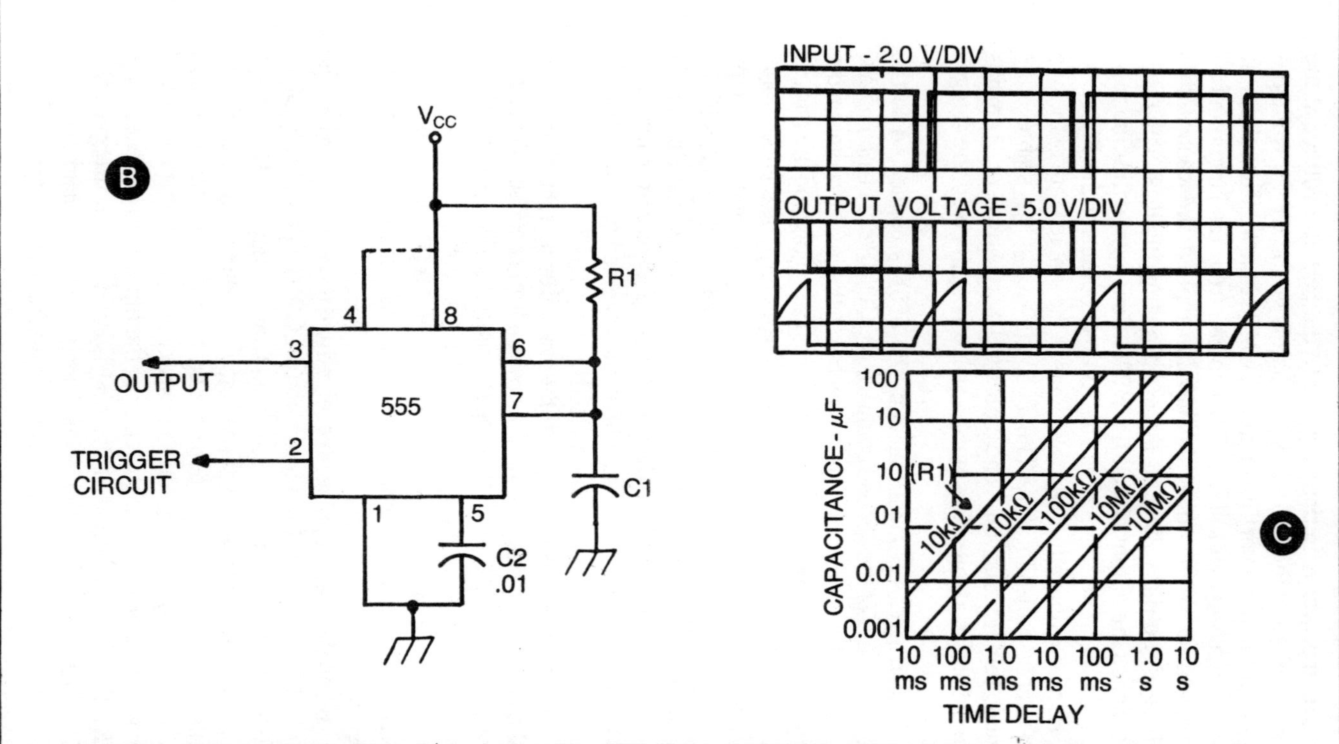

Fig. 6-1. Monostable circuit using the 555: A) Block form; B) Circuit diagram form; C) Timing waveforms; D) Duration Chart.

When power is initially applied to the 555, the voltage at the inverting input of the 555 will immediately go to ⅔(V+), while the voltage at the noninverting input of COMP1 is low. The voltage at the noninverting input of COMP1 is taken from the junction of an RC network. When power is initially applied the capacitor is uncharged, so the voltage at the noninverting input is near zero. This makes the output of COMP1 HIGH, causing the flip-flop to be reset. In the reset condition, the NOT-Q output of the RS FF is HIGH, so the 555 output is LOW. This is the stable state of the 555 monostable multivibrator. The 555 will go to its stable state immediately at turn-on. The 555 one-shot, however, remains dormant in its stable state until a pulse is applied to the trigger input.

The trigger terminal (pin 2 of the 555) is connected to the inverting input of the second comparator, COMP2. The noninverting input of this comparator is connected to the ⅓(V+) point on the resistor voltage divider. This means that the voltage on the trigger input must drop to a point less than ⅓(V+) before the comparator is activated. But, when the trigger pulse is applied and the voltage drops below the required point, the output of COMP2 snaps HIGH and causes the set input of the RS FF to be activated. In the set condition of the RS FF, we find that the NOT-Q output of the RS is LOW. Since the output of the 555 is the inverted output of the RS FF, the output of the 555 is now HIGH.

During the dormant period, the capacitor is kept discharged by the action of transistor Q1. This transistor is connected directly across the capacitor so it will keep the capacitor discharged as long as the transistor is forward biased. In the dormant period, the NOT-Q output of the RS FF (the source of the Q1 base voltage) is HIGH so this transistor is on, keeping the capacitor at ground potential.

When the trigger pulse is received and the RS FF goes into the set condition, the NOT-Q output drops LOW (while the 555 output is HIGH). This removes the bias from transistor Q1 so the discharge transistor is now turned off. This allows the capacitor to begin charging at a rate determined by the RC time constant, R1C1. When the voltage on the capacitor reaches ⅔(V+) it will activate comparator COMP1, causing the flip-flop to be reset. This forces the NOT-Q output of the RS FF HIGH again, dropping the 555 output LOW. The 555 is once again in the dormant state and will remain in that condition unless an additional trigger pulse is received.

There is a reset input on the 555 (pin 4) that serves to force the RS flip-flop into the condition in which NOT-Q is HIGH and the 555 output is LOW. This terminal can be used to terminate an output pulse (something that you might want to do in some instrumentation circuits). If the reset terminal is not needed, it is to be tied to the V+ (also called V_{cc}).

The control voltage (which forms the bias on the inverting input of COMP1) is stabilized by capacitor C2. The value of this capacitor is not too critical, and anything from 0.001 μF to 0.05 μF will prove satisfactory.

The output period of the 555 monostable multivibrator is set by the time constant of R1 and C1. The equation is:

$$T = 1.1\, R_1\, C_1$$

where:

T is the time the output is HIGH, in seconds
R_1 is the resistance of R1 in ohms
C_1 is the capacitance of C1 in farads

EXAMPLE

Find the duration of the output pulse of a 555 monostable multivibrator if a 22 kohm resistor is connected to a 0.15 μF capacitor in the timing network.

Solution:

$$\begin{aligned} T &= 1.1 R_1 C_1 \\ &= (1.1)(2.2 \times 10^4 \text{ ohms})(1.5 \times 10^{-7}\text{F}) \\ &= 0.0036 \text{ seconds} = 3.6 \text{ milliseconds} \end{aligned}$$

The graph in Fig. 6-1C shows the solution to $T = 1.1\, R_1 C_1$ for times from 0.01 seconds (10 ms) to 10 seconds, allowing you to select resistor and capacitor values suitable for your application. Note that it is usually the practice to select a capacitor value to meet one of the standard available values and then calculate the required resistance to achieve the time desired. This is because it is easier to derive oddball resistor values than oddball capacitor values. Of course, if there is no need for precision, then select the nearest values that do the job.

Precision in any RC circuit is difficult to achieve because the component values tend to change in temperature. This can be overcome by using metal film or wirewound resistors, and capacitor types that are known to be stable (silver mica, polyethylene, etc.).

SOME 555 ONE-SHOT APPLICATIONS AND PROJECTS

The 555 is one of the easiest ICs to learn to apply. In this section we will discuss some of the types of projects and applications for the 555 one-shot circuit. If you learn well the theory of operation, then you will be able to design circuits for your own applications.

Bounceless Pushbutton

Ordinary mechanical switches suffer from *bounce*. When the leaves of the contact electrodes come together, they do not mate solidly at the instant of first contact; they bounce together before coming to a nice solid contact. The result is that there will be a series of pulses produced by the switch as it makes contact. If you connect a circuit such that the pulses from the bouncing are input to an event counter, then you will see an accumulation of pulses every time the switch is closed. This should tell you that bouncing switch contacts can increment or decrement binary counters, toggle flip-flops, and pass through gates as if they were valid digital pulses. In short, these spurious pulses raise hob with digital circuits.

Whenever we want to use a pushbutton (or other type of switch) in digital circuits we must condition the contacts to eliminate the bounce. This is done by using the pushbutton to activate a one-shot multivibrator that has a period long enough to allow the bounce to damp out. The digital circuit then sees the one-shot output pulse as the switch closure, and it is a clean signal. The first bounce pulse from the pushbutton switch will trigger the one-shot. This means that the subsequent pulses (which die out prior to the time-out of the one-shot pulse) cannot pass through the circuit because they will not retrigger the one-shot.

We can use the one-shot circuit of Fig. 6-1B to form a bounceless pushbutton by using the triggering arrangement of Fig. 6-2. In the circuit of Fig. 6-1B, use values of R1 = 56 kohms and C1 = 0.01 μF (for a time of 600 milliseconds). The trigger circuit (Fig. 6-2) uses 10 kohm pull-up resistors and a 0.1 μF capacitor. When the circuit is first turned on, both ends of the capacitor are connected to the V_{cc} (+) terminal and have the same potential. But when pushbutton S1 is closed momentarily, one end of C3 is suddenly grounded. This means that there is a path for current to flow and charge C3. But at the instant the switch is closed the potential is zero, so the potential on the trigger input of the 555 (pin 2) is also zero. The 555 wants a potential of less than ⅓(V+) in

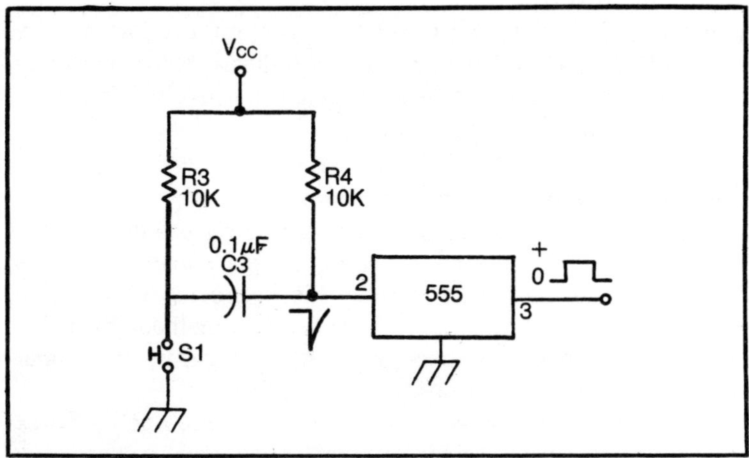

Fig. 6-2. Manual triggering of the 555.

order to trigger, so this condition will trigger the 555 and cause the output to go HIGH for 600 ms. The voltage at pin 2 will rapidly build back towards Vcc as the capacitor charges through the 10 kohm resistor. If you want to design time constants for other applications, then keep in mind that the time constant of R4 and C3 must be considerably less than R1C1 (Fig. 6-1). Otherwise, the 555 will remain in the trigger condition and will retrigger when the period (600 ms) times out.

Positively Triggered One-Shot (I)

The 555 normally requires a negative-going transition to trigger the operation of the one-shot. In many digital circuits, however, it is necessary to use a positive-going trigger pulse. The circuit in Fig. 6-3 is designed to allow positive triggering of the 555 monostable multivibrator.

Most of the circuit is the ordinary 555 one-shot circuit shown in the first section of this chapter. The timer period is determined by the RC time constant and is equal to $T = 1.1\, R_1\, C_1$. The control voltage is stabilized by capacitor C2, a 0.01 µF unit. The trigger terminal is normally held HIGH by the 10 kohm pull-up resistor (R2) connected to the V+ power supply. The triggering is accomplished by a transistor (Q1) connected from the trigger terminal to ground. When Q1 is forward biased it will drop pin 2 of the 555 to ground, thereby triggering the one-shot. Since this transistor is an NPN (almost any small-signal NPN unit could be used), it requires a positive forward bias voltage—our positive-going pulse, which will forward bias transistor Q1 and trigger the one-shot. This circuit operates as an ordinary 555 monostable multivibrator, except that the triggering is positive-going instead of negative-going.

Positively Triggered One-Shot (II and III)

In circuits where you are already using various digital integrated circuits it might be easier to use spare gates (wired as inverters) or sections of hex inverter chips to provide the inversion of the trigger pulse. In Fig. 6-4 we see two methods for using digital IC sections to produce the needed trigger inversion that allows the 555 to fire from positive-going pulses.

The circuit in Fig. 6-4A uses either the 7405 or 7406 hex inverter circuit. The 7405 and 7406 devices are open-collector hex inverters. The output stage of each section is a single transistor without a collector load. The user provides an external collector load to V+. The 7405 device will allow collector voltages up to +15 volts, while the 7406 device accommodates collector potentials up to +30 volts. In each case, the actual IC V+ (pin 14) must be connected to the +5 volt TTL power supply, even though the collector load resistor may be connected to potentials as high as 30 VDC.

The pull-up resistor to V+ will have different values depending upon the collector voltage. In general, use 2.7 kohm for

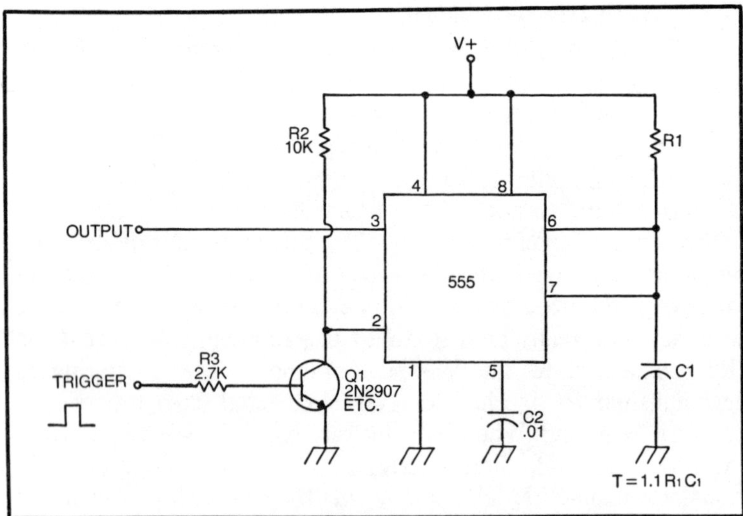

Fig. 6-3. Inverted triggering of the 555 using a bipolar transistor.

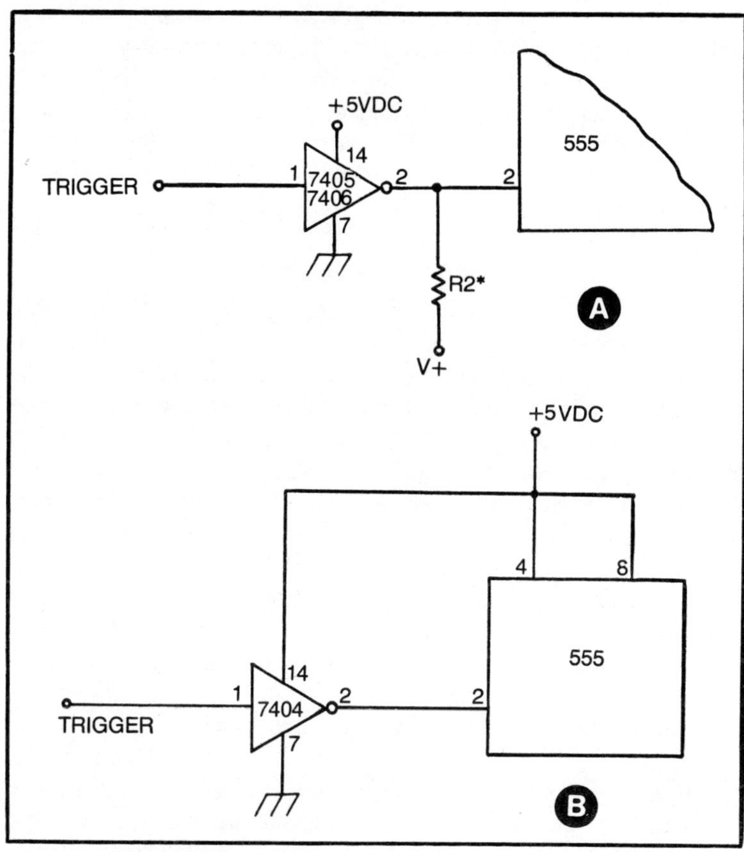

Fig. 6-4. Inverted triggering of the 555; A) Using a TTL inverter with open-collector outputs; B) Using regular TTL inverter.

potentials up to 7 volts, 10 kohm up to 12 volts, and 20 kohm up to terminal maximum. It is a general rule to keep the collector current in the IC output transistor below 25 mA.

When the trigger signal (i.e., the inverter input signal) is LOW, then the inverter output is HIGH. This level is ensured by the pull-up resistor to V+. The trigger input of the 555 (pin 2) is then in the dormant state and nothing happens. But when the trigger signal snaps HIGH, then the inverter output drops LOW and triggers the 555. Recall that the criterion for triggering a 555 is to cause the pin 2 voltage to drop below ⅓(V+).

A slightly different version is shown in Fig. 6-4B. In this circuit, we consider the 555 TTL-compatible because it is operated from +5 volts DC. The TTL inverter 7404 is a regular output version and needs no pull-up resistor. We can connect the 7404 in the same manner as in version II, but deleting the pull-up resistor. Note that TTL inverter stages can easily be made from NAND and NOR gate sections by tying all unused inputs together. In the 7400 and 7402 devices, for example, each section has two inputs. If these inputs are strapped together, then the IC section will operate as an inverter instead of a NAND or NOR gate.

Retriggerable Monostable Multivibrator (555)

The one-shot circuits discussed thus far are not retriggerable until the predetermined period has timed out. In some one-shot circuits based on bipolar transistors there is an additional time called the refractory period during which the one-shot cannot be retriggered. This is the reason why such circuits are used for pushbutton debouncing. The extra pulses produced by the bouncing pushbutton contacts do not affect the circuit, so are eliminated. In that case, the "switch contact closure" seen by the digital circuitry is the one-shot output signal. But sometimes we want to be able to retrigger a one-shot while it is still in the output sequence, before it times out. We may, for example, want a signal HIGH as long as some process is going on, and will generate a one-shot pulse each time the event occurs. The retriggerable monostable multivibrator of Fig. 6-5 will fill that bill.

The circuit for the retriggerable one-shot is shown in Fig. 6-5A, while the timing diagram is shown in Fig. 6-5B. The basic circuit is that of an ordinary 555 one-shot circuit, and the time of the output pulse (T) is given by the familiar $1.1R_1C_1$. This circuit, however, has been modified to accommodate the retriggering by adding two transistors. One transistor (Q1) is the same type of circuit as used in Fig. 6-3 to make a positive-triggered one-shot (version I). A 10 kohm pull-up resistor from the 555 trigger input (pin 2) to the V+ power supply keeps the trigger terminal HIGH. It will remain HIGH until transistor Q1 is turned on by a positive trigger pulse applied to the input resistor R3. A second transistor (Q2) is used as an extra discharge transistor across capacitor C1. When the trigger pulse is applied to the input (R3), it will also turn on transistor Q2 causing it to dump the charge in capacitor C1 and thereby restart the timing sequence.

Figure 6-5B shows two complete timing sequences for the circuit of Fig. 6-5A. The first sequence (T_1 to T_2) is an ordinary one-shot cycle. A trigger pulse is applied to the circuit causing pin 2 to drop LOW (note that the trigger pulse applied to the input

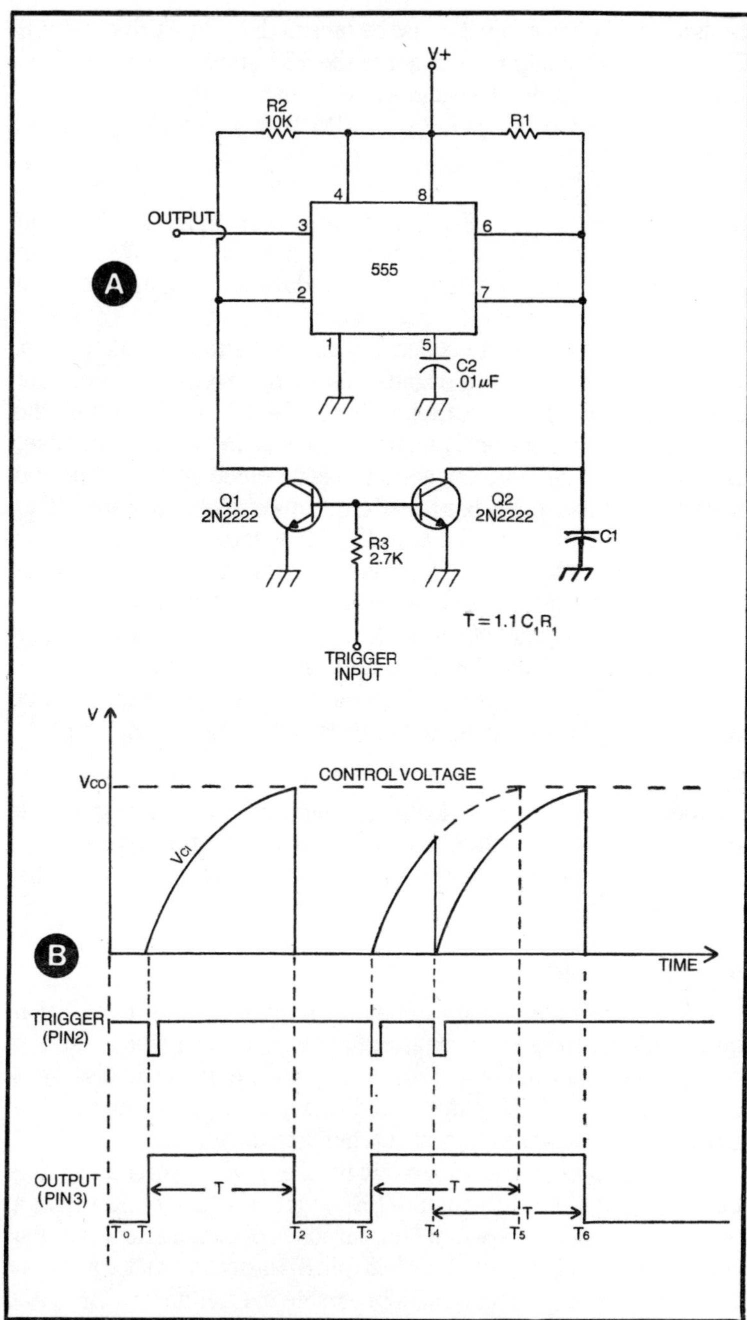

Fig. 6-5. Retriggerable monostable multivibrator using the 555.

resistor, rather than pin 2, must be inverted; i.e., positive-going). This causes the output terminal of the 555 (pin 3) to snap HIGH, and voltage V_{C1} to begin rising. Recall from the last chapter that the 555 contains two comparators. COMP1 is biased to a control voltage (V_{co}) that is equal to ⅔(V+). The voltage across C1 will rise until it is equal to V_{co}, at which time the comparator activates and causes the internal discharge transistor to dump the charge in C1 to zero. This also causes the output to drop LOW, and the one-shot has timed out. It will not produce another output pulse until another trigger pulse is received.

The second timing sequence illustrates the consequences of retriggering. At time T_3 a trigger pulse is received by the dormant one-shot, causing the output to snap HIGH. At this time, the voltage across capacitor C1 begins to rise, as in the previous case. But, at time T_4, another trigger pulse is received before the period from the previous pulse has timed out. This causes transistor Q2 in Fig. 6-5A to dump the charge in C1. (The dotted line shows how V_{C1} would have increased had this new pulse not been received.) The capacitor voltage is dropped back to zero and will begin its sequence over again. When the time period dictated by the R1C1 time constant expires, then the output drops back to LOW.

The total time of the retriggerable monostable is *not* 2T, but rather it is T plus the unexpired portion of the first period, i.e.:

$$T^1 = T + (T_4 - T_3)$$

Theoretically, we could keep the output of the retriggerable multivibrator HIGH as long as we continue to hit the trigger input with fresh trigger pulses prior to the expiration of the period of the one-shot.

One-Second Timer

There are many applications for a one-second timer. If you build a digital frequency counter, for example, you would use the one-second timer to control the display update line. The display is kept stable by only displaying "old" data, already measured. The LEDs do not show the count as it is being totalized.

In the one-second timer circuit of Fig. 6-6 we connect the output to a light emitting diode (LED) that will blink on and off at a one-second rate. This will do for purposes of explanation, but you could connect the output of the 555 (pin 3) to any line that you wish. The circuit is an ordinary 555 monostable multivibrator. The time constant (R1 + R2) C1 is adjusted to give the one-second period required. Repetitive blinking requires repetitive triggering.

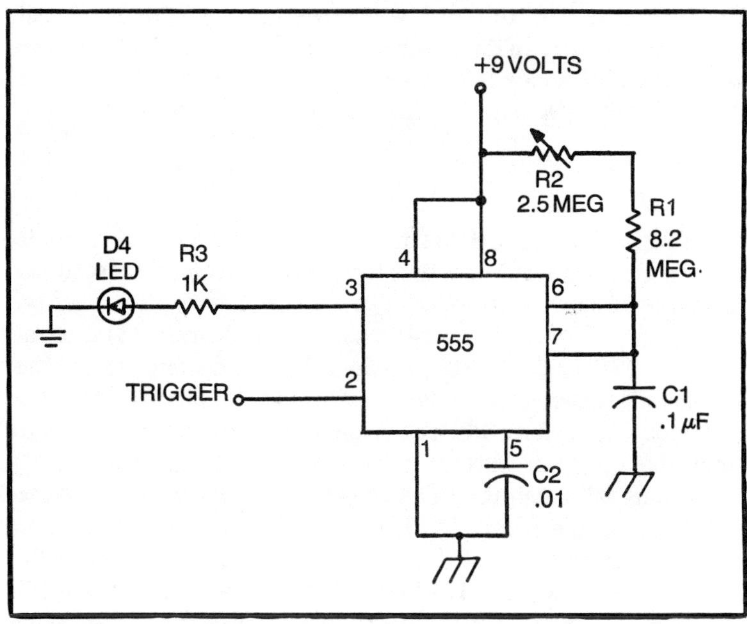

Fig. 6-6. One-second timer.

Note that the resistance of the circuit is split between two resistors, while the capacitance of the timing circuit is in a single capacitor. It is easier to obtain relatively precise values of capacitance, *provided* that you will accept one of the *standard* capacitor values. There are only a few standard capacitor values, so we usually select a likely capacitor value and then calculate the needed resistance. Variable resistors (in the form of potentiometers) are easily obtained in precision components, so we make the resistance the variable arm of the circuit. By connecting a fixed resistor (which has most of the circuit resistance) in series with a variable resistance, we can make any resistance needed within the range of the combination. This feature allows us to: 1) adjust the combination to oddball values of resistance; and 2) adjust for any errors in the value of the capacitor (remember, it is the combination of the resistance and capacitance—i.e., the *product* R × C—that is important to the timing of the circuit).

TTL-Compatible 555 Output

It is sometimes necessary to use the 555 timer to drive TTL devices. The 555 output will either sink or source up to 200 mA of current, so it will be directly compatible with TTL devices *if* the

555 is operated from a +5 volt power supply. No additional circuitry is needed in this ideal situation. But what about the case where the 555 device is operated at power supply voltages other than +5 volts? Remember, the 555 will allow operation at any potential from +4.5 to +15 volts. In the case where the V+ (also called V_{CC} is *not* +5 volts, then use one of the circuits shown in Fig. 6-7.

Method I. Figure 6-7A shows the use of a bipolar transistor to make a TTL-compatible output for the 555. Note that the output from the transistor is inverted from the normal 555 polarity. Almost any NPN small-signal transistor will work for Q1. (e.g., 2N2907, 2N3393, 2N2222, or 2N3904). The collector of the transistor is connected to the +5 volt (not V_{CC} source through a collector load resistor R2. The value of this resistor is selected to limit the collector current to a value within the transistor's limitations (although the value shown is nearly universal). When the output of the 555 is LOW, the transistor is unbiased and no collector current will flow. This makes the collector terminal of the transistor +5 volts (or TTL HIGH). But when the output of the 555 is HIGH, the transistor is forward biased. The base resistor is selected to permit a base current that drives the transistor into complete saturation. This situation drives the collector LOW, i.e., to below the $V_{ce(sat)}$ of the transistor.

Method II. Practical experience with a lot of 555 devices reveals a basic problem: the output is not very well isolated from the triggering circuitry inside. A pulse applied to the output terminal (pin 3) will cause triggering of the 555 (not too cool). We can isolate the output in circuits such as Fig. 6-7A by connecting several diodes. (Fig. 6-7B) that have contact potentials that must be overcome before any external pulses reach the 555.

Method III. The 555 supports pretty much the same range of operating potentials as the CMOS line of logic. Because CMOS input impedance is very high, the 555 is not taxed at all driving CMOS devices.

There are two CMOS devices (of immediate interest here) that are CMOS input but TTL output. The 4049 device is a set of inverters (six of 'em) in a single package. The output stage of each inverter is a TTL stage. The 4050 device is similar, except that it is a hex buffer (the same thing as an inverter but without the inverter action—HIGH in = HIGH out, LOW in = LOW out). We can use either of these devices to interface 555 timer circuits to TTL logic circuits. An example is shown in Fig. 6-7C. One section of a 4050

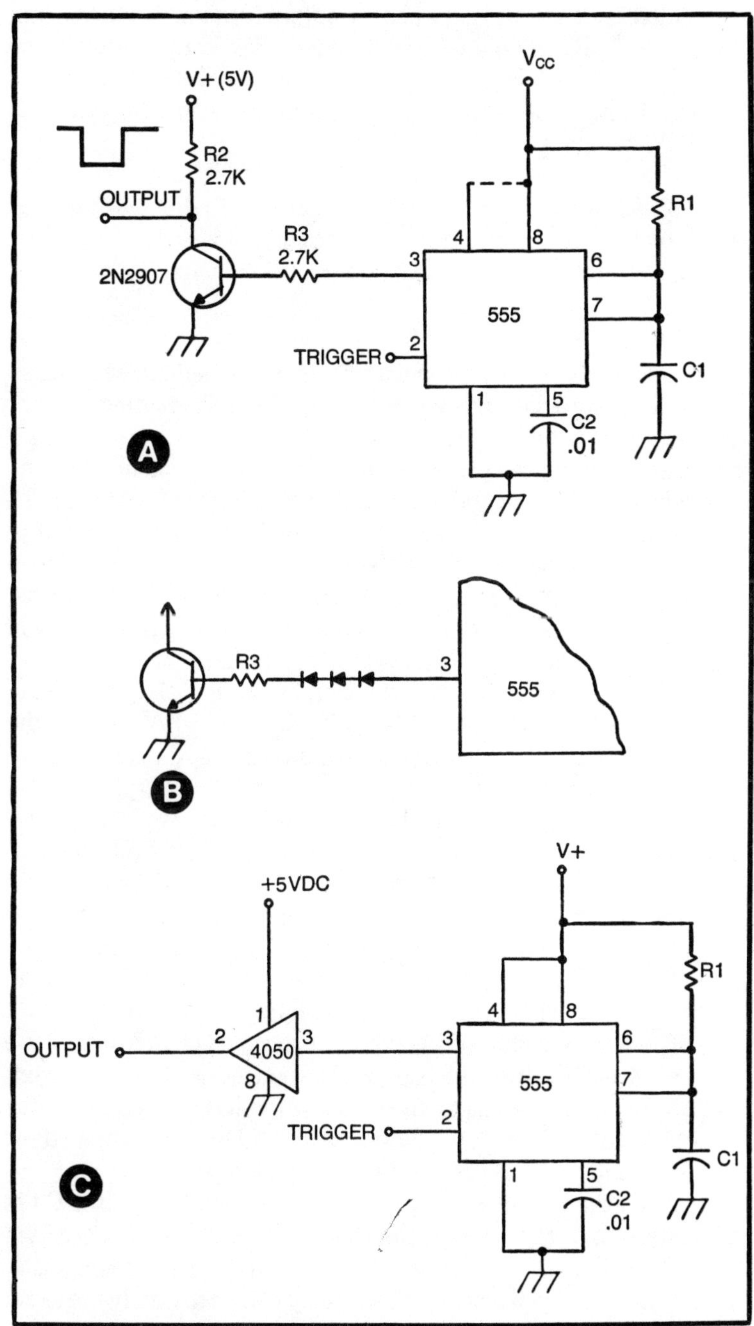

Fig. 6-7. 555 one-shot with inverted output.

hex buffer is connected between the output terminal of the 555 (pin 3) and the load. When the 555 output goes LOW, the output of the 4050 is also LOW. Similarly, when the output of the 555 is HIGH the 4050 output is also HIGH. The difference is that the output is not TTL-compatible.

The use of the 4049 device makes the circuit behave similarly, except that the output is inverted compared with the 555 output. A LOW on the 555 output makes the 4949 output HIGH; and a HIGH on the 555 output makes the 4049 output LOW.

Pulse Delay Line

A delay line produces an output signal that is delayed in phase (or time) compared with the input signal. In alternating current circuits we can build delay lines from transmission line segments, or from LC networks designed to behave like transmission line segments. But delay for pulses is a little harder to accomplish cheaply. The problem is that analog methods produce different transmission velocities for different frequencies. The pulse is always made up of a fundamental frequency and a load of harmonic frequencies. This situation means that the shape of the pulse will be distorted at the output of the typical analog delay line.

A digital pulse delay line can be built using a pair of monostable multivibrators (Fig. 6-8). In Fig. 6-8A we see the timing diagram for a pulse delay line, and in Fig. 6-8B the circuit that will produce this delay.

Consider the situation shown in Fig. 6-8A. We receive a 1 millisecond (1 ms) pulse at the trigger input of the first one-shot (OS1). This trips OS1 into operation. The period for the first one-shot is set to give the desired amount of delay (in this case 100 ms). When the first one-shot times out, it will trigger the second one-shot (OS2), and this will produce a new pulse with the same duration as the original (if we do our arithmetic correctly).

The first one-shot (OS1) is an ordinary 555 circuit of the type discussed earlier in this chapter. The period is set to approximately 100 milliseconds by the combination R1C1 (e.g., 10^7 ohms x 10^{-8} farads = 100 ms). The output of OS1 (pin 3) snaps HIGH when the 1 ms pulse is received at the trigger input (pin 2).

When the output of OS1 is LOW, one end of capacitor C3 is grounded while the other is connected to the +12 volt source via pull-up resistor R3. But when the output of OS1 suddenly snaps HIGH, the grounded plate of C3 suddenly comes to nearly the same potential as the other plate. The capacitor is suddenly discharged through R2 and R3. The RC time constant of this circuit is selected to differentiate the pulse applied to the capacitor. This means that there will be a positive spike as the OS1 output snaps HIGH and a negative spike as the OS1 output drops LOW again. The diode (D1) clips the positive spike to 0.6 to 0.7 volts, but it is reverse biased for the negative spike. It is the negative spike that drops below ⅓(V+) and triggers OS2. The time constant of OS2 is approximately 1 ms, closely matching the input pulse that triggered OS1.

You can design circuits such as Fig. 6-8 for almost any delay. In fact, a variable-delay line can be built by making the period of OS1 variable. The variable two-stage monostable circuit can also be used as a pulse positioner circuit.

Pulse Positioner

This project grew out of a visit by a student to my lab when I worked at a university on the east coast. The student wanted to

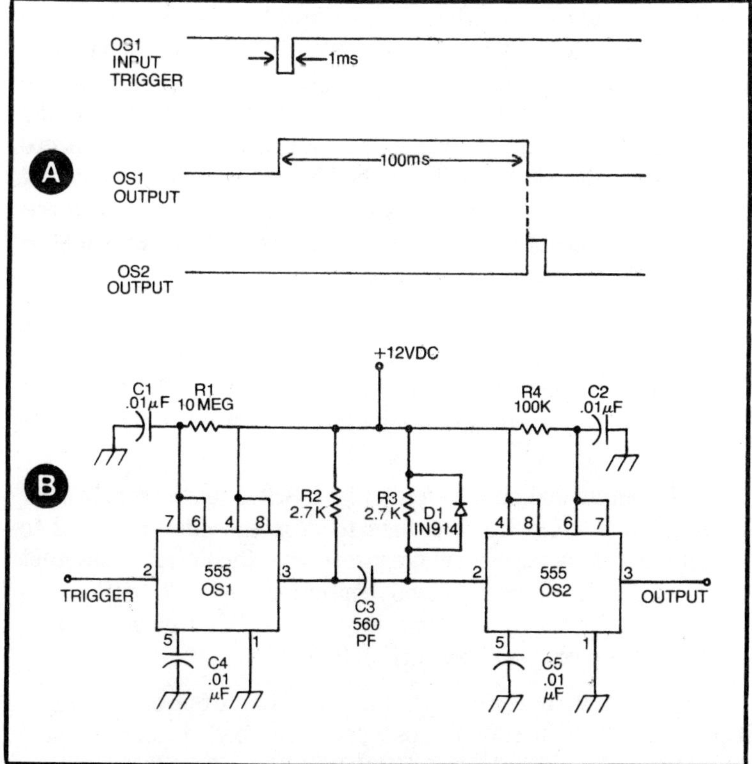

Fig. 6-8. Pulse delay line: A) Operation; B) Pulse delayer.

trigger a "sample and hold" circuit close to the peaks of a 60 Hz sinewave derived from the AC power line. It is relatively easy to design a circuit that will indicate the zero-crossings of a sinewave but a little more difficult (not by much, though) to make it synchronize with the peak. The peak is not a sharply defined event (as is the zero-crossing) but is broadened (see Fig. 6-9A). The problem, then, is to use some easily discriminated event such as the zero-crossing to position a pulse at the point in time when the peak occurs. The solution is the circuit of Fig. 6-8B modified according to Fig. 6-9B. By modifying the input one-shot (OS1) to trigger on zero-crossings, we can then vary the period of OS1 to position the pulse from the output of OS2 at any point within the range "delta-T" (in Fig. 6-9A).

The input of OS1 is modified to trigger from a zero-crossing detector, IC3. This IC is an LM-311 voltage comparator. Recall that a voltage comparator will issue an output that tells whether two voltages are equal. When the two input voltages are equal, the output of the comparator drops LOW (i.e., zero), otherwise it remains HIGH. In the circuit of Fig. 6-9B, we ground the noninverting input (pin 3), and connect the sinewave to the inverting input pin 2. As long as the sinewave voltage is nonzero the output of the 311 is HIGH, so the 555 one-shot is dormant. But when the AC sinewave voltages crosses zero, the output of the comparator drops LOW for a brief instant and will trigger one-shot OS1.

The position of the output pulse (which is generated when OS1 times out) is set by varying resistor R1A (a potentiometer). This will adjust the period of the one-shot over a small range that is less than the 8.35 ms period of half a 60 Hz sinewave (total period is 16.7 ms).

The output one-shot (refer to Fig. 6-8B) is set to 1 ms. In most cases such as this, we will want to drop the period of OS2 to something that is very small compared with the period of interest (i.e., 8.35 ms). I recommend approximately 100 to 800 microseconds for the output one-shot period. The pulse-position nature of this circuit is shown in Fig. 6-9C.

Note that there is a limit to the period of the 555 output pulse. If periods less than 100 microseconds are needed, then perhaps one of the TTL one-shots (e.g., 74121, 74122, or 74123) would be a better selection.

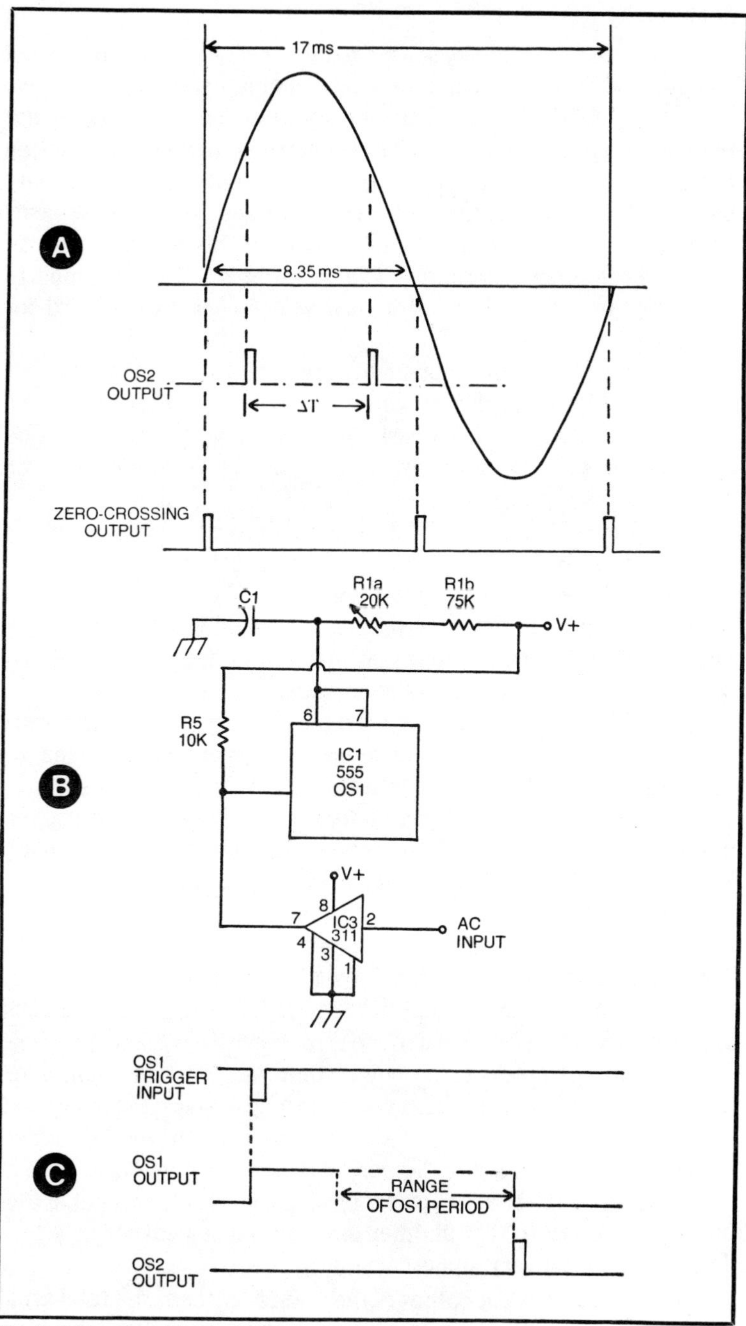

Fig. 6-9. Pulse positioner: A) Operation; B) Circuit; C) Timing.

Tachometry

The word tachometry is used to designate the measurement of a repetition rate. In the automotive tachometer, for example, we count the pulses produced by the ignition coil to measure the engine speed (in *rpm*). In medical electronic devices, they often measure factors such as heart rate or respiration rate electronically using tachometer circuitry. A heart rate meter measures the heart rate in beats per minute (*BPM*) using the signal from an electrocardiograph amplifier. The respiration rate is similarly measured in a device called a *pneumotachometer* ($10 word for breathing rate meter).

There is a certain commonality among non-digital tachometer circuits. It doesn't matter whether you are measuring audio frequency, engine rpm, heart rate, or any other rate. The difference seems to be in the component values of very similar circuits.

Figure 6-10 shows the basic tachometer circuit in block diagram form. Not all of the stages will be present in all instances, but some of them are basic to the problem. The AC amplifier and Schmitt trigger will, for example, only be used when needed. The one-shot and integrator, however, are always needed for they are fundamental to the measurement process.

The essential idea is to convert a frequency or repetition rate to an analog voltage. This is done by first converting the signal to pulse form. The AC input amplifier is used only if it is necessary to scale the input signal to a level where it will drive a Schmitt trigger or other squaring circuit. The purpose of the next stage, as we have just revealed, is to produce a square wave output signal at the same frequency as the input signal.

We are striving mightily to get the signal into shape to time-average it in the integrator. We must try to achieve a state in which the only variable is repetition rate. The integrator produces an output that is proportional to the *area under the input signal*. The squaring circuit produces an output that has constant amplitude, but its *duration* may vary. This variation may obscure the results.

The output pulse of a one-shot stage will have a constant amplitude and, because of the period of time-out, will also have a constant duration. The area under each pulse from the output of the one-shot is equal to that of other one-shot pulses earlier or later. The area does not vary, but is constant.

If we integrate the output of the one-shot, then, the total area is proportional to the repetition rate of the signal. We will then

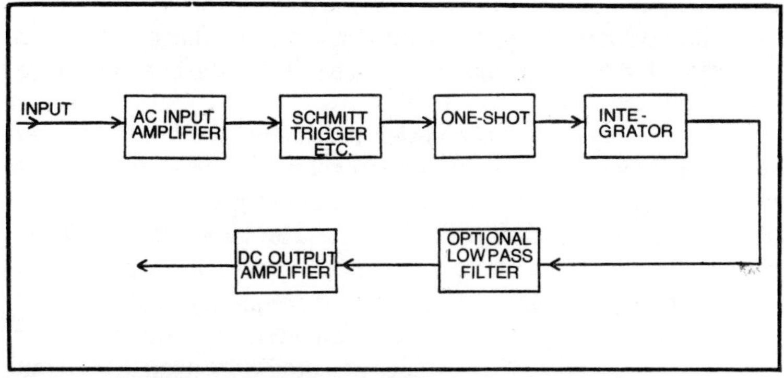

Fig. 6-10. Basic tachometer circuit.

have a voltage that is proportional to the frequency of the input signal.

We could then pass the signal through optional stages such as a low-pass filter (often unnecessary because the integrator is a low-pass filter itself) or DC amplifier (used to scale the output signal).

Telemetry Demodulator

One application of the tachometer principle is to make an FM telemetry subcarrier decoder. In some data telemetry systems the data signal is too low in frequency for direct transmission over telephone lines or radio channels. Typical communications circuits have a low-end frequency response of 300 Hz. The human ECG signal, on the other hand, has a maximum frequency content (highest harmonic in the Fourier series) of approximately 100 Hz. Clearly, an ECG signal cannot be directly transmitted along a telephone line or over a radio channel. Similarly a low-frequency signal such as the ECG cannot be tape-recorded directly. Again, low-frequency response of the medium is the problem.

The answer is to use an FM subcarrier. The subcarrier will have an unmodulated carrier frequency within the audio range and will be caused to deviate above and below this frequency by the data signal. The total signal, then , is a frequency-modulated audio subcarrier that is compatible with telephone lines, radio channels, and tape recorders. The tachometer circuit of Fig. 6-11A is used to demodulate the FM subcarrier when it is received.

The audio input signal (200 mV or more) is applied to a zero-crossing detector (IC1) similar to that in the previous circuit. This produces a chain of squarewaves at pin 7 of the LM311 that has the same frequency as the audio input signal. These pulses are differentiated by R2C2 and then used to trigger the one-shot stage (IC1, a 555).

The output of IC2 is a series of *constant amplitude-constant duration* pulses that have the same frequency, or repetition rate, as the AC input signal. These pulses are integrated in a three-stage RC integrator (C4/C5/C6/R5/R6/R7). The time constant of the integrator must be carefully selected to be very short with respect to the frequency of the data signal that frequency-modulates the subcarrier, yet long enough to act as an integrator/low-pass filter for the one-shot pulses. The values selected were used to transmit 1 Hz triangle waves with little distortion. The graph in Fig. 6-11B shows the output signal at point A as a function of carrier frequency.

Note in Fig. 6-11B that there is an offset voltage of approximately 0.5 volts. When the frequency is near-zero, then we would expect a zero output. But we have an offset potential instead. We use amplifier IC3 (an operational amplifier) to cancel the offset. A cancellation current is input to the op-amp inverting input through resistor R12. This current is adjusted by potentiometer R13.

The second operational amplifier reinverts the signal (for a net zero phase shift), and additionally produces some output scaling through gain variation (potentiometer R11). If the output potential is too small, incidentally, then use a larger value for R11 according to the ordinary rules for an operational amplifier inverting follower circuit.

The circuit in Fig. 6-11A is reminiscent of many demodulator circuits used in data telemetry and similar applications. The circuit has even been used (in highly modified form) as an FM demodulator in communications radio receivers.

Audio Frequency Meter

The same basic idea presented above can also be used to form an audio-frequency meter (Fig. 6-12) that uses an ordinary analog meter movement as the output indicator. We will vary the period of the one-shot to produce different ranges. Change resistor R1 according to Table 6-1.

The integration in this circuit is two-fold. One integration is performed by the single-section RC filter R4/C4, while the second is the damping time constant of the meter movement. This meter is

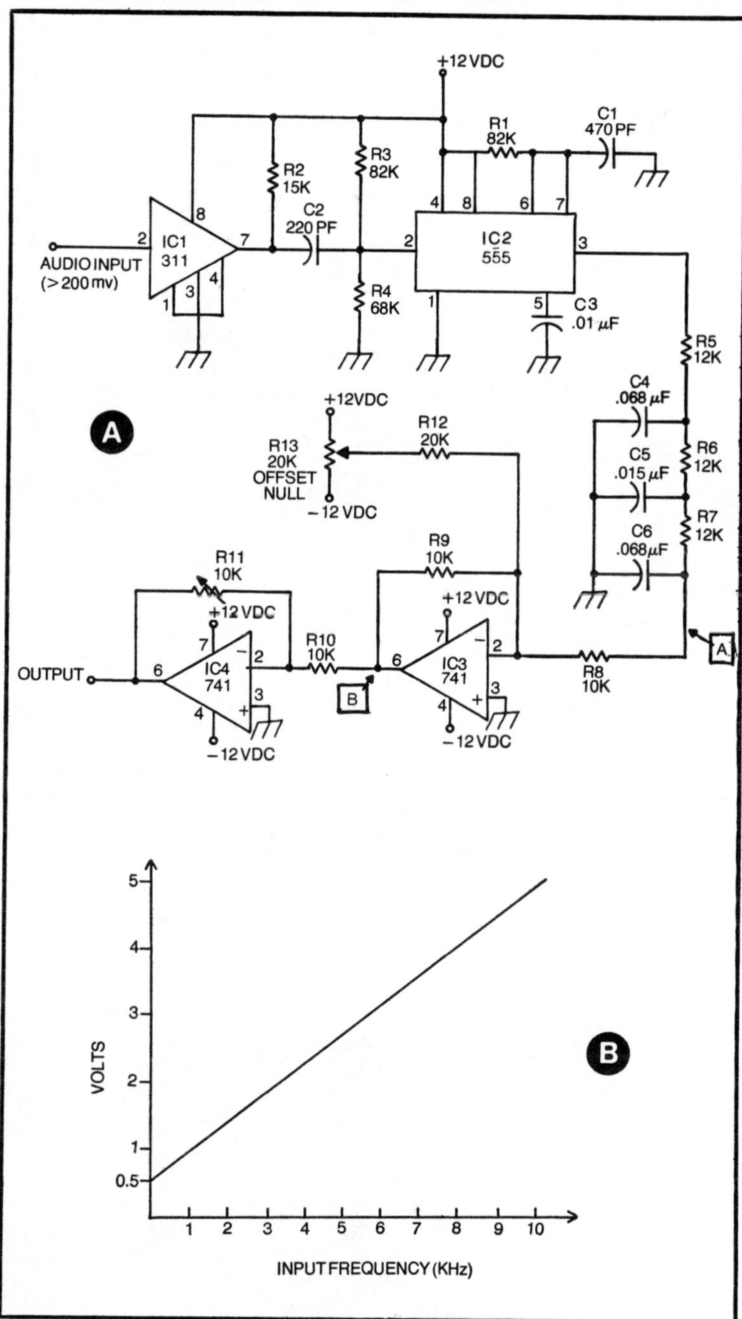

Fig. 6-11. Low frequency tachometer: A) Circuit; B) Output function.

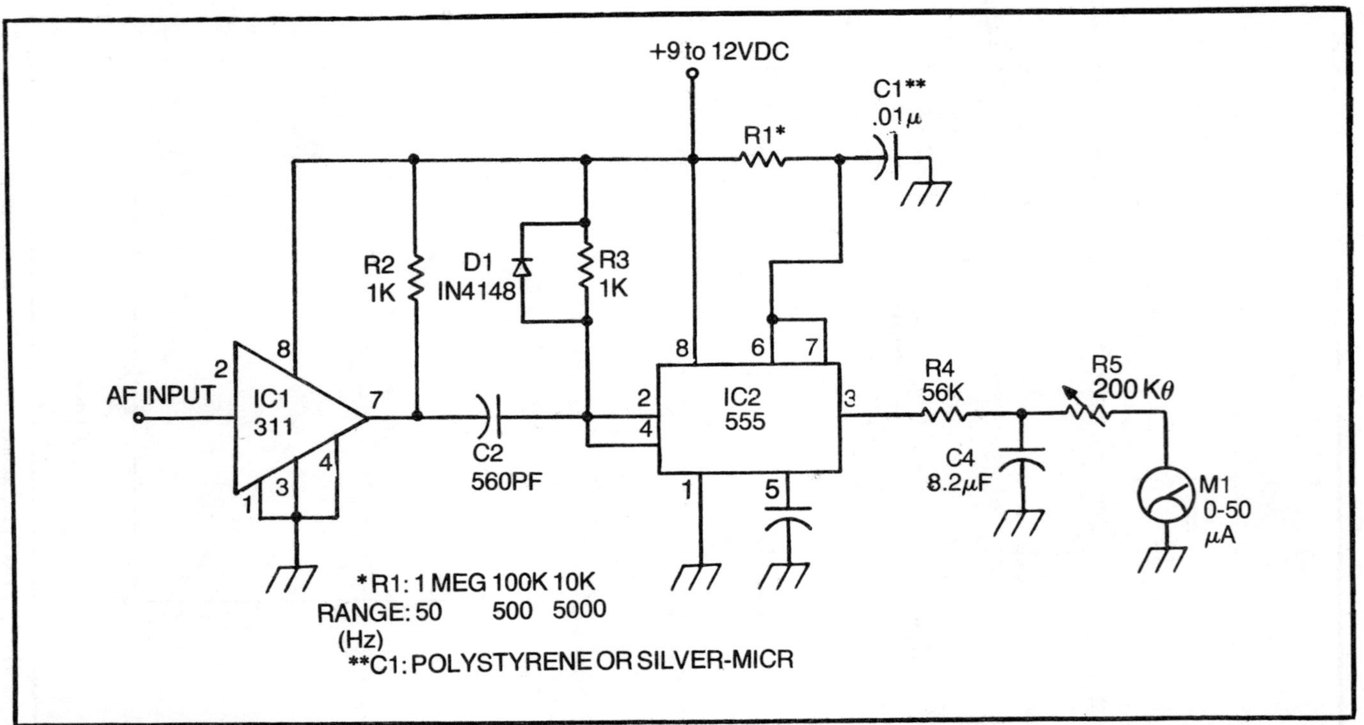

Fig. 6-12. Audio frequency meter.

Range (Hz)	50	500	5000
R1	1 meg	100 k	10k

Table 6-1. Resistor Changes.

a 0 to 50 μA current meter converted to a voltmeter by the action of series multiplier resistor R5. Adjust the resistor until the meter reads "50" when a maximum range AC signal (50, 500, or 5000 Hz) is applied to the input. It is important to the accuracy of the frequency meter to use precision resistors for R1 and R5 and a polystyrene capacitor for C1.

Chapter 7

Astable Operation of the 555

An astable multivibrator has *no* stable states. The output will flip back and forth between the two possible states, being happy in neither. The output of the astable multivibrator, then, is a square wave with a frequency set by the rate at which the output states change. We sometimes think of the astable multivibrator as a monostable that continually retriggers itself.

EXAMPLE

An example of a 555 used as an astable multivibrator is shown in Fig. 7-1. In Fig. 7-1A we see the block diagram form of the circuit, while Fig. 7-1B shows the circuit as it would normally appear in schematic diagrams. The timing waveform is shown in Fig. 7-1C, and a chart of the frequency/component-values is shown in Fig. 7-1D. The capacitor waveform and its relationship to the output waveform is shown in Fig. 7-1E.

We can easily understand the operation of this circuit by examining Fig. 7-1A. The circuit is similar to the monostable version but with some important differences. The two comparators (COMP1 and COMP2) are still biased by an internal resistor network such that COMP1 is biased to ⅔ (V_{cc}) and COMP2 is biased to ⅓ (V_{cc}). But note the treatment of the other two inputs to the comparators; here is a point of difference between the monostable and astable circuits. In the astable version the remaining inputs of the two comparators are strapped together and are connected to the capacitor in the timing circuit.

Under the initial conditions, when power is first applied to the circuit the voltage along the internal voltage divider comes up almost immediately. But the voltage across the capacitor rises much more slowly at a rate that is determined by the time constant (R1 + R2)C1. The trigger input at this instant sees a potential (the capacitor potential) that is less than ⅓ (V_{CC}, so the output of the comparator (COMP2) goes HIGH forcing the internal RS flip-flop to set (i.e., go to the condition where Q = HIGH, NOT-Q = LOW). The output of amplifier A1 is now HIGH, and transistor Q1 is turned off. Capacitor C1 will charge according to the usual time constant rules until the voltage reaches ⅔ (V_{CC}). At this point the output of COMP1 goes HIGH forcing the internal flip-flop to reset. This will cause the NOT-Q output of the FF to go HIGH, causing the a1 output to go LOW and turning on transistor Q1. The transistor is connected to the junction of resistors R1 and R2, so the charge on the capacitor is dumped through R2. The rate of decay of this capacitor voltage is set by time constant R2C1.

The voltage across C1 continues to decay until it drops below ⅓ (V_{CC}). At that point comparator COMP2 triggers, and forces the internal flip-flop into the set condition again. Note that the voltage across C1 (Fig. 7-1E) varies between ⅔ (V_{CC}) and ⅓ (V_{CC}), and this causes the output to flip back and forth between HIGH and LOW states.

There is a difference between the HIGH and LOW times. Capacitor C1 will charge through both resistors (R1 and R2) but will discharge through R2 alone. This means that the HIGH and LOW times will not be the same (see Fig. 7-1C). The HIGH time, T_1, is given by the expression:

$$T_1 = 0.693(R1 + R2)C1$$

while the LOW time is set by

$$T_2 = 0.693(R2)C1$$

The total period of the output signal is the sum of the HIGH and LOW times for each cycle, or

$$T = T_1 + T_2$$

$$T = 0.693(R_1 + 2R_2)C_1$$

The frequency of oscillation in any type of oscillator circuit is always the reciprocal of the period (i.e., the period divided into 1), so

$$F_{Hz} = 1/T_{sec} = \frac{1}{0.693\,(R_1 + 2R_2)C_1} = \frac{1.44}{(R_1 + 2R_2)C_1}$$

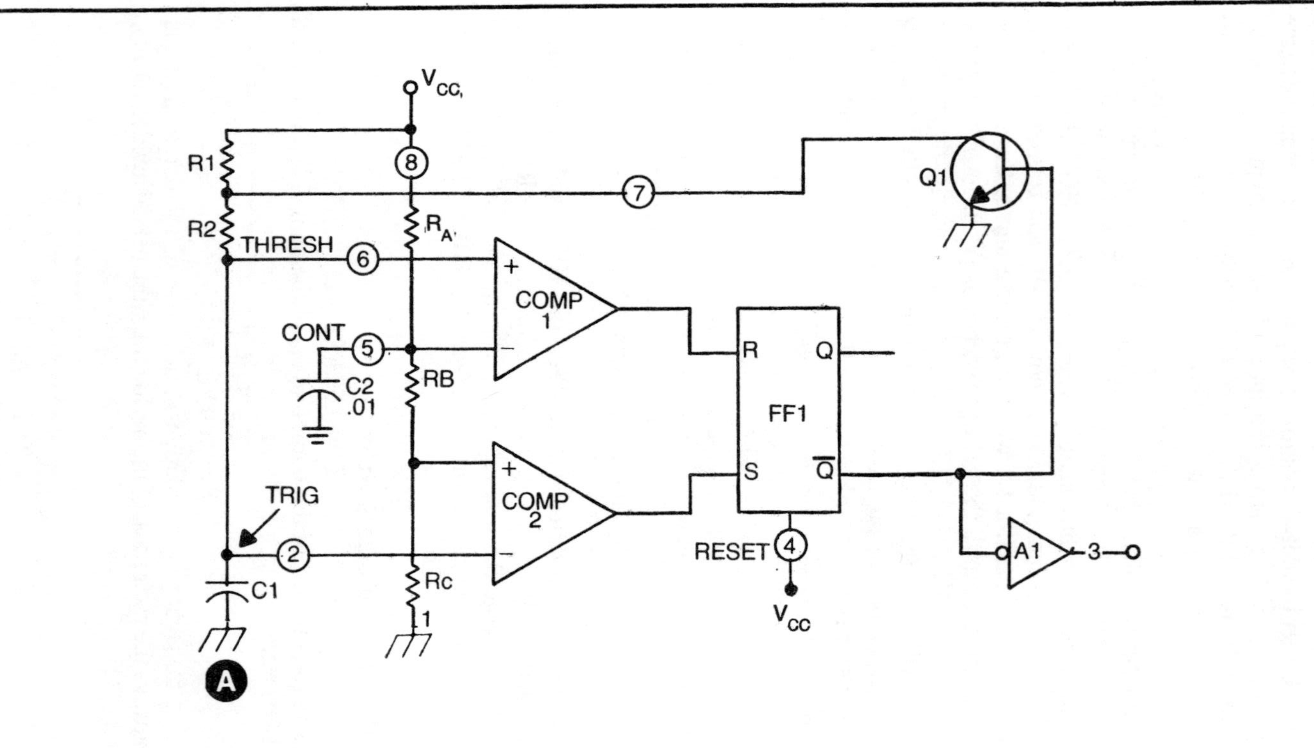

V_{CC}
R1
8
7
Q1
R2
R_A
THRESH
6
COMP 1
CONT
5
C2
.01
RB
R
Q
FF1
S
$\overline{Q}$
COMP 2
TRIG
2
RESET
4
A1
3
C1
Rc
1
V_{CC}
A

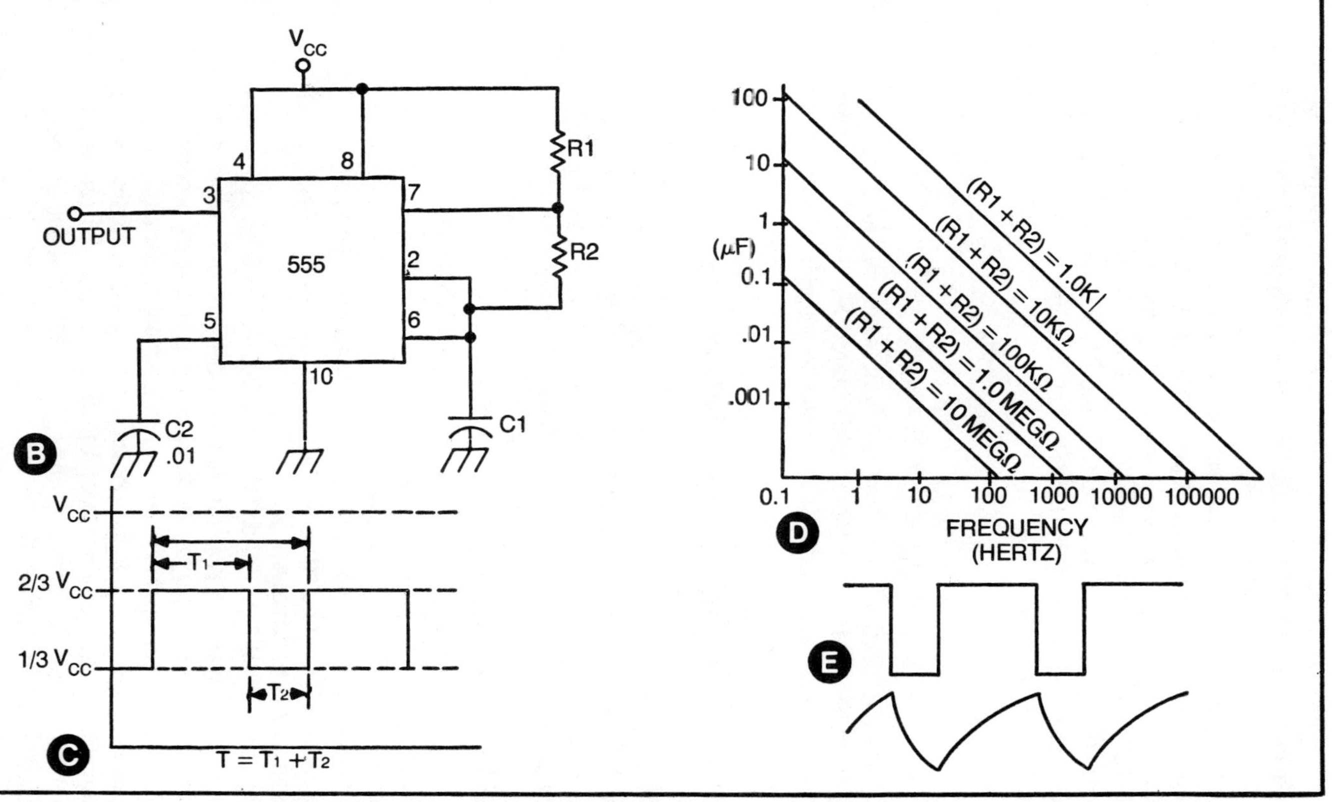

Fig. 7-1. Astable operation of the 555: A) Block diagram circuit; B) Circuit diagram; C) Duty cycle; D) Timing chart; E) Timing waveforms.

Figure 7-1D shows the above relationship graphed for frequencies from 0.1 Hz to 100 kHz, using "standard" component values.

The duty cycle of an astable multivibrator is the ratio of the HIGH to the LOW times. In a perfectly symmetrical squarewave generator the output is HIGH for as many milliseconds as it is LOW. The duty cycle, then, is 50 percent. In the 555 astable multivibrator, however, the HIGH and LOW times are unequal. The duty cycle, or duty factor as it is sometimes called, is expressed by

$$DF = \frac{T_1}{T_1 + T_2}$$

or,

$$DF = R2/(R1 + R2)$$

In the section to follow we will examine some of the applications of the 555 astable multivibrator. Some of the circuits are strictly astable, while at least one combines the astable and monostable circuits.

555 ASTABLE MULTIVIBRATOR APPLICATIONS

In this section we will consider some sample applications for the 555 astable multivibrator. These projects are designed to be built as-is, but they should best serve as a guide to circuit design. You are encouraged to make your own circuit designs to fit the actual need at hand, rather than aping the designs of the book.

1000 Hertz Oscillator

Figure 7-2 shows the circuit for a 1000 Hertz 555 oscillator that will drive a small loudspeaker. The circuit is a classic 555 astable multivibrator with component values selected to create an oscillation frequency of approximately 1000 hertz. The 555 output terminal will either source or sink up to 200 milliamperes, so it has ample power to drive a speaker.

If you want to use the 1000 hertz oscillator for troubleshooting, then delete the resistor and loudspeaker, and output pin 3 directly to the circuit under test. Since this circuit produces a squarewave signal, it will be rich in harmonics. Because 1-kHz square waves produce harmonics well into the medium-wave frequency range, they are useful to troubleshoot AM radios and ham radio equipment on frequencies in the 80-, 75-, and 40-meter bands. The signals will appear every kilohertz through the spectrum.

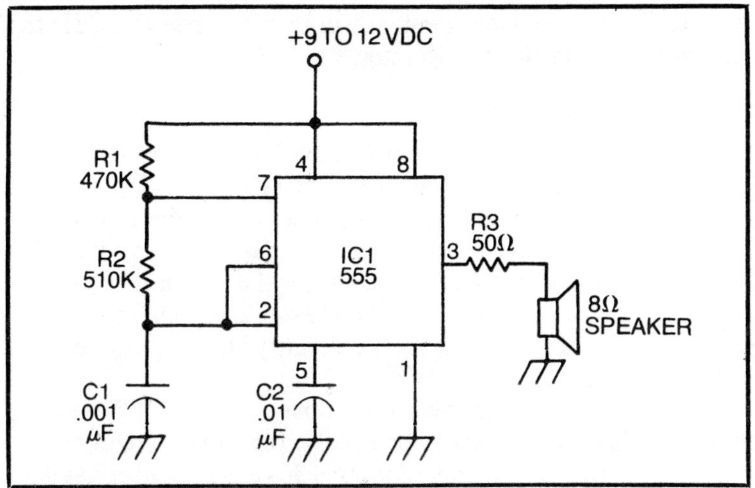

Fig. 7-2. 1000 Hz oscillator.

Water Level Detector (I)

Water level detectors are alarm circuits that are designed to alert a human or machine when the liquid level in a tank or other holding apparatus exceeds a certain level. This may be a means of detecting flooding conditions or to automatically control the filling process.

Several different circuits are possible for liquid level detection. One version is shown in Fig. 7-3A. This circuit is a 555 astable multivibrator that is connected to a sensing circuit that turns on the astable when the electrodes are shorted by water.

Transistor Q1 is connected in parallel with the discharge transistor inside the 555 timer IC. When transistor Q1 is turned on, the capacitor will discharge through resistor R2. If Q1 remains turned on, then the capacitor voltage will remain discharged, and the circuit will not oscillate. But if the forward bias is removed from Q1, then the transistor is off and capacitor C1 will work as ordinarily anticipated.

The sensor consists of two electrodes set into the container being filled. When the liquid level exceeds the point where the electrodes are shorted, then the forward bias of transistor Q1 is shorted to ground. This will cause Q1 to cut off, and the circuit begins to oscillate.

We have two options for the frequency of this astable. The high frequency version will turn on the LED and drive an audio output. The low frequency option will merely cause the LED to turn on and off, alerting the operator.

Water Level Detector (II)

There are times when we will want to control appliances or devices (e.g., pumps) that operate from 115 volts AC power in response to rising liquid level. Most solid-state components will not tolerate 115 volt AC power, so we must resort to relay control. An example of a relay-controlled water level detector is shown in Fig. 7-3B. This circuit is similar in all respects, except the output circuit, to Fig. 7-3A. The rest of the circuitry has been deleted for sake of simplicity.

The relay driver, in this case, is one section of a 7406 hex inverter. This TTL IC has an open-collector output circuit, so it can interface with devices operating to +30 volts instead of just +5 volts normally expected of TTL devices. The relay coil is connected as the collector load of the inverter output transistor, so will turn on when the inverter goes LOW. Normally, when the water level is low, the output of the 555 is LOW so the inverter output is HIGH. This keeps the relay turned off. But when the water level is high, then the 555 output is also HIGH and this forces the inverter output LOW (turning on relay K1).

The diode shunting the relay coil is used to damp any high voltage spikes caused by the "inductive kick" seen when the relay turns off. Such spikes can ruin solid-state electronic circuits, so must be suppressed. The diode can be almost any rectifier diode, and I recommend the 1N4007 device.

The load is controlled by relay contacts A1, A2, and A3. Contact A2 is the armature and is common to the two circuits. Contact A1 mates with A2 in the normally closed (i.e., coil de-energized) mode, while contact A3 is mated with A2 in the energized mode.

We can trigger this circuit either with the system of Fig. 7-3B, or we can use the method of Fig. 7-3C. By connecting a pull-up resistor to V+ from the trigger input, and then connecting the electrodes between the trigger input and ground, we can trigger the 555 when the water level rises to the trip point.

There is still one other version that is useful, and this is shown in Fig. 7-3D. The previous two circuits have been designed to trip the relay only during times when the water level is above the set-point of the electrodes. The version of the circuit in Fig. 7-3D, however, will latch and hold once tripped. The circuit works in the same manner as the other two, except that now we have two additional relay contacts. When the output of the 7406 inverter goes LOW, the "cold" end of the relay is grounded. Current will flow to ground from the coil through both the inverter output transistor and the relay contacts B2-B3. But when the inverter output tries to go HIGH again it sees the grounded relay contacts, and the short keeps the relay self-latched. The only way to remove the latched condition is to either turn off the power to the circuit or interrupt the current path to the relay contacts. This is the function of reset switch S1. When this normally-closed pushbutton is pressed, the path is opened and current flow is interrupted. This will de-energize the relay and turn off the alarm. The alternate contacts (A1 through A3) are used to control the pump, sound the alarm, or whatever you need for the application at hand.

Frequency-Synched Astable

It is often desirable to synchronize an oscillator to some external frequency source. Your television, for example, uses both vertical and horizontal oscillators that are synched via sync pulses to the TV transmitter at the station. This is done to ensure that the TV receiver scans the same raster as the TV camera.

We can synchronize the 555 astable multivibrator by using a circuit such as Fig. 7-4. This circuit uses the coincidence of two pulses to trigger the 555 astable multivibrator. The gate is a TTL NAND gate. Recall that these gates will produce a HIGH output if either input is LOW and will produce a LOW output only if *both* inputs are HIGH. As long as the 555 output is HIGH, then pulses from the sync-clock frequency F_s can pass through the gate to the differentiating network that triggers the 555 device. If the sync-clock frequency disappears, then the astable will free-run on its resonant frequency. This frequency must be close to the sync frequency or have a harmonic relationship to the sync frequency.

One-Resistor Astable

We can simplify the design of the astable multivibrator and eliminate one of the two resistors (Fig. 7-5A). This mode of operation is perhaps, a little far from the method intended by the 555 designers, but it works. When power is initially turned on the output is HIGH. This will cause the capacitor to begin charging through resistor R1. When the capacitor voltage exceeds the magic figure, ⅔(V+), the output of the 555 will drop LOW causing C1 to

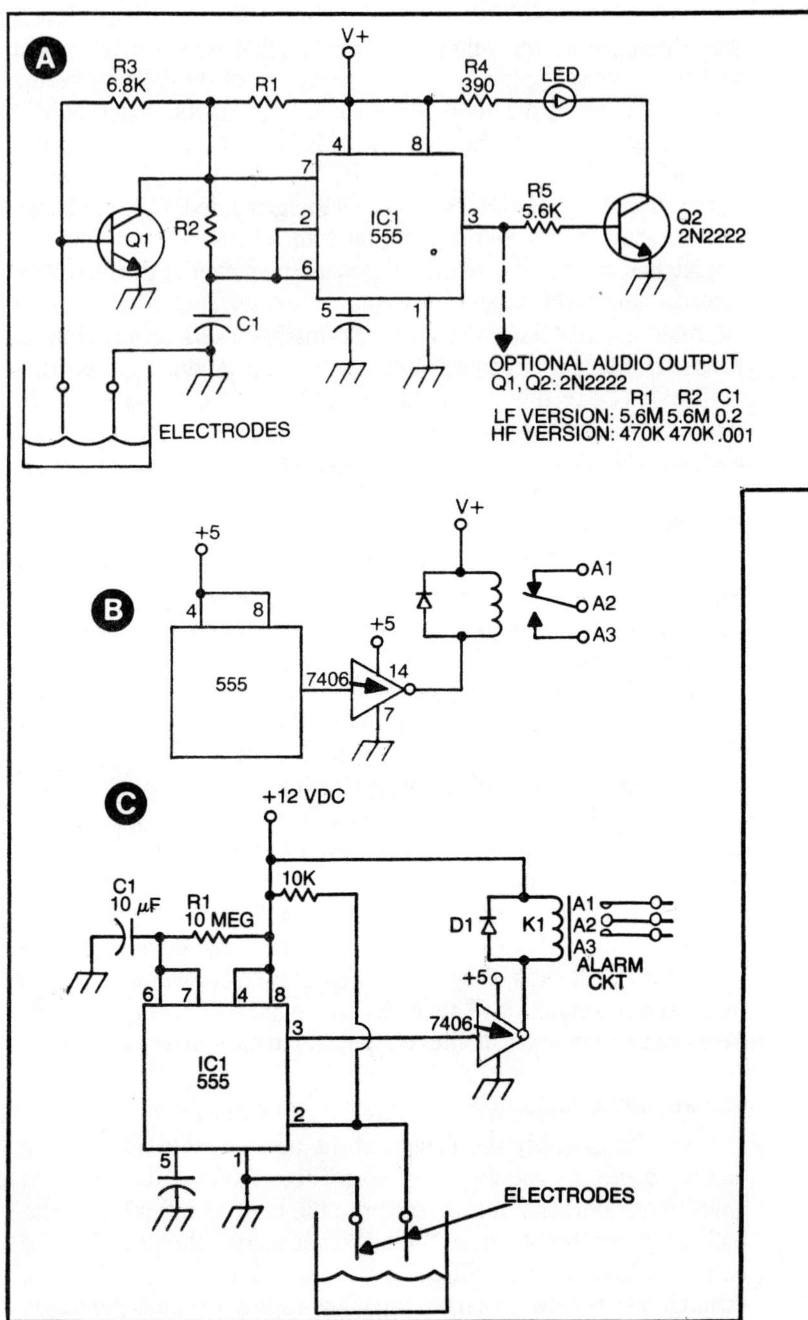

Fig. 7-3. Water level detector.

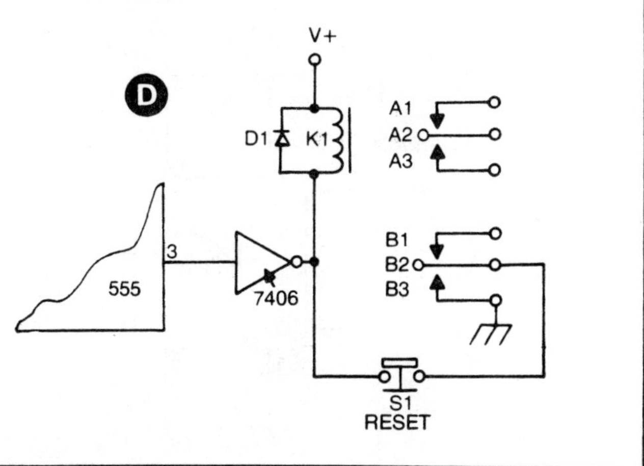

discharge through resistor R1. The approximate frequency of oscillation is given by

$$F = 0.7R_1C_1$$

We can also make this circuit variable, allowing the output frequency to be trimmed to the desired frequency. This is shown in Fig. 7-5B. In this case, we have split the resistance of R1 into two portions. One is a fixed resistor while the other is a potentiometer.

Variable Duty Cycle Astable Multivibrator

The duty cycle, or duty factor, of an astable multivibrator relates the total HIGH time to the total LOW time. In the 555 astable circuit, the duty cycle is set by the two resistors in the frequency-setting circuit. The relationship is:

$$DF = R2/(R1 + R2)$$

So, by varying the ratio of the two resistors, we can also vary the duty factor. This is done in the circuit of Fig. 7-6. Resistors R1 and R2 are connected to either end of a potentiometer. The wiper of the potentiometer serves as the "junction" between the two resistances and is connected to pin 7 (i.e., the collector of the internal discharge transistor).

Code Practice Oscillator

Learning the International Radiotelegraph (Morse) Code requires practice. We must listen to competently transmitted code to learn to receive, and we must transmit code in order to learn to send. One of the tools needed to properly learn the code is a *code*

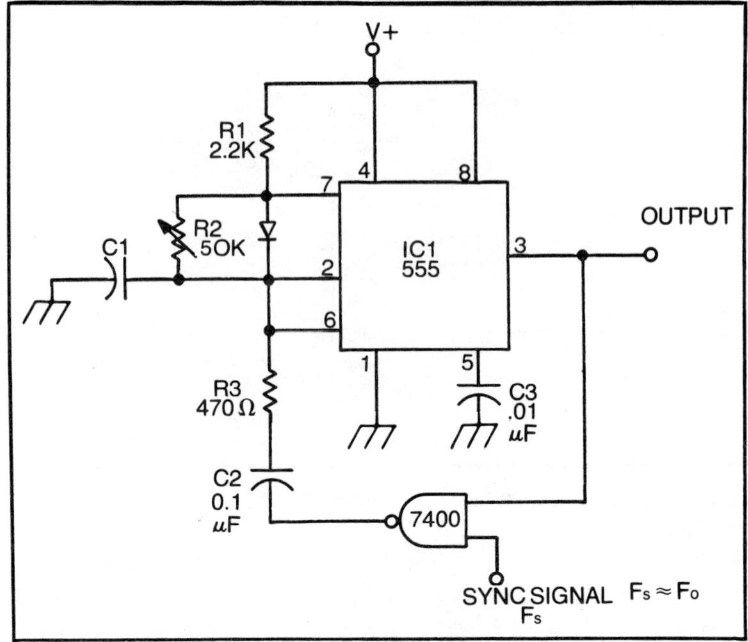

Fig. 7-4. 555 synchronized oscillator.

practice oscillator (CPO). The CPO is an audio oscillator (300 to 1200 hertz) that can be keyed on and off by a telegraph key.

An example of a code practice oscillator circuit is shown in Fig. 7-7. The oscillator is a 555 astable multivibrator with component values selected to produce a frequency on the order of 600 hertz. The 555 is capable of driving a small loudspeaker or pair of high-impedance earphones. The earphones are connected to a 5000 ohm volume control (R3) and to the series-wired telegraph key jack (J1). The volume control is adjusted to produce a comfortable level in the earphones. The operator can then turn the tone on and off with the telegraph key.

Tone Generator

A tone generator is used to produce a nearly sinusoidal audio tone for purposes of testing, signaling, etc. The 555 astable multivibrator will produce the audio tone, but it is in the form of a square wave. Figure 7-8 shows how to arrange the output circuit for an astable multivibrator to filter out the harmonics of the squarewave and thereby produce a sinusoidal (or nearly so) output waveform.

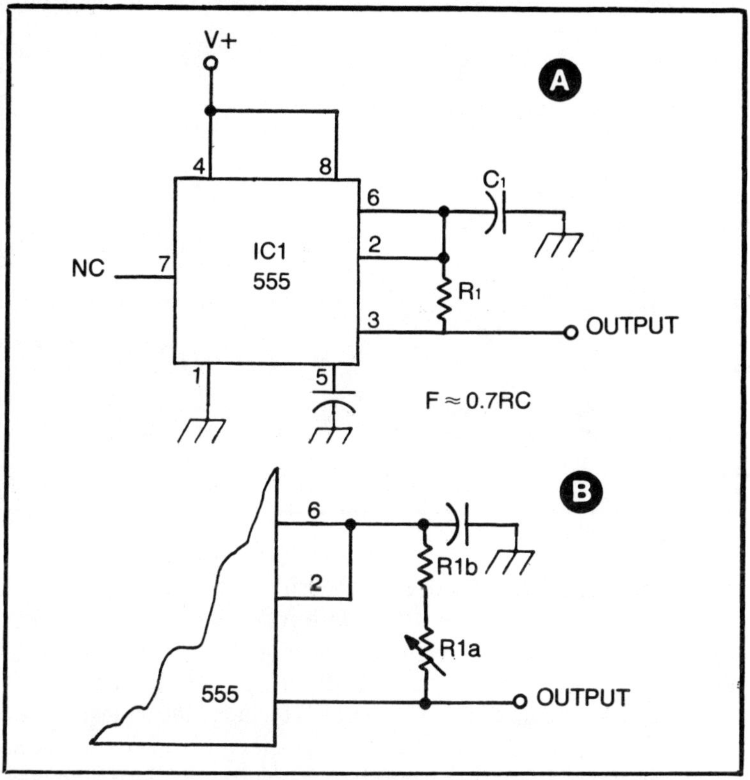

Fig. 7-5. 555 astable multivibrator: A) Simplified; B) Variable frequency.

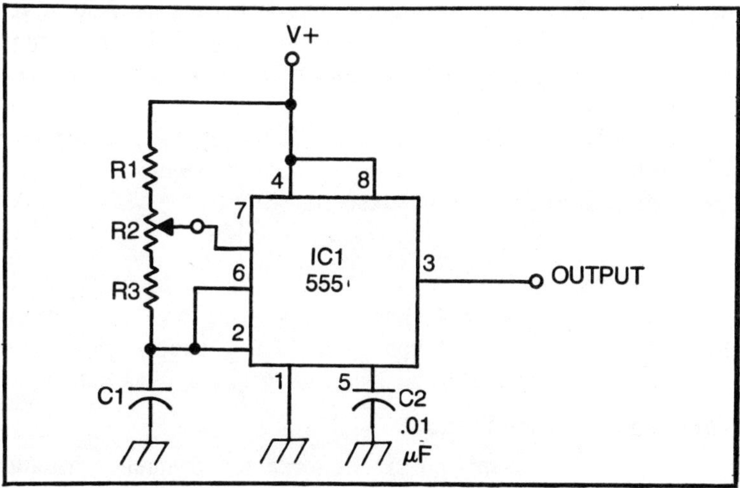

Fig. 7-6. Variable duty cycle 555 astable multivibrator.

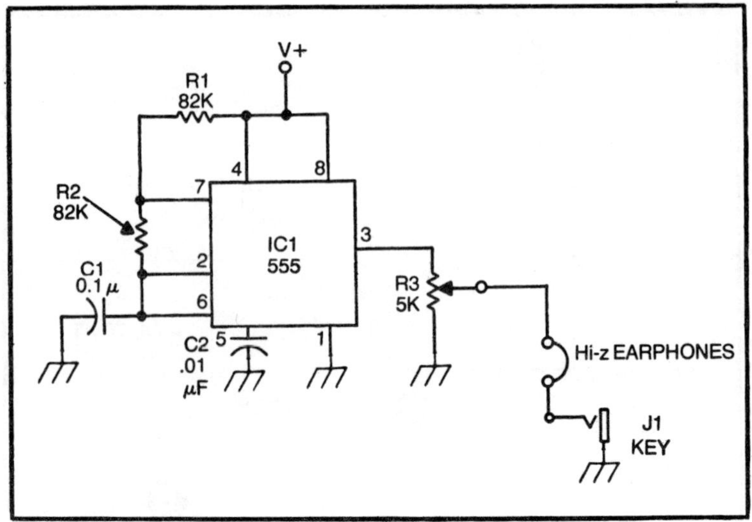

Fig. 7-7. Code practice oscillator.

The 555 astable section of the circuit produces an output signal consisting of squarewaves with a frequency set by timing resistors R1 and R2 and by timing capacitor C1. The output (pin 3) is applied to the input of a two-stage RC integrator. The function of an integrator, in this case, is low-pass filtering of the square wave. The components of this network are selected to have a cutoff frequency that is slightly higher than the fundamental frequency of the 555 astable multivibrator. This will cut off most of the harmonics and severely attenuate the lower harmonics. Additional sections of filtering will provide a more nearly sinusoidal output but will cost something in amplitude. If the amplitude is too low, then an operational amplifier output stage could be provided (see TAB Book No., 787, *Op-Amp Circuit Design and Application*, by J.J. Carr) to boost the output signal. The component values of the integrator stage can be selected according to

$$F = \frac{1}{2\pi RC}$$

A volume control is shown in this circuit, but it is optional. Whether you use it will depend upon your own application. Resistor R5 is used as a voltage divider to produce low-level outputs, and (again) it is optional depending upon application.

One Hertz Astable Multivibrator

Figure 7-9 shows the circuit for a low-frequency astable multivibrator. This circuit will produce an output signal of

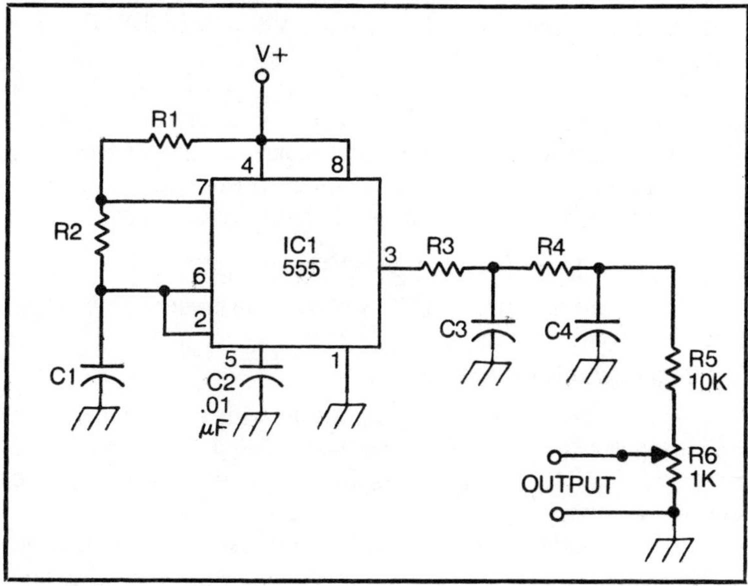

Fig. 7-8. Tone generator.

approximately 1 hertz so can be used as a timer or as a display clock for a digital frequency counter.

In cases where the circuit is used as a timer, little modification is needed. The circuit will drive most counters nicely. If TTL counters are used, then set the V+ potential to +5 volts DC. If

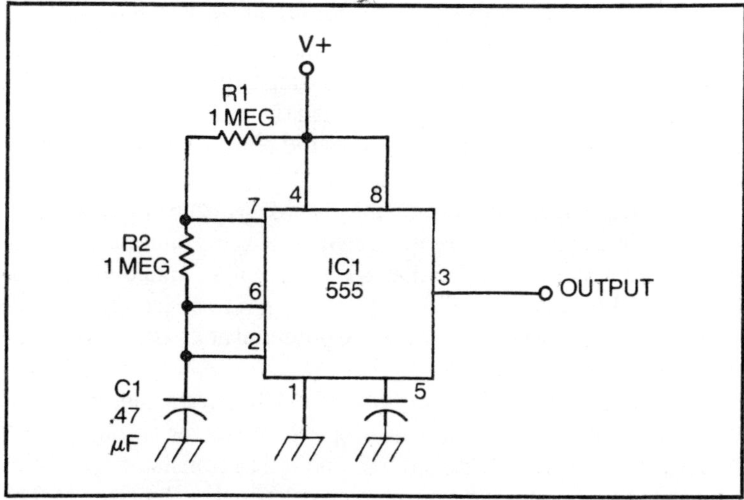

Fig. 7-9. One-Hertz astable multivibrator.

CMOS devices are used, then set the V+ potential to the V+ potential of the CMOS circuitry.

Using the circuit as a display clock may require some adjustment of the duty cycle. Most such applications prefer a small duty cycle, so it would be necessary to adjust R1 and R2 according to the formula given earlier in this chapter. The frequency accuracy will be maintained, however, if the rule below is followed:

$$R_1 + 2R_2 = \frac{1.44}{(1\,\text{Hz})\,(4.7 \times 10^{-7}\,\text{farad})}$$

Preservation of this equality will result in preservation of the frequency accuracy of the circuit.

TTL Output Astable Multivibrator (I)

The 555 is not necessarily compatible with TTL devices. The 555 is bipolar and uses an NPN transistor output stage. But it will not be compatible with TTL (despite the ability to sink or source 200 mA of current) unless the V+ supply potential is +5 volts DC.

Figure 7-10A shows a method for interfacing the 555 astable multivibrator to TTL digital ICs. The timing network (R1, R2, and C1) sets the operating frequency of the circuit. The output signal amplitude, however, depends upon supply potential, V+.

We can use any V+ potential we want because the output stage that actually drives the TTL circuit operates from +5 volts DC. We could set a separate +5 volt supply for transistors Q1 and Q2, or we may use the zener diode method shown in Fig. 7-10A. The 4.7-volt zener diode and series-limiting resistor R7 serve to drop the V+ to a regulated +5 volts that is used in TTL circuits.

If we could tolerate an inverted output (often the case with clock signals) then we could eliminate transistor Q2 plus resistors R5 and R6. The use of the two transistor serves to provide an output signal phase that is the same as the 555 phase applied to the base of Q1.

When the output of the 555 (pin 3) is LOW, transistor Q1 is cut off. This means that there is no current flowing in collector resistor R4, so the collector potential at Q1 is +5 volts. This potential is applied to the base of transistor Q2, causing it to turn on hard. The effect of saturating Q2 is to cause the potential at its collector to be very nearly zero (i.e., a LOW).

When the output of the 555 goes HIGH, transistor Q1 is turned on hard (saturated). This drops the collector voltage of Q1 to zero, thereby removing the bias applied to transistor Q2. This causes Q2 to cut off, and its collector voltage thereby rises to +5 volts (i.e., HIGH).

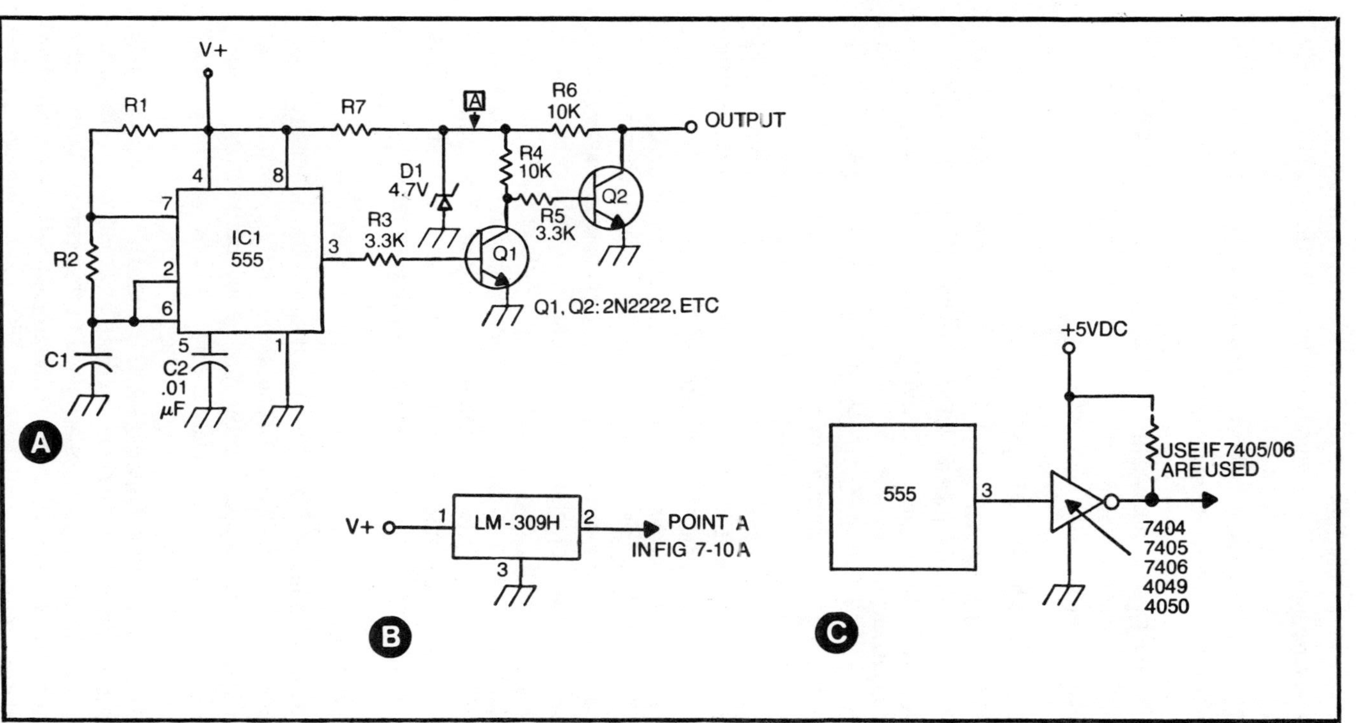

Fig. 7-10. TTL compatibility.

We can modify this circuit somewhat to eliminate the zener diode and resistor R7. This modification serves to bring the supply potential up to +5 volts, instead of the nominally +5 volts represented by using a 4.7-volt zener diode. The regulation and voltage drop in this case is provided by a small LM-309H +5 volt (100 mA) three-terminal IC voltage regulator. The H-series LM-309 devices are housed in TO-5 or plastic transistor cases and are limited to providing 100 mA of current (instead of 1000 to 1500 mA of the TO-3 LM-309K devices).

TTL-Output Astable Multivibrator (II)

The transistor circuit of Fig. 7-10A will provide proper operation, especially if we do not have a spare TTL gate or inverter section available. But, in most digital projects, we will have a spare gate or inverter section available. Devices such as the 7400 (NAND gate) or 7402 (NOR gate), plus several related gates, have multiple sections and not all may be used in any given application. Similarly, the inverter IC packages are *hex inverters*, meaning that they contain six inverter stages. The 7404, 7405, and 7406 devices are examples. In the CMOS line, we have the 4049 inverter and the 4050 hex buffer (noninverting). Both of these devices will provide TTL output of the package supply potential is +5 volts D.C.

The use of an inverter to provide TTL output is shown in Fig. 7-10C. The circuit does not show any pinouts because any of several inverters or gates can be used. The pull-up resistor is needed if any of the open-collector (e.g., 7405 or 7406) inverters are used. The value of the pull-up resistor should be 1.5 kohm to 4.7 kohms. If the NAND or NOR gates are used, then connect the two inputs of the single section together to form an inverter. This use was discussed in the chapter on digital electronics.

Warble Siren

Have you heard those electronic sirens the police are using these days? They don't make a continuous wailing, such as the old mechanical sirens, but warble, alternating between two tones (BeeeeeBooooo). The circuit in Fig. 7-11 is a bitonic audio generator that can be used to form a warble siren.

The main siren section is formed by IC2, which is a 555 astable multivibrator that will oscillate at either of two frequencies. Two timing capacitors are used in this circuit. When relay contacts 1-2 are open, then only capacitor C4 is in the circuit. The oscillation frequency will then be 550 hertz. But when relay

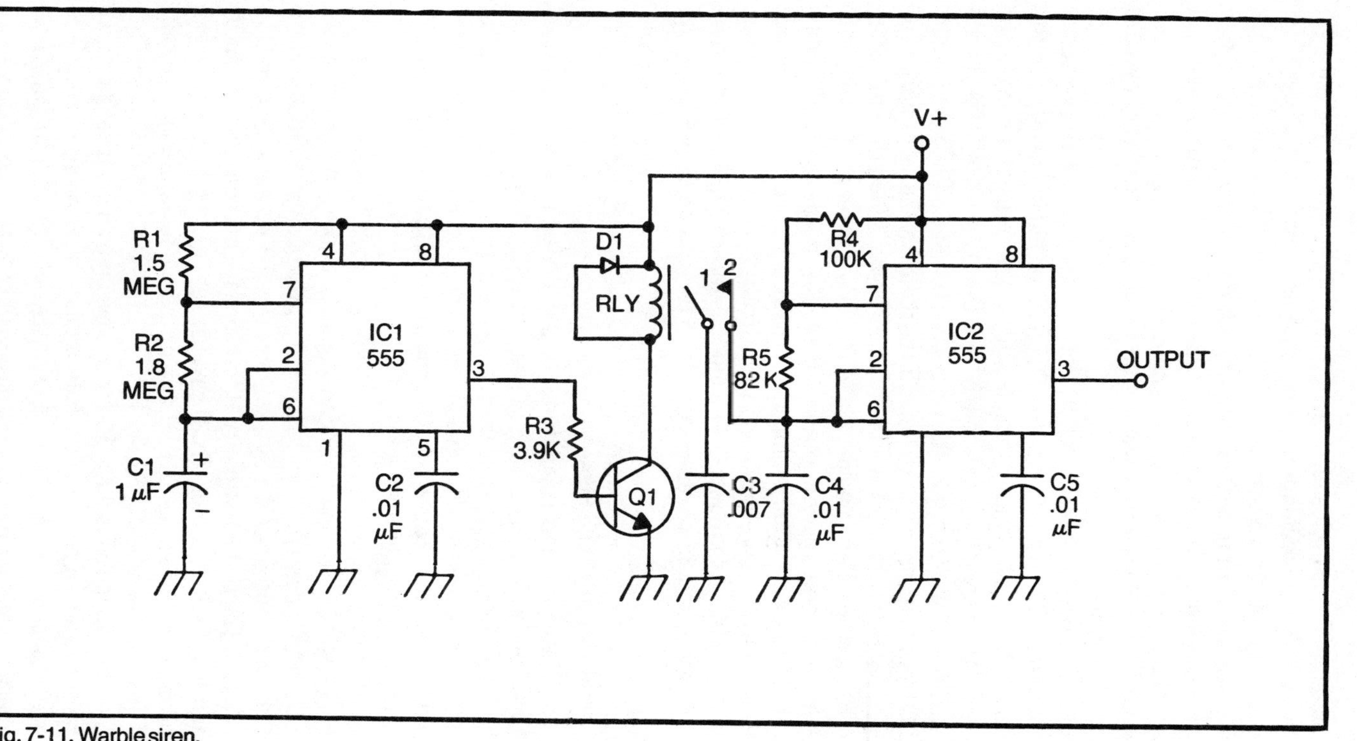

Fig. 7-11. Warble siren.

contacts 1-2 are closed, both C3 and C4 are in the circuit (in parallel), so the oscillation frequency is 350 hertz.

Control of relay switching is the function of IC1. This circuit is also an astable multivibrator (555) but operates at a much lower frequency (i.e., approximately 0.28 hertz). This frequency will allow the tones of the siren to cycle on and off at approximately 3.5 seconds (the tone times will not be equal, due to the duty cycle of the 555 output signal).

The output signal from IC1 drives the base of relay control transistor Q1. This transistor (2N2222 or similar) will turn the relay on and off which is connected as the transistor's collector load. When the output of IC1 is HIGH, then transistor Q1 is turned on hard and will energize the relay coil. But when the 555 output is LOW, then the transistor is cut off and the relay thereby energized.

Diode D1 is used as a spike suppressor. When the coil of the relay is energized, a magnetic field builds up around its windings. When the coil current ceases, then this magnetic field dumps current back into the circuit and causes a voltage spike that can damage transistors and integrated circuits; and it can raise havoc with all forms of circuitry. The diode is forward biased for normal potentials, but is reverse biased for the *inductive kick* potential spike.

Ten-Minute Timer for Amateur Radio ID

The rules and regulations of the Federal Communications Commission *(FCC)* require an amateur radio operator to transmit his call sign as part of the identification of the transmission at intervals not more than ten minutes apart. The amateur engaged in a roundtable, or long-winded ragchew, must, therefore, keep track of the time in order to ID legally.

The circuit in Fig. 7-12 provides a brief "beeeep" every 10 minutes. The timer operation depends upon a 0.1 hertz (i.e., period 10 seconds) clock (IC1) and a series of dividers. IC2 divides the 0.1 Hz output of IC1 by a factor of ten. This makes the output of IC2 0.01 Hz, or 1 pulse every 100 seconds. This pulse is counted by another 7490 counter (IC3), which is designed to produce a binary-coded decimal (BCD) output that reflects the input count. The output states of IC3 follows the A-B-C-D (1-2-4-8) system.

Ten minutes is 60 × 10, or 600 seconds. This means that we want to recognize the sixth 100-second pulse in our circuit. The sixth count will produce an output (ABCD) of 0110 from IC3. This signal is applied to one set of inputs of the 7485.

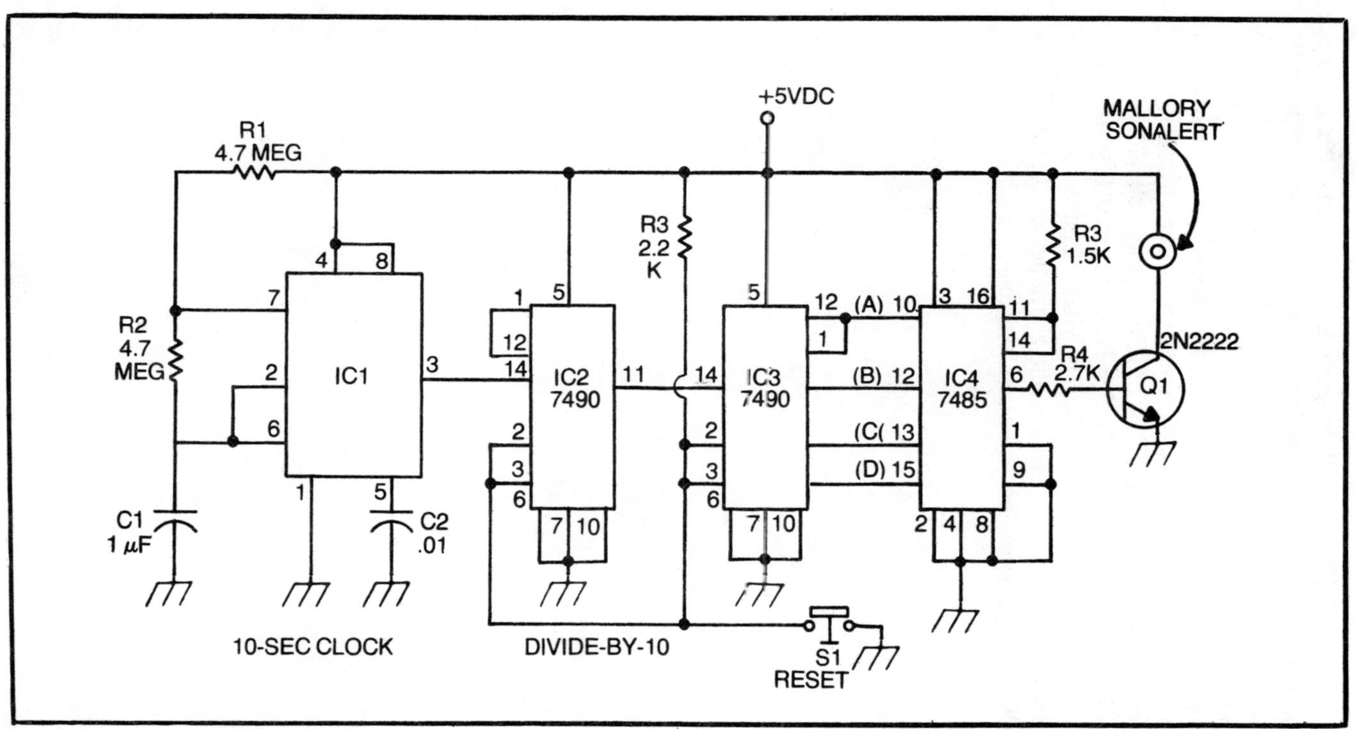

Fig. 7-12. Ten-minute Ham radio timer.

The 7485 IC is a TTL magnitude comparator. It compares two, four-bit words ("A" and "B") and issues an output that indicates whether A is less than B, A is greater than B, or A equals B. The output of IC3 is applied to the four A inputs of the 7485. The B inputs of the 7485 are connected by hard-wired logic to V+ or ground, as needed to produce the 0110 pattern that indicates "6." When the output of IC3 reaches 0110 (6_{10}), then the 7485 sees an equality situation, so causes the A = B output (pin 6) to go HIGH.

When the output of the 7485 is HIGH, then transistor Q1 is forward biased and this turns on the alarm device. In this case, a Mallory *Sonalert* is selected for the alarm tone.

Chapter 8

Exar XR-2240

The RC timing scheme introduced in the chapter on the 555 has been extended in the Exar XR-2240 timer. This timer uses an eight-bit binary counter to produce times of 1 to 256 times the clock period. The internal circuitry of the XR-2240 is shown in Fig. 8-1. The version in Fig. 8-1A shows the internal block diagram, while Fig. 8-1B shows the actual circuitry in simplified form.

The XR-2240 will operate over a supply voltage range of +4.5 volts to + 18 volts DC. These voltages are also used for CMOS devices, but the XR-2240 is a bipolar integrated circuit, not CMOS. The timebase section is a clock circuit based on the same principles as the 555 timer circuit (Fig. 8-1B). One main difference between the timebase portion of the XR-2240 and the 555 is in the relative reference levels created by the internal resistor voltage dividers (R1, R2, and R3 in Fig. 8-1B). In the 555 timer all three resistors had the same value, giving comparator bias levels of ⅓(V +) and ⅔(V +). In the Exar XR-2240, on the other hand, the reference levels biasing the comparators are 0.27V + and 0.73V +, respectively. One result of this (and probably the intended result) is simplification of the equation that gives the period of the output waveform:

$$T = RC$$

Where:

T is the time in seconds of the monostable circuit
R is the resistance in Ohms
C is the capacitance in farads

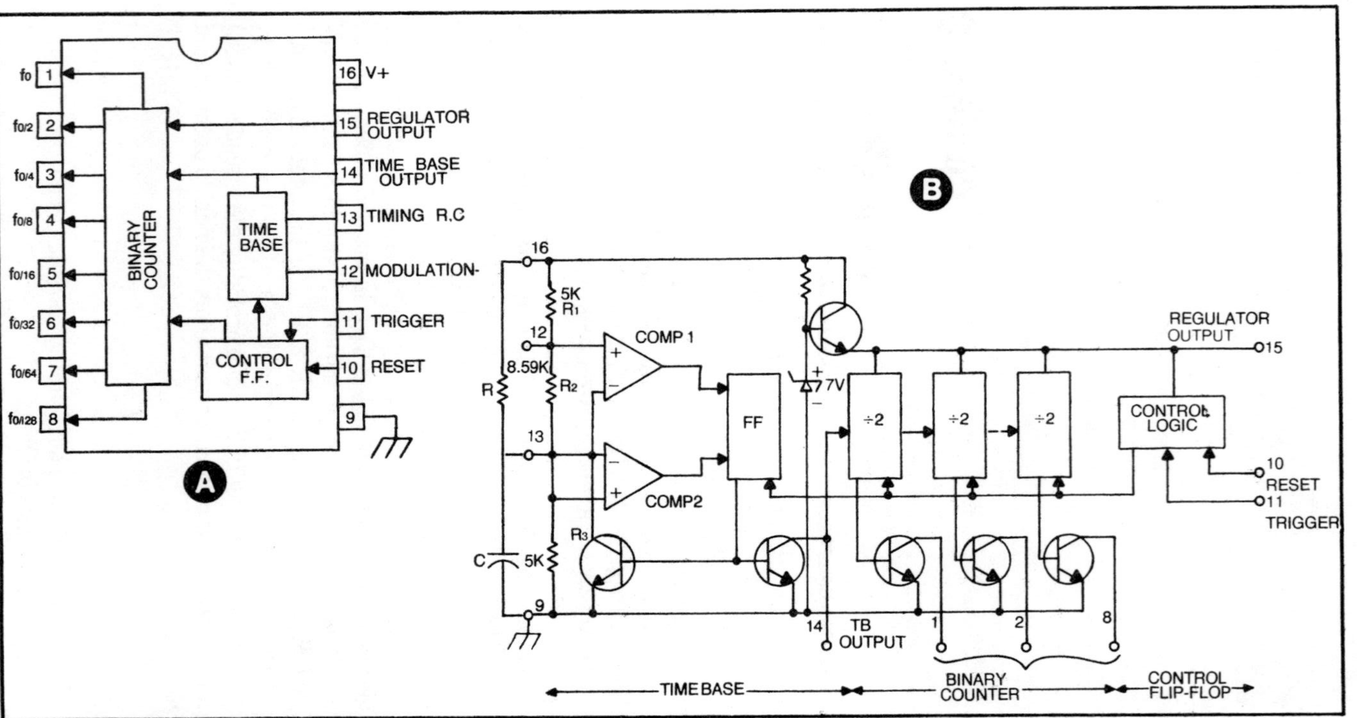

Fig. 8-1. XR-2240 timer: A) Pinout; B) Block diagram.

The binary counter section contains a cascade chain of J-K flip-flops connected in the standard manner for binary frequency division. Each stage in the binary counter functions as a divide-by-two counter. The binary counter chain is connected to the output of the timebase section through NPN open-collector transistors. The transistor collector of the timebase section is also connected to IC pin 14 (timebase output) so that a 20 kohm pull-up resistor can be connected between the collector and the output of the internal regulated power supply (pin 15).

Digital outputs from this counter are, in the usual fashion, given as voltage levels at the IC pins. Each output bit is delivered to a specific terminal of the IC package where it is connected to a pull-up resistor similar to that of the timebase out connection. The output terminals will generate a LOW condition when active. This may seem to be opposite the usually accepted arrangement (except in certain microprocessor chips), but there is a method to this madness. The binary counter is capable of producing a very stable long-duration timer.

Figure 8-2 shows the basic operating circuit for the XR-2240 in monostable and astable multivibrator modes. This chip proves interesting in that the sole difference between astable and monostable modes is feedback connection consisting of a 51 kohm resistor. This resistor provides for automatic resetting of the chip

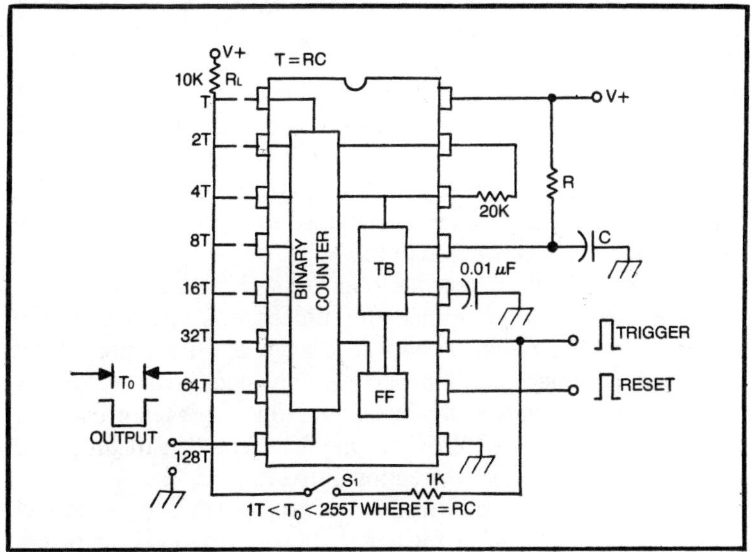

Fig. 8-2. Normal circuit for XR-2240.

by linking the wired-OR output terminals with the reset terminal of the chip. The timer is set into operation by applying a positive-going pulse to trigger input pin 11. This pulse is routed to the control logic and has several jobs to perform simultaneously: resetting the binary counter flip-flops, driving all output low, and enabling the timebase circuit. As was true in the 555 IC timer, this timer works by charging capacitor C1 through resistor R1 from the positive voltage source V_{cc}. The period of the output waveform (monostable) is given by T = RC.

The pulses generated in the timebase section are counted by the binary counter section, and the output stages change states to reflect the current count. This process will continue until a positive-going pulse is applied to the reset terminal.

Figure 8-3 shows the relationship among the timing pulses in the XR-2240. The reason for the open-collector arrangement is to allow the user to wire a permanent-OR output so that the actual output pulse LOW duration can be programmed. Each binary output is wired in the usual power-of-2 sequence: 1-2-4-8-16-32-64-128. If these are wired together the output will remain LOW as long as any one output is LOW. This allows us to program the output duration from 1T to 255T (where T is the RC time constant of the timebase section) by connecting together those outputs which sum to the desired period. For example, let us design a 57 second timer using a 1 second timebase (RC=1). From the binary powers of the outputs we know that

$$1 + 8 + 16 + 32 = 57$$

so, we will connect together in wired-OR fashion those pins. These times are available on IC pins 1, 4, 5, and 6. So, to make our 57 second timer, we must set RC = 1 second (i.e., 1 megohm and 1 μF) and then strap together pins 1, 4, 5, and 6 of the XR-2240. The output terminals selected are connected to a pull-up resistor (10 kohm) that is also connected to the V_{cc} power supply. The output will remain LOW for 57 seconds following each trigger pulse.

We could change either the timebase frequency or the wired-OR configuration or both to change the timer period. Of course, if the timebase frequency were doubled, then the counter would reach the desired state in half the time. This feature allows programming of the XR-2240 to time durations that might prove difficult to achieve using conventional circuitry.

Each output must be wired to V_{cc} through a pull-up resistor of 10 kohm unless, of course, the wired-OR configuration is selected. In that case, a single pull-up resistor is used.

Current through each output terminal must be limited to 5 milliamperes, or less. This value can serve as a general rule for the selection of appropriate pull-up resistor values if 10 kohms is not desired.

The amplitude of the trigger and reset pulses must be at least two PN junction voltage drops (2×0.7 V = 1.4 volts). In most practical applications it might prove wise to use values greater than 4 volts amplitude, or standard TTL levels, in order to guard against the possibility that any particular chip may be a little difficult to trigger near minimum values, or that outside factors conspire to reduce pulse amplitude at some critical moment.

Synchronization to an external timebase or modulation of the pulse width is possible by manipulation of pin 12. In normal operation this pin, which is the inverting input of comparator 1, is bypassed to ground through a 0.01 μF capacitor so that noise signals will not interfere with operation of the chip. A voltage applied to pin 12 will vary the pulsewidth of the signal generated by the timebase section. This voltage should be between +2 volts and +5 volts for a timebase change factor of 0.4 to 2.25, respectively.

If it is desired to synchronize the internal timebase to an external reference, connect a series RC network consisting of a 0.1 μF capacitor and a 5.1 kohm resistor (see Fig. 8-4) to pin 12. This network differentiates input pulses, which should have an amplitude of at least 3 volts and periods of 0.3T to 0.8T.

Another way to link the XR-2240 to another timebase is to use an external timebase. This signal may be applied directly to pin 14.

Each Exar XR-2240 has its own internal voltage regulator circuit to hold the DC potentials applied to the binary counters to a

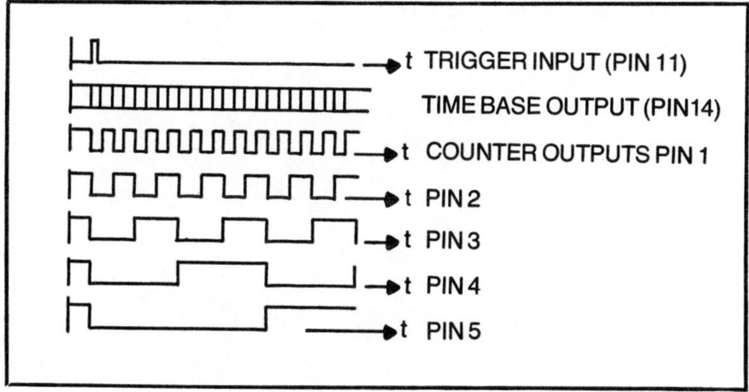

Fig. 8-3. Timing waveforms.

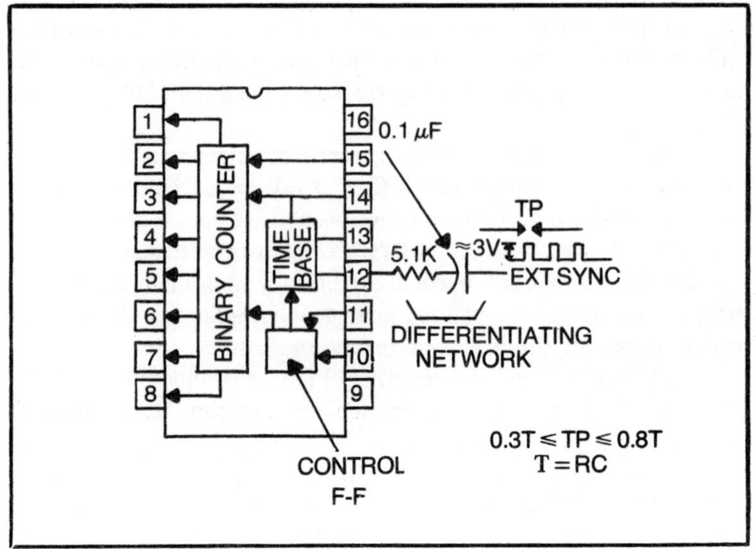

Fig. 8-4. External synchronization.

constant level. This regulator consists of a series-pass transistor which has its base held to a constant potential by a zener diode. If operation below 4.5 volts DC is anticipated, then it becomes necessary to strap the regulator terminal (pin 15) to V_{cc} (pin 16). The regulator terminal can be used to source up to 10 mA to external circuitry or to another XR-2240.

The XR-2240 timer is second-sourced by Intersil under the designation 8240. The timer is also available to hobbyists through Calectro, under the part number designation J4-1214. A pair of related timers, once made under the designations XR-2250 and XR-2260 (now Intersil 8250 and 8260), are equipped with BCD outputs.

Chapter 9

XR-2240 Projects

The XR-2240 (and related chips) can be used in any application where a monostable or astable multivibrator is called for, and the frequency of operation is compatible. But the XR-2240 also contains a binary counter, and this single fact extends the range of possible applications over that of the other single-chip IC timers. In this chapter, we will discuss some of the many applications for the XR-2240. Also note that the chapter on long-duration timers will contain XR-2240 projects.

STAIRCASE OR RAMP GENERATOR

A staircase generator will create an analog output waveform consisting of increasing or decreasing steps. It is similar to a ramp waveform, except that the voltage output levels can take on only certain specific values. The ramp is continuous and has an infinite range of possible values. Figure 9-1 shows how to use the XR-2240 to create a staircase waveform. We are pressing the XR-2240 into service in this application as a digital-to-analog converter with clock input. The binary-ramp type of DAC will use a binary counter to sequentially select resistors in a weighted network.

In Fig. 9-1 the binary counter outputs of the XR-2240 are connected to the resistor ladder. In this case, the resistor values are R, 2R, 4R, 8R, 16R, 32R, 64R, and 128R; where R is the resistance of the lowest value resistor (e.g., 1000 ohms in most cases). The outputs of these resistors are summed at the inverting input of a 741 operational amplifier. The feedback resistor around the negative feedback loop of the operational amplifier sets the gain of the stage, and thereby serves as a scale factor adjustment.

The slope of the ramp is dependent upon the gain of the operational amplifier and the speed with which the binary counter increments through its range. This latter parameter is set by the timing of the resistor and capacitor in the timebase section.

Every time the XR-2240 is triggered the output of the operational amplifier will ramp downward from maximum (fullscale) to zero. Resetting the XR-2240 returns the output to zero. We can disable the staircase generator by causing the cathode of the diode to drop LOW. When pin 14 is clamped less than 1.4 volts DC, then the circuit is inhibited.

Now, here's the $64 question. The outputs of the XR-2240 are open-collector, and require a pull-up resistor in normal operation. Right? Then how come these resistors, wired as they are in Fig. 9-1, will cause a current to flow to the input of the operational amplifier? The answer to this question will test your fundamental knowledge of the properties of operational amplifiers. There is a fundamental property of the operation amplifier that tells us that applying a voltage to one of the two inputs requires that we treat the other input as if it had the same voltage applied. This is the origin of the concept of the so-called "virtual ground" (hideous term). If we apply a voltage, such as V_{ref} in Fig. 9-1, to the noninverting input of an operational amplifier, then we must treat the other input as if the same voltage were applied there also. This is not some illusory theory: if the voltage at the inverting input in Fig. 9-1 is actually measured, then it would be found that V_{ref} is on that terminal of the operational amplifier. The currents come from the fact that the XR-2240 outputs ground when active. This will cause the resistors to be grounded at one end and connected to a V_{ref} source at the other. For example, let's say that the pin 3 output is active, so is LOW. This means that the 32R resistor is grounded, so a current of $(V_{ref})/(32R)$ flows in that XR-2240 output.

We adjust the slope and frequency of the staircase waveform by adjusting the clock rate. We can adjust the maximum amplitude of the reset output by using the scale-adjust potentiometer in the feedback loop of the operational amplifier.

ANALOG-TO-DIGITAL CONVERTER

An analog-to-digital converter (ADC) will produce a binary output word that is proportional to an analog input voltage (or, in

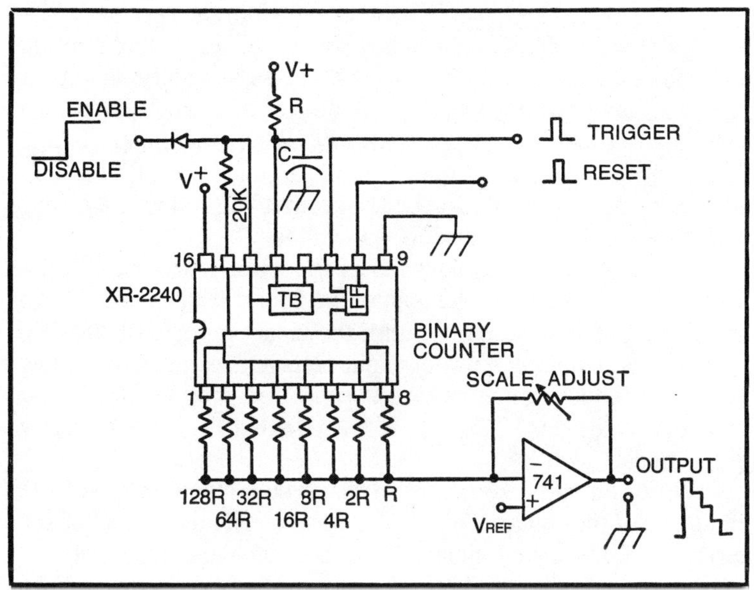

Fig. 9-1. Staircase generator.

some cases, a current). The ADC is used in digital instruments, digital voltmeters, and digital computers to provide data concerning the analog input voltage applied. An example of a binary ramp ADC is shown in Fig. 9-2.

The XR-2240 is used, similar to the previous circuit except that in this case the staircase waveform is applied to a voltage comparator to compare with the input signal. The binary outputs of the XR-2240 are used as the binary word to the computer (or digital display).

A voltage comparator is basically an amplifier with too much gain, and it is used to indicate the equality or inequality of two voltages. An operational amplifier, minus the feedback resistor, serves admirably as a comparator. The gain of the operational amplifier is so high that the output will saturate HIGH when the differential input voltage is more than a couple millivolts. This means that the output is HIGH when the two input signals are not equal and LOW when the two input voltages are exactly equal.

In the circuit of Fig. 9-2, the unknown analog input voltage is applied to the noninverting input of the comparator amplifier. The output of the staircase generator (a voltage) is applied to the noninverting input of the comparator. The output of the comparator controls an RS flip-flop by applying a signal to the *set* input.

When a strobe pulse (start conversion) is received, the XR-2240 timer will cause the binary counter to begin incrementing and will reset the RS flip-flop. The staircase waveform begins at full scale and drops one incremental step for each clock pulse from the timebase section of the XR-2240. When the ramp voltage finally equals the input voltage, the comparator *sets* the RS flip-flop and the count is stopped. The binary counter will reflect the binary word that represents the analog input voltage.

The scaling is accomplished in the same manner as in the previous case: adjusting the gain of the operational amplifier. We adjust the gain potentiometer until the binary output is 00000000 when a maximum-value analog input voltage is applied. We know that the maximum value voltage is represented by 00000000 and the zero value is 11111111. Values in between are directly proportional.

We could add a zero control that would allow us to set the other end of the range. This can be done by adding a small offset current (in the usual manner) to the inverting input of the comparator. This would help us null out any devious little offset potentials that conspire to make zero look like something other than zero.

DIGITAL SAMPLE-AND-HOLD CIRCUIT

A sample-and-hole circuit is designed to take a sample of a voltage level, and then hold it for future processing. There are many applications for S&H circuits in electronic instrumentation. One application is to reduce the slew error in an analog-to-digital converter. Such errors occur because of changes in the analog input voltage while the conversion is taking place. In other cases, the S&H will be used to hold a voltage sample for purposes of comparision. In many instruments, the circuit will perform such comparisons. These circuits are often ratiometric, meaning that the idea is to examine the *ratio* between two voltage values rather than the actual voltages. This technique cuts down markedly on the error of measurement caused by factors such as power supply changes.

The classic sample-and-hold circuit consists of a switch (either electronic or relay) that closes long enough to store the voltage sample in a small value capacitor. These circuits work well for very short-duration hold periods, but they fail miserably in the long-duration hold mode. The reason for this is the normal leakage in any capacitor. The digital sample-and-hold circuit does not suffer

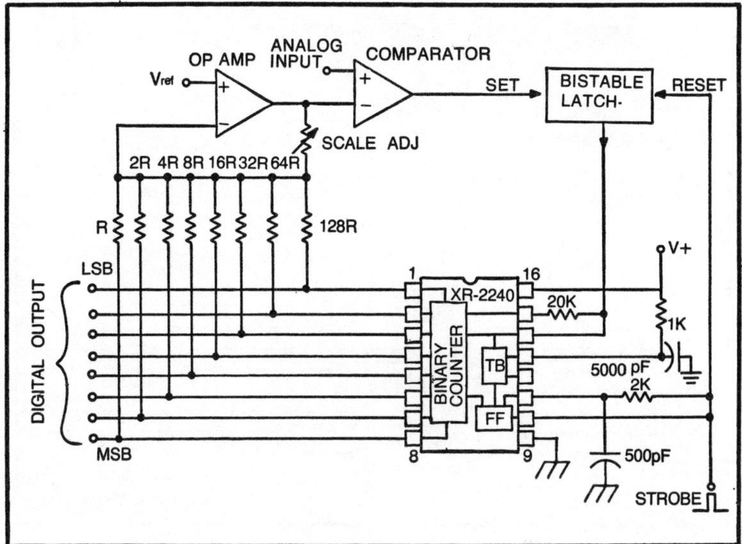

Fig. 9-2. A/D converter.

from the "droop" caused by capacitor leakage. An example of such a circuit, based on the Exar XR-2240 timer IC is shown in Fig. 9-3.

The digital sample-and-hold circuit in Fig. 9-3 is almost identical to the ADC circuit using the same integrated circuits. When a strobe pulse is received the circuit will sample the analog input waveform. The XR-2240 binary counter will ramp downwards until it reaches the analog input voltage, at which point the comparator sets the RS flip-flop and stops the counter. The output of the staircase amplifier (the operational amplifier) is the same as the analog input voltage. The analog voltage can then change, but the sampled output V_0 remains the same. We could also use the binary outputs of the XR-2240 to form the binary equivalent to the sampled voltage, if that were desired (in which case, the circuit would be an ADC).

There will be a significant error in the sampled voltage if the XR-2240 counter increments too slowly. This will allow the analog voltage to change too much during the sample period. The solution to the problem for this circuit is to make the sample fast by keeping the timebase frequency fast (e.g., approximately 200 kHz).

PROGRAMMABLE ASTABLE MULTIVIBRATOR

There are two methods for "programming" the output of the XR-2240. One is to change the RC time constant of the timebase

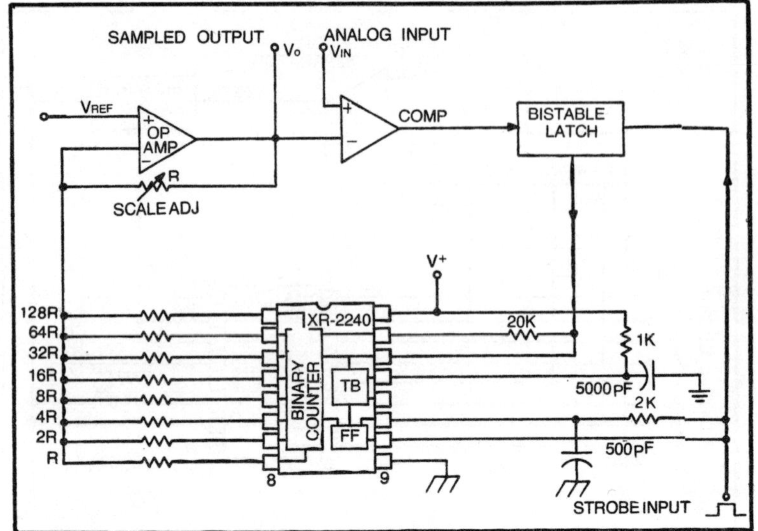

Fig. 9-3. Digital sample-and-hold.

section, while the other is to vary the selection of the outputs that are wired-OR together. In Fig. 9-4, both methods are used. The capacitor is not usually amenable to varying the frequency because the values normally required are not available in variable form. The resistance, however, can be easily made variable (in the form of a potentiometer). The time constant of the timebase circuit is (R1A + R1B)C1.

Switches are used to select the outputs that are used in the wired-OR configuration. These switches are weighted 1-2-4-8-16-32-64-128, so we can make a time constant that is the sum of the selected output weights. Either monostable or astable configurations can be used.

TEN-MINUTE TIMER FOR AMATEUR RADIO ID

The circuit shown in Fig. 9-5 is another version of the ten-minute timer. These timers are used by amateur radio operators to warn them of the ten-minute deadline for transmitting call sign identification. A period of ten minutes is 600 seconds, and this circuit "time out" in 572 seconds, just in time for the amateur to transmit the required ID without running into the red.

All outputs of the XR-2240 are wired together in a wired-OR configuration. This means that the output duration will be 255T. The value of T, in this case, is approximately 2.24 seconds, so the output period will be

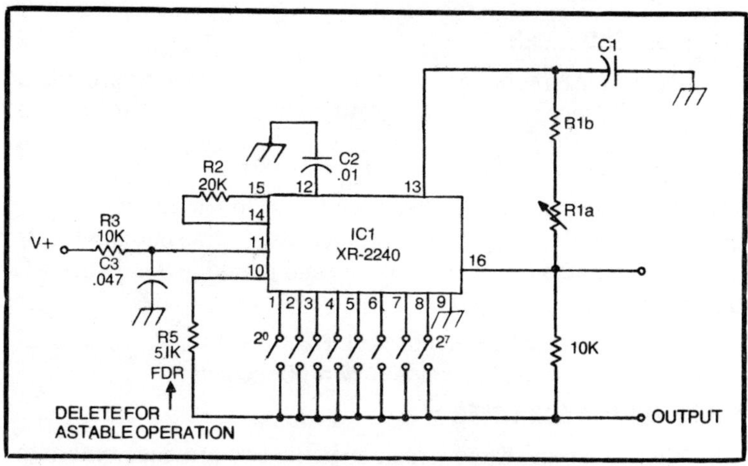

Fig. 9-4. Programmable monostable.

$$T = (2.24) \times (255)$$
$$= 572 \text{ seconds}$$

The circuit is a one-shot multivibrator. Applying a positive level to the trigger input (pin 11) by pressing the switch will start the timer sequence. The output will drop LOW for 572 seconds. When it goes HIGH again a signal is applied to pin 10 that resets the counter to its dormant state. A circuit that will be presented a little later will serve to produce an audible alarm output for this circuit.

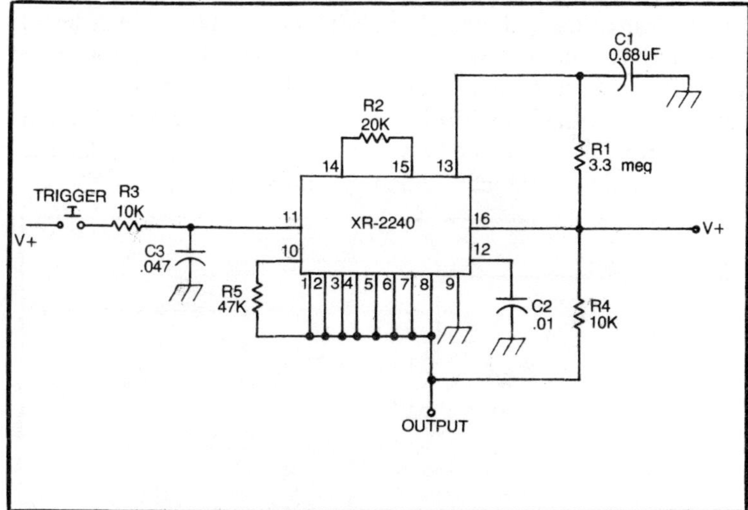

Fig. 9-5. Ten-minute timer.

SIXTY-SECOND TIMER

The timer circuit shown in Fig. 9-6 is designed to provide a period of 60 seconds. The XR-2240 circuit is similar to the ten minute timer of the previous example, but the output strapping is changed to select the shorter timer period. Note that Fig. 9-6 is a simplified drawing, and all other circuitry is the same as in Fig. 9-5. In this case, we want a period of 60 seconds so strap outputs 1,2,8, and 16. When used with the 2.24 second period given by the RC timer network, the period is:

$$T = 2.24 \text{ seconds} \times (1 + 2 + 8 + 16)$$
$$= 2.24 \text{ seconds} \times 27 = 60 \text{ seconds}$$

THREE-MINUTE EGG TIMER

We can use the same circuit as in the two previous timers to make a three-minute timer for such applications as cooking your egg in the morning. Figure 9-7 shows the correct output strapping (when the timebase is 2.24 seconds) for the three minute, or 180 second, timer. In this case, we strap together in wired-OR fashion the 16 and 64 outputs, so the output time is

$$T = 2.24 \text{ seconds} \times (16 + 64)$$
$$= 2.24 \text{ seconds} \times (80) = 180 \text{ seconds}$$

TIMER AUDIO OUTPUT CIRCUIT

The timers in the three previous examples lack an audible output. Most timers, however, require such an output in order to be effective. The sound source in this circuit is a 555 astable multivibrator set to a frequency (see Fig. 9-8) that you feel comfortable with. The 555 is turned on by a silicon-controlled rectifier (SCR1) connected between the V+ power supply and the V+ connections of the 555. When the timer output is HIGH (i.e.,

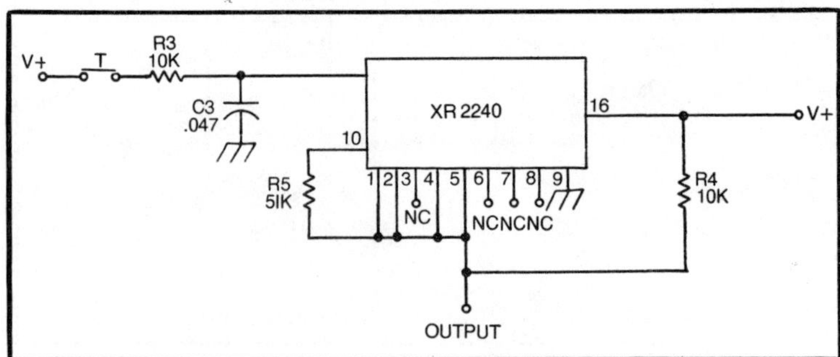

Fig. 9-6. Sixty-second timer.

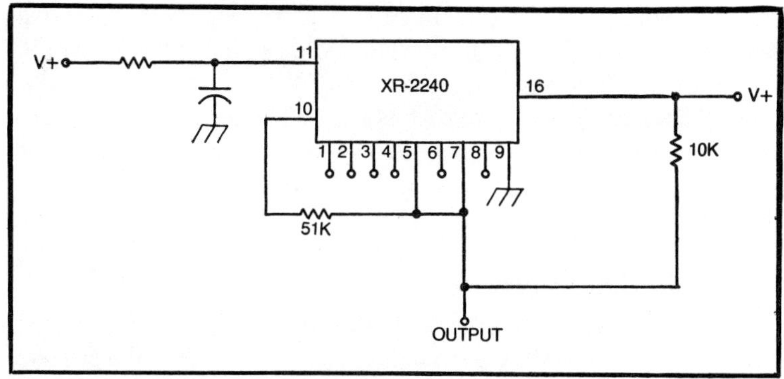

Fig. 9-7. Three-minute timer.

dormant), the voltage applied to the 2.7 kohm resistor sets up a current that turns on the gate of the SCR. The alarm sounds. But when the timer output is LOW, indicating that it is performing its timing function, then the SCR will not turn on.

The action of the output alarm is controlled by switch S1. This switch can be either an SPDT switch or a dual pushbutton switch (as shown) in which one set of contacts is normally open (N.O.) and the other set is normally closed (N.C.). This type of switch is shown in Fig. 9-8, and is functionally equivalent (the way it is wired) to the SPDT switch. Switch S1 serves the *reset* function for the alarm. When S1 is pressed, section S1A opens, and thereby commutates (turns off) the SCR (silencing the alarm). At the same time, section S1B is closed, so the trigger level is applied to the input of the XR-2240 timer IC.

A loudspeaker is used at the output of this stage. Note that there is a resistor in series with the loudspeaker. The value of the

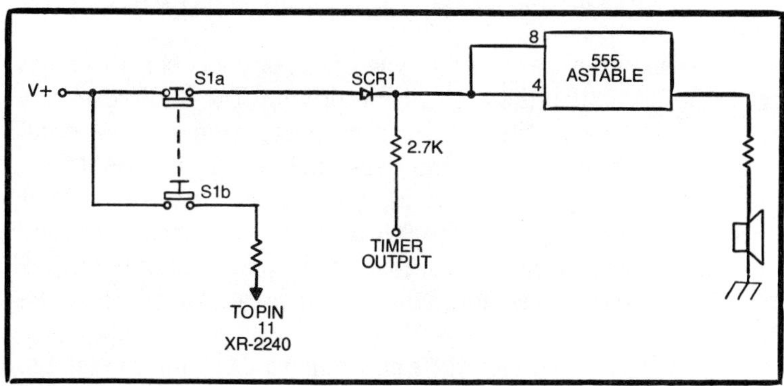

Fig. 9-8. Alarm output.

resistor will depend upon several factors and will be found experimentally. The factors involved are the power supply voltage, the loudness desired, and the impedance of the loudspeaker selected. Note that use of a 200 ohm speaker (hard to get, but not impossible) will allow you to delete the resistor *provided* that the speaker can handle the power level and you want the full volume!

BINARY ADDRESS SEQUENCE

The Exar XR-2240 timer IC is almost unique in that it contains a timebase section followed by a binary counter circuit. We can use this eight-bit counter to make a binary address sequencer. These circuits are used in microprocessor checkout and other digital experiments to manually step through the operation of a circuit. We are incapable of following the action of a digital circuit with a one megahertz (or more!) clock, so need some way to slow the action down to a dull blur when troubleshooting or designing such circuits. Such a circuit is also useful when building a microcomputer front panel. It will allow us to manually step through memory address locations to insert the desired program by hand (don't try that on a large program unless you have a strong hand—otherwise you will learn why they call the process "fingerboning").

The circuit shown in Fig. 9-9 is a binary address sequencer based on the XR-2240. The circuit is connected as a one-shot multivibrator with a duration of 10 microseconds. Every time the one-shot is triggered, a 10 μS pulse is generated that increments the binary counter by one state.

The output connections of the binary counter are wired independently, each with its own pull-up resistor to the V+ line. The outputs are also brought out to form the binary output of the address sequencer.

We can either tie the reset line LOW, or connect it to a reset line in the circuit that the sequencer is driving. The LOW reset line will forbid us to reset the circuit, so we must increment through all of the remaining possible 256 address combinations in order to get back to 00000000 or 11111111. We would, therefore, usually want to use the reset input of the XR-2240 in the normal manner. Connect it through a resistor and a normally open switch to the V+ line. When the switch is closed the reset input sees a HIGH, so the IC is reset.

The trigger circuit is a little different from the other versions. In this case, a resistor (R2) is connected permanently between the

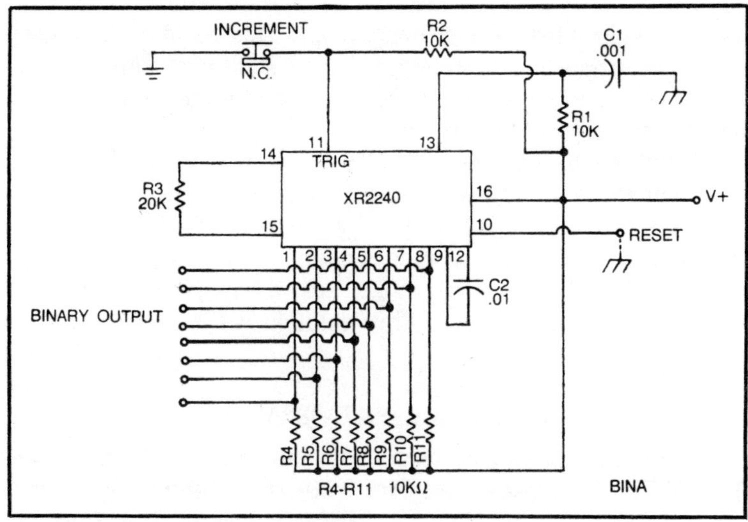

Fig. 9-9. Binary address sequencer.

trigger input of the XR-2240 and the V+ power supply. A normally-closed pushbutton switch is connected to short out the trigger input. This will prevent noise transients from accidentally triggering the circuit. When the user desires to operate the circuit, incrementing the counter by one, then the button is pressed. This action will remove the short circuit, causing the trigger input to rise to a HIGH—triggering the XR-2240.

Once triggered, the XR-2240 will generate a single 10-μS pulse that increments the internal binary counter by one. Note that the outputs, in this case, are *inverted*, so the count may look a little strange (e.g., 1110 instead of 0001 for the first count). This is actually an advantage when using the circuit for microcomputer front panel design because it will be necessary to pass the signal through a buffer or line driver in order to drive more than a few "K" worth of memory devices. These drivers are often inverters, or will have optional inverting outputs. In fact, some circuits use ordinary high-power TTL inverters for this service.

A potential problem with some versions of this circuit is *contact bounce* in the switch. Some switches are notorious for producing a large number of output counts each time they are pressed. This is because the contacts don't "make" immediately and solidly, but will bounce. Each bounce produces an additional trigger for the timebase section of the XR-2240. The duration of the output pulse in this circuit is short enough to expire before the spurious pulses from the switch die out. This could easily cause additional triggering thereby producing additional pulses. If a circuit seems to increment more than once for each press of the button, then suspect this problem.

Solutions to the contact bounce problem are easy. One is to simply increase the RC time constant of the timebase section until a point is reached that allows the bounce pulses to die out before the timer times out. Another solution is to use a one-shot multivibrator (e.g., a 555) with a long duration to trigger the address sequencer circuit. Either solution will take care of the problem.

PROGRAMMABLE MONOSTABLE MULTIVIBRATOR

We can use a four-bit binary comparator circuit to make the XR-2240 programmable (see Fig. 9-10). Two 7485 comparators are needed with the eight-bit XR-2240 device. The 7485 is designed to compare two input binary words (designated A and B) and issue outputs that tells us whether A is greater than, less than, or equal to B. In this case, we use the A=B output (pin 6) as the output for our circuit.

The XR-2240 can be connected in the monostable or astable modes, depending upon your application. The eight outputs of the

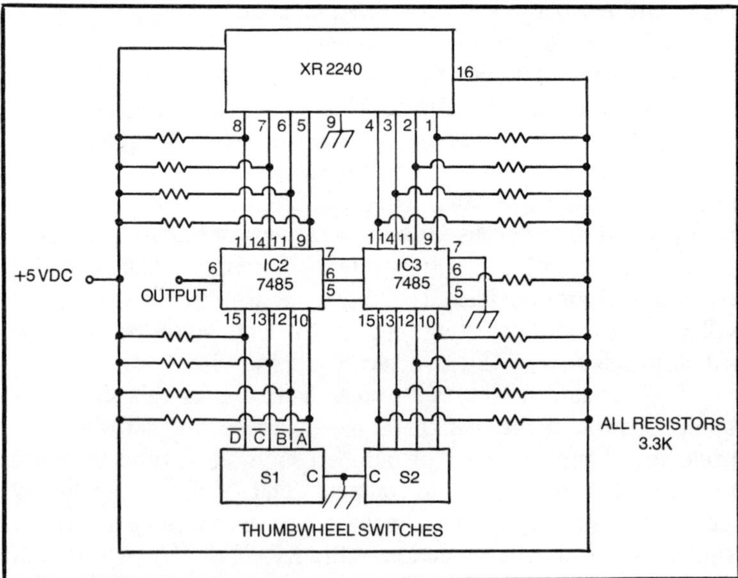

Fig. 9-10. Switch-programmable monostable.

internal binary counter are connected 4×4 to the inputs of the two 7485 comparators. The alternate inputs of the 7485 devices are connected to the outputs of four-bit binary thumbwheel switches. Be sure to use a switch that has the inverted outputs ($\overline{A}$, $\overline{B}$, $\overline{C}$ and $\overline{D}$) so that the correct bits will be LOW when the XR-2240 has a LOW on the corresponding output. It is also necessary to make sure the switch is a hexadecimal model, and not BCD (binary coded decimal). The BCD switch will only go as high as 10 and therefore will miss the remaining states of the counter (base-16).

The output of IC2 is the output for the circuit. It will remain LOW until the count of the XR-2240 is equal to the inverted binary word applied to the alternate 7485 inputs from the thumbwheel switches. When the two counts are equal, the output goes HIGH. This HIGH can be used by itself or can be used to either retrigger or reset the XR-2240, depending upon the application.

Chapter 10

CMOS Timer Circuits

The CMOS (complementary metal oxide semiconductor) family of IC digital logic devices presents some interesting possibilities in circuit design. The CMOS input is a very high impedance, so will require very little driving power. This is in contrast to the TTL input which is a current source. The TTL input requires a considerable amount of power compared with the CMOS input. The output of the CMOS device consists of a pair of resistors in series across the power supplies. In the HIGH condition the output (the junction of the two resistors) sees a low resistance (200 ohms) to the V_{dd} positive power supply, and a very high resistance to ground, or V_{ss} power supply. In the LOW condition, the relative resistances change. In this condition, the output sees a very high resistance to the V_{dd} positive power supply and a low resistance to the V_{ss} negative power supply, or ground (the V_{ss} power supply terminal is often connected to ground). An implication of the output circuit behavior is that the series connection is always a combination of a high resistance and a low resistance. As far as the power supplies are concerned, therefore, the resistance is always high. This fact means that very little current is drawn by the CMOS device! The only time that any significant amount of current is drawn is for a brief instant during the *transition* of the output terminal.

The low power requirement is in contrast to TTL devices. All TTL devices, even the so-called "low-power" types, require substantial amounts of electrical power. There are good scientific reasons why the high-speed TTL requires so much power, but that does not help us much when we are trying to fit a small digital timer into a portable configuration that does not eat batteries alive. Those modern calculators, having liquid crystal displays and operating for two years on a pair of hearing-aid batteries, use CMOS devices.

It is, therefore, rather obvious that CMOS devices are the choice to be made in any application where the power consumption is important. But there are also other reasons why the CMOS device should be selected over TTL whenever the superior speed of the TTL is not needed. One of these reasons is that the CMOS device can operate from unregulated power supplies up to ± 15 volts. The advantage of this, quite apart from the savings in power supply cost, is that the CMOS device operated from power supplies greater than 5 volts DC has a greater noise immunity than the TTL. This factor can become important in environments where there is a lot of electrical noise interference. It requires relatively low-level noise pulses to trigger a TTL device. Remember, the LOW condition is any potential less than 0.8 volts, while the HIGH condition is any potential greater than 1.4 volts. The difference between these levels is only 600 millivolts. In a worst case situation, a noise spike needs only 600 to 700 millivolts of amplitude in order to falsely trigger a TTL gate circuit, counter, or register (sigh). The CMOS device sees a transition between HIGH/LOW states when the input voltage passes a point that is defined as one-half the difference between V_{dd} (+) and V_{ss} (−) power supply levels. In the case where the V_{ss} supply is zero (i.e., the V_{ss} terminal on the CMOS device is grounded), then the transition point is ½ (V_{dd}). If the V_{dd} power supply is +12 volts, therefore, the trip point will be ½ (12), or 6 volts. This voltage is considerably higher than the trip point in TTL, so the CMOS device is less sensitive to noise pulses.

But the use of CMOS devices is not totally "for free." Thus far, I have painted a relatively bright picture of the CMOS device. There are, however, a couple of disadvantages to the CMOS device. One of them is the lack of speed inherent in the CMOS process. The high resistances that reduce the power consumption of the device also increases the time required to react, and that decreases the speed of the device. Low-frequency operation, then, is inevitable when using CMOS devices. But what does this mean? What is low speed and how does it affect timer circuits? In point of fact, the low speed of CMOS devices is a lot faster than anything needed for the timer circuits in this book. Most garden-variety CMOS devices will operate up to 4 or 5 MHz, with some devices capable of operating close to 11 MHz. Thus, we are not terribly affected by the CMOS device speed. Most of the circuits we will consider are very low speed.

Another difference between CMOS and the other logic lines is sensitivity to static electricity charges. The CMOS (and, to some extent, the "LS" series of TTL) devices will blow out if improperly handled. It seems that the static charge that normally builds up on your body, tools, and the chassis of the device that you are working with will burn out the thin metal oxide layer between the metallic gate structure and the semiconductor channel of the MOSFET transistors used in the CMOS device. Most of these devices will burn out if less than 100 volts is placed on them! In fact, the maximum gate-channel voltage on most of these devices is around 80 volts, yet a static charge can easily reach into the kilovolts range.

The problem of static burn-out is more pronounced with the A-series CMOS devices (i.e., those marked with only the "4xxx" number or those marked "4xxxA"). The B-series CMOS devices have built-in zener diode protection that serves to shunt the high-potential static charges around the delicate gate insulation.

The main thing that affects your success in using CMOS devices (as well as MOSFET transistors) is the handling procedure. We must use methods of handling that keeps all pins of the CMOS device at the same potential. One tactic is to keep the black conductive foam that CMOS devices are shipped on permanently in place on the pins of the device until we are ready to insert it into the circuit. A little curiosity can be expensive! Also, some shipping tubes and shipping bags are "anti-static" devices, meaning that they are slightly conductive. These carrying devices serve to keep the pins of the device at the same potential, but work only if you can stifle the desire to take the IC out and play with it. Try that on a $35 frequency counter, and you'll learn to keep the foam in place and the device in its carrier. My own first lesson in this matter was costly.

It also helps to keep the circuit board that contains CMOS devices on a metal "cookie sheet" work surface. Some professional technicians who work on CMOS devices all the time use standard size sheets of "do it yourself" sheet aluminum on the work bench. This sheet should be connected to ground through a 1 megohm, 2 watt resistor. The resistor allows the static charge to drain off harmlessly, yet places sufficient resistance in the line to prevent a catastrophe should the worker (you!) come between a source of 115 volts AC and the grounded cookie sheet. Any time the cookie sheet is directly grounded, there is a potential for disaster!

Some technicians also prefer to place an alligator-clip lead between a metallic ID or watch band on their wrist and the cookie sheet. The wire will ground you to the metal work surface, thereby neutralizing any potential differences that could hurt the CMOS/MOSFET devices. But, BE ABSOLUTELY SURE THAT THERE IS NOT DIRECT CONNECTION BETWEEN YOU AND GROUND! The 1 megohm resistor should protect you, but don't forget the power line ground (i.e., "third wire") on the test equipment or power supply being used! This often overlooked source of a ground could easily wipe you out in the event of a short circuit.

Now that we have recapped the advantages and disadvantages of the CMOS integrated circuit, let's get busy and look at time practical circuits. The first circuits will be a group of monostable multivibrators.

But first, let's recap the definition of the monostable multivibrator, or "one-shot," circuit. The nickname "one-shot" is very descriptive of the operation of this device. The one-shot will produce one, and only one, output pulse every time that it is triggered. When a trigger pulse is applied to the trigger input of the one-shot, the output of the one-shot goes HIGH for a specified, predetermined, period of time. After the time expires, the output drops LOW again and the circuit becomes dormant.

This description is another way of saying that the one-shot circuit has but *one stable state* (i.e., monostable). The dormant state is the stable state, and the monostable multivibrator remains in this state until a trigger pulse is applied. Following the trigger pulse, however, the output snaps to the unstable state for a predetermined length of time. In the example of the last paragraph, the dormant state was output-LOW, and the unstable transient state was output HIGH. There are monostable circuits with both logic levels as the dormant state (i.e., dormant HIGH, transient LOW).

CMOS 4528 DEVICE

There are fewer standard one-shot ICs in the CMOS line than there are in the TTL line. One device, however, is the 4528, a dual monostable (see Chapter 3). Figure 10-1A and 10-1B are two configurations using the 4528. The circuit in Fig. 10-1A is a positive-edge triggered one-shot, while that of Fig. 10-1B is a negative-edge triggered one-shot.

Application of the 4528 is simple and straightforward. The duration of the output pulse is set by resistor-capacitor network R1/C1. In this case, the RC time constant is approximately 10 microseconds. The accuracy of the output duration is enhanced markedly if the capacitor is a polyethylene or silver-mica type with a close tolerance. Similarly, the resistor should be a precision model with a low temperature coefficient.

Each section of the 4528 has two trigger inputs, labeled positive and negative (or A and B). If we want positive-edge triggering, then we would use the positive, or "A" input; and connect the negative, or "B" input, to V_{dd}. If, on the other hand, we wanted to use negative-edge triggering, then we would use the negative input, and ground the positive input.

In Fig. 10-1A we are seeking positive-edge triggering, so the positive input (pin 4 for the section selected) is used, and the negative input (pin 5 in the section selected) is tied to +9 volts DC.

In Fig. 10-1B, we are seeking negative-edge triggering, so the negative input (pin 5) is used; and the positive input (pin 4) is grounded.

The 4528 is made more useful by the fact that we can use either positive-or negative-edge triggering without external components. All that is necessary is to reconfigure two terminals to obtain the two different modes of operation. Both sections of the 4528 are independent, so we can wire one section to be a positive-edge triggered device, and the other as a negative-edge triggered device.

4013 AS A ONE-SHOT

The CMOS device numbered 4013 is a dual type-D flip-flop that has direct set and direct clear inputs. Recall the rules of the Type-D flip-flop in the clocked mode: the logic level on the D input will be transferred to the Q output only when the clock line is active. In this case, the clock is a positive-edge triggered terminal.

The direct set and direct clear lines are used in unclocked, or direct, operation of the flip-flop. These inputs are active-HIGH. This means that a positive HIGH applied to the set input will force the Q output HIGH and the NOT-Q output LOW. Similarly, a positive HIGH applied to the clear input will force the Q output LOW and the NOT-Q output HIGH.

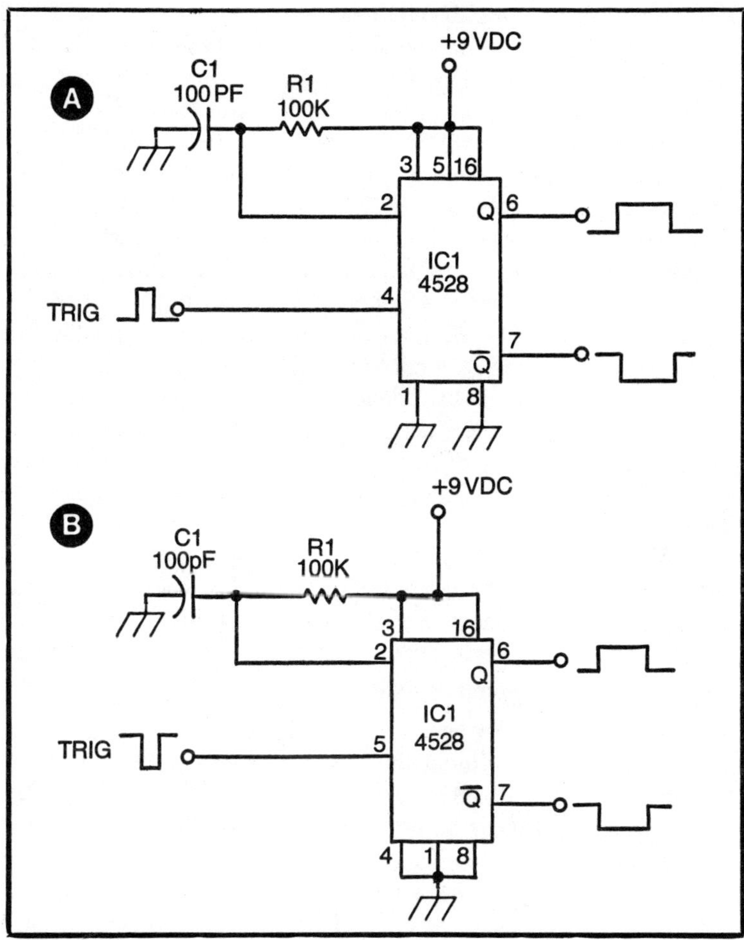

Fig. 10-1. 4528 CMOS one-shot; A) Positive triggered; B) Negative triggered.

Positive Edge-Triggered 4013 One-shot

The circuit in Fig. 10-2A uses one section of the 4013 device as a monostable multivibrator. The D input of the flip-flop is connected permanently HIGH. When the clock line goes HIGH when a trigger pulse is received, this HIGH condition is passed to the Q output. The Q output is used as the output of the one-shot if you want an active-HIGH output. If, on the other hand, you need an active-LOW output, then use the NOT-Q output of the 4013 as the output of the one-shot.

During the dormant period, when the output of the one-shot is LOW, diode D1 will cause any charge on capacitor C1 to drain off.

After a very brief time, the charge on capacitor C1 will drop to the forward bias potential of the diode. In the case of the 1N4148 device shown, this potential will be 0.6 to 0.7 volts. If a germanium diode, such as the 1N60, is used instead, then the potential across the capacitor will be only 0.2 to 0.3 volts.

When a trigger pulse is applied, the Q output snaps HIGH effectively transferring the HIGH on the D input to the Q output. This condition places a potential across RC network R1/C1, causing C1 to begin charging. Capacitor C1 is connected to the clear input of the Type-D flip-flop. When the voltage across the capacitor reaches ½(V+), or about 6 volts in this case, the flip-flop will clear, causing the Q output to drop LOW again.

The time between the trigger pulse (causing the Q output to snap HIGH) and the clear occurring is set by the time constant of the RC network. This time constant sets the period required for the voltage at the clear input to build up to the trigger point of the clear input.

After the Q output drops LOW again, the capacitor is discharged through the diode (D1) and the LOW output terminal (which is now at ground potential). The time constant of this circuit is set to 10 microseconds, but you can lengthen that time if needed.

Negative-edge Triggered One-shot

It is not easy to use the 4013 device as a negative-edge triggered one-shot without the use of the external components. The circuit in Fig. 10-2B shows how a single NOR gate (or NAND gate, or NOT gate) can be used to make the circuit of Fig. 10-2A operate on the negative edge. The 4001 NOR gate is connected as an inverter (NOT gate) because one of the two inputs is grounded. A HIGH placed on the other input of the NOR gate causes the input of the flip-flop to be LOW, and a LOW placed on the NOR gate input causes the flip-flop to be HIGH. We can, therefore, obtain negative-edge triggering.

BOUNCELESS PUSHBUTTONS

An ordinary electrical switch is not suitable to be used in most digital electronic circuits. When the two contacts come together, they do not "make" perfectly but will *bounce* several times, causing several "glitches" (i.e., spurious pulses) to be generated. These glitches are capable of resetting counters, incrementing counters and shift registers; or they can be propagated through gates to raise all kinds of havoc with digital circuits. Pushbutton and toggle switches are especially afflicted with this problem.

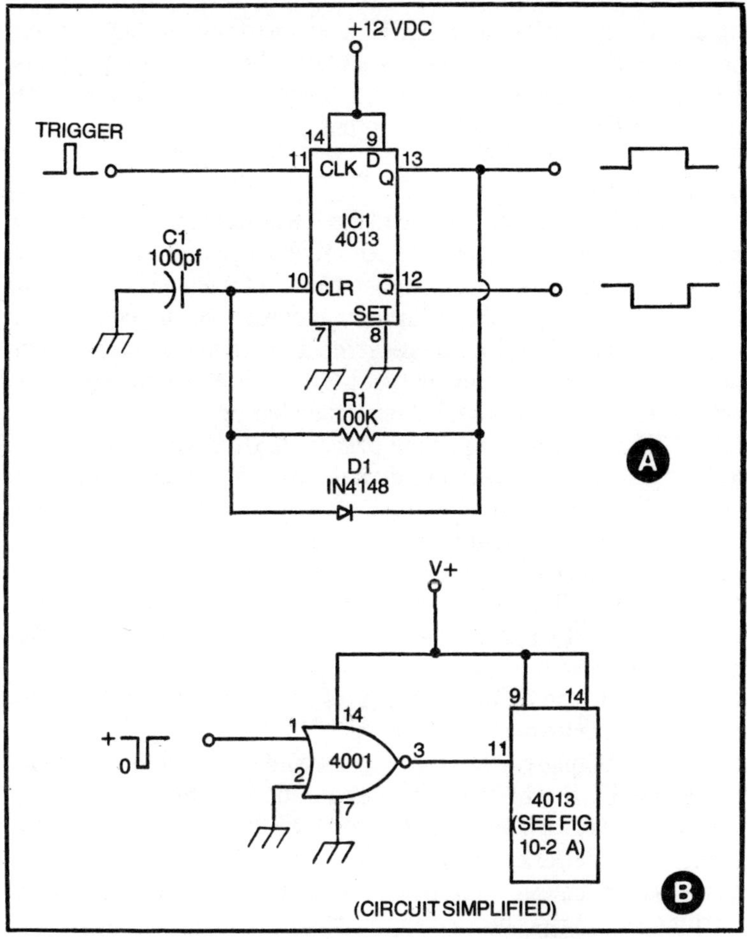

Fig. 10-2. CMOS 4013 one-shot: A) Positive triggered; B) Negative triggered.

The problem of pushbutton bounce is serious enough in digital equipment to make the designers use "bounceless pushbutton" circuits. These are circuits that use digital logic levels as switch closures, but generate the logic levels using a switch and some additional circuitry.

Figure 10-3 shows three forms of bounceless pushbutton circuit. The version shown in Fig. 10-3A uses a CMOS RS flip-flop to accomplish the dirty deed. The 4044 is a quad NAND-logic RS flip-flop (only one section is used here). There is no NOT-Q output on this chip, so we must be satisfied with an active-HIGH output condition. The use of NAND logic tells us that the *set* and *reset* inputs are normally held HIGH, and are brought LOW when activated. If the set input is made LOW, then the Q output goes HIGH and remains HIGH (this is a bistable circuit). If the reset input is made LOW, then the Q output goes LOW and stays LOW. The inputs (R and S) are able to operate even if brought momentarily LOW.

The R and S inputs are tied to the V+ power supply through 10 kohm pull-up resistors in Fig. 10-3A. A switch S1 is connected such that terminal A is connected to the reset input and terminal B is brought to the set input. In normal operation, the common is grounded and set to short the reset input to ground. This condition holds the RS flip-flop in the output LOW state. When the switch is actuated, however, the set input is grounded, and the reset input is HIGH. This causes the Q output to snap HIGH, a condition seen as a "switch closure" by the digital circuits which follow the "bounceless pushbutton."

There is a problem with the circuit of Fig. 10-3A. It is still possible for switch transients to actuate the RS flip-flop in some circumstances. This would mean that the pushbutton is no longer "bounceless," it just bounces in a more sophisticated manner. We could slow down the operation by placing 0.1 uFd capacitors from each input (R and S) to ground. This means that the reset input would take a finite period of time to go HIGH following actuation of the switch. Similarly, the reset input would require a short time to rise to the HIGH condition after the operation was finished.

The use of capacitors in the circuit of Fig. 10-3A is not terribly appealing. It lacks elegance as a solution. We would make much better use of the capacitor in a half-monostable or full-monostable circuit operating as a bounceless pushbutton.

Figure 10-3B shows the use of the 4528 monostable multivibrator IC as a bounceless pushbutton. This circuit is the ordinary 4528 one-shot configuration set to have a pulse duration of 10 milliseconds. This may seem like a long period, but it actually might be too short! The idea is to have the period of the one-shot *longer* than the bouncing of the switch contacts. The contacts of most switches may tend to bounce for several milliseconds. For most cases, the 10 millisecond period is sufficient, but for some it might be wise to use a 0.1 μF capacitor at C1 in order to produce a 100 millisecond time constant.

The 4528 in Fig. 10-3 is wired in the negative-edge triggered configuration. The trigger input (pin 5) is kept normally HIGH during the dormant period by pull-up resistor R2. When the

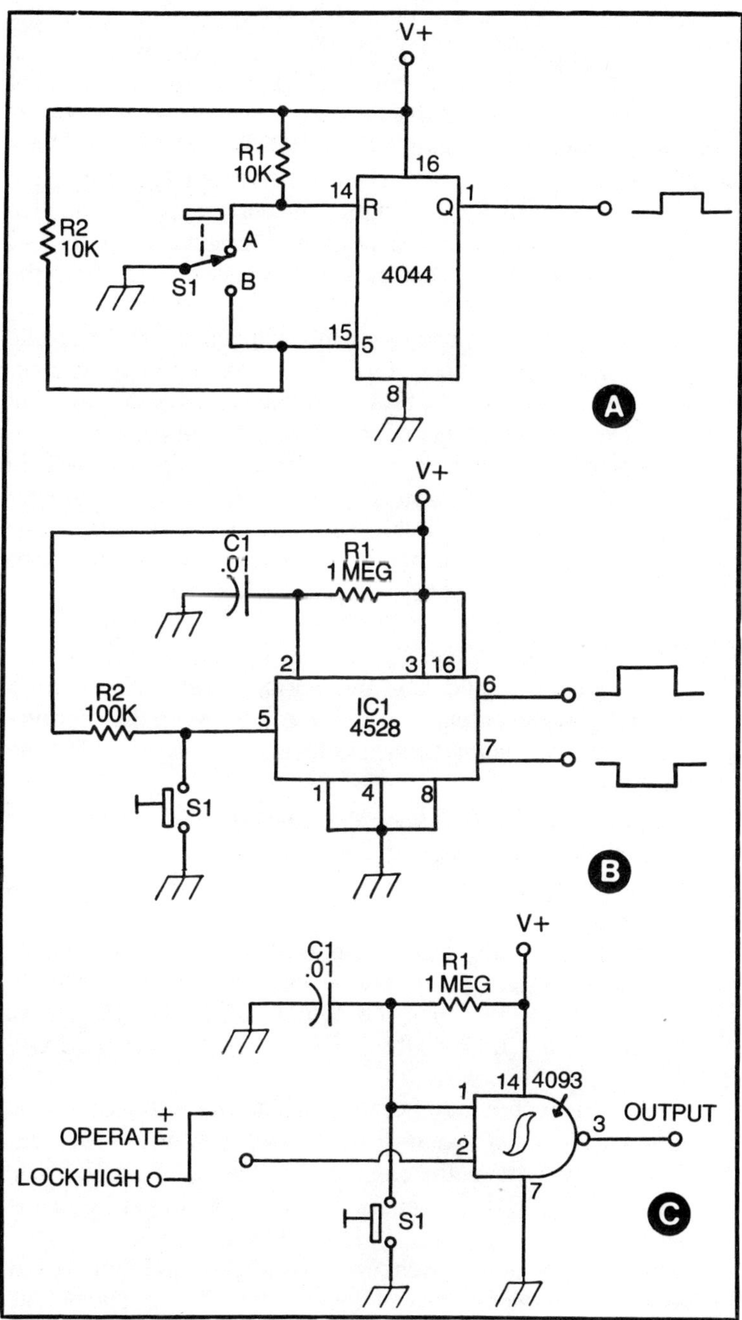

Fig. 10-3. Bounceless pushbutton circuits.

operator wishes to operate the circuit, he or she will press the button thereby closing switch S1. This drops the trigger input LOW for the duration of the period that S1 is closed. The one-shot outputs will then become active for the period of time set by R1C1. Because the 4528 has both Q and NOT-Q outputs, we have complementary outputs that can be used for a variety of purposes. The Q output is normally LOW during the dormant period and then snaps HIGH during the transient period. Alternatively, the NOT-Q output is HIGH during the dormant period and drops LOW for the transient period.

Some particularly sloppy switches will bounce on both make and break, although the bounce on the make operation is substantially worse. In those cases, then make the period of the one-shot several hundred milliseconds or possibly even a full second.

Figure 10-3C shows the use of a Schmitt trigger one-shot in order to make a bounceless pushbutton. What is a *Schmitt trigger*? The S.T. is a special circuit that has a lot of interesting applications, most of them in the area of cleaning up data pulses that are noisy. The schmitt trigger output will snap HIGH when the input potential passes a certain threshold, V_1, and will remain HIGH until the input potential decreases past a second threshold, V_2. An irregularly shaped input waveform, therefore, becomes a squared pulse at the output. The difference between the positive-going and negative-going threshold levels ($V_1 - V_2$) is called the *hysteresis* of the device.

The 4093 device is a special CMOS IC that contains four two-input NAND logic Schmitt triggers. The device can be used as an ordinary NAND gate or as a Schmitt trigger circuit. On a five-volt supply, the positive-going trigger point is +2.9 volts, while the negative-going trigger point is 2.3 volts. At higher supply potentials, the trigger points are set higher. If either, or both, inputs are made LOW, then the output will be HIGH (normal NAND operation). If both inputs are HIGH, then the output is LOW (also normal NAND operation).

In the circuit of Fig. 10-3C we are using one of the inputs (pin 2) as a lock-in/lock-out line. If we make pin 2 LOW, then the circuit will not operate because the output will be permanently HIGH. If we make pin 2 HIGH, however, the circuit will operate in the normal manner.

The active input in this circuit is pin 1. This pin is held HIGH by the V+ power supply through resistor R1. The voltage on this input is the voltage across capacitor C1.

Normally, capacitor C1 is charged so the voltage at pin 1 is close to the supply potential. This situation means that both inputs are HIGH, so the output will be LOW. Pressing switch S1 will short capacitor C1, causing its charge to dump to ground. It also places the input of the Schmitt trigger at ground potential, thereby causing the output to snap HIGH. The capacitor will begin to charge as soon as switch S1 is open again, but it will not allow the 4093 input terminal to see a HIGH condition until the voltage across the capacitor rises to the positive-going threshold. The time required to reach this potential is a function of the RC time constant (R_1C_1) and the supply voltage. The total time that the output is HIGH will depend also upon the time that the operator keeps the switch closed. This method is a little superior to the others because of this feature; we don't worry too much about the one-shot timing out before all of the bouncing ceases.

WATER LEVEL DETECTOR

A water level detector will warn you when the level of the water in a tank, a swimming pool, or some other receptacle is high enough—or too high. Although the digital logic following the water level detector will differ from one application, depending upon just what you want done when the water level reaches the mark. The circuit shown in Fig. 10-4 makes a nice, low power consumption, high impedance water level detector *PROVIDED* that the V+ power supply is *isolated* from the 115 volt AC mains wiring! Otherwise, it could be a very dangerous circuit! I recommend that you operate the alarm from batteries, and either replace them, or recharge them—away from the water—when they go dead. Note that the use of CMOS logic means that almost any practical-size battery will seem to last forever—at least a very long time. The circuit will draw microamperes, so it is not at all unlikely that a battery will last for a year or more.

The CMOS device used in this circuit is the 4049 inverter. This device is actually a hex inverter, so there will be five additional inverters in the package to use in other parts of the circuit. The 4049 device, like its noninverting counterpart the 4050, is capable of driving TTL inputs if the V+ power supply is held to +5 volts DC.

The input of the 4049 inverter is held HIGH by the V+ power supply operating through resistor R1. A pair of electrodes in the fluid tank will be shorted out when the water level reaches a certain point, and that will cause the input of the 4049 to go LOW. When

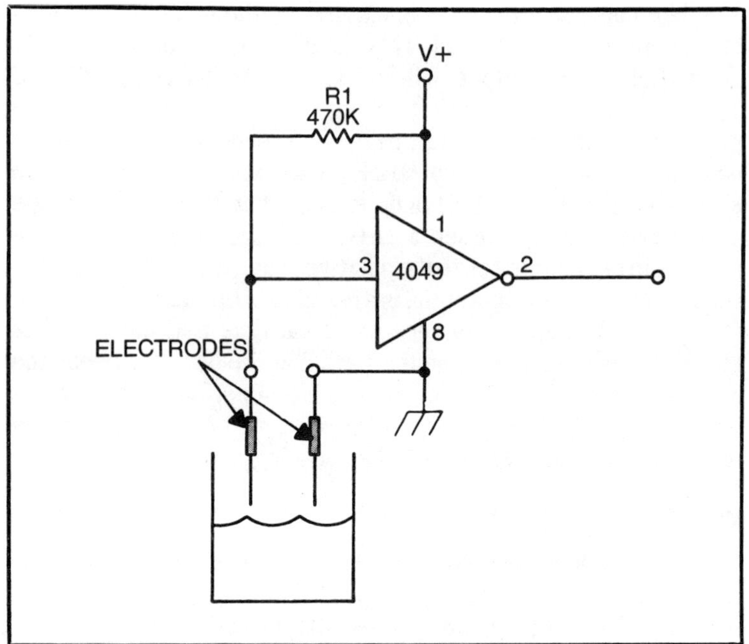

Fig. 10-4. CMOS water level detector.

the input is LOW the output will be HIGH, so the alarm will be triggered.

Another word of caution in the use of this circuit: don't use it on flammable liquids or on electrically active liquids. In the former case, there is always the possibility of fire, while in the latter the reaction is unknown to me—so be careful (some fluids emit "un-nice" fumes when an electrical current is passed through—so use the circuit on *water only* unless you personally are an expert in the chemistry of the other fluid).

TOUCH-PLATE SWITCHES

I recently got on an elevator when the mechanic was making some repairs. The floor-call switches in that system were of a type that you simply touch, and then they light up and call the elevator. At first I though that these were *thermally* operated (in a fire?!?!), but the mechanic told me that they were *touch* sensitive.

Early attempts at touch switch construction used an LC oscillator in which the touch plate of the switch was part of the capacitor in the LC network. An output discriminator wanted to see a certain frequency all of the time; if the frequency varied, then the phase detector would emit a DC level that could be used to close a relay or perform some other wonderful task.

But those early models were not too good. The finger of the operator would surely mistune the oscillator tank circuit, but so would almost everything else in close proximity to the damn thing. Also, the circuits were easily affected by stray electromagnetic interference fields—of which there are zillions.

Touch-plate switches have improved over the years. Today, we can use the high input impedance of the CMOS device to make a touch switch for us that is relatively trouble free. The circuit in Fig. 10-5A shows how to make use of the stray AC power mains field that permeates all space around inhabited dwellings. We are capable of acting as an "antenna" that picks up *several volts* of 60 Hz energy whenever we are indoors. If you doubt this, then take hold of the open probe of an oscilloscope and observe the 60 Hz waveform!

When a person touches the touch plate of Fig. 10-5A, a 60 Hz signal is induced into the input of the 4011 NAND gate. Since this

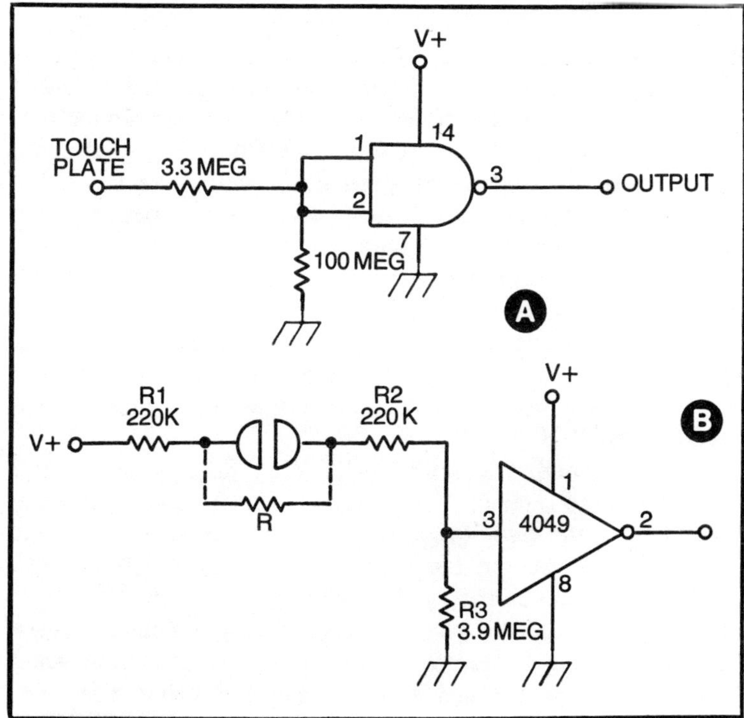

Fig. 10-5. Touch-plate switches.

gate is inverter wired, the output will be a chain of 60 Hz squarewaves that can be used to trigger additional circuits "downstream." This circuit works only when the V+ power supply potential is low, i.e., 5 volts or so.

Figure 10-5B shows another form of touch plate. Again, we must insist that this circuit be operated from an isolated power supply, such as batteries. This circuit uses a 4049 CMOS inverter. This device is the one that will provide TTL outputs if the V+ is kept at +5 volts DC.

The output is normally HIGH in this device because the input is effectively grounded through R3 (even though "grounded" means a 3.9 megohm path!). The touch plate is *split* into two halves. When a finger is placed on the touch plate, the circuit is completed, and the voltage from the V+ power supply is applied to the input of the 4049 device through R1, R2 and the resistance of the finger (R).

ASTABLE MULTIVIBRATORS

An astable multivibrator has no stable states: the output terminal will flip and flop back and forth between HIGH and LOW, creating a wavetrain of squarewaves. The astable multivibrator is used as a clock in timers and other forms of digital circuits. In this section, we will consider five different types of astable multivibrator, of which only three are *fully* CMOS devices. All of them, however, are capable of being used in CMOS circuits as clocks.

The circuit in Fig. 10-6A uses a 4093 Schmitt trigger device (described earlier in this chapter) to operate as an RC-tuned astable multivibrator. We could tie input 2 of this NAND logic Schmitt trigger HIGH, and make the circuit operate all the time. Alternatively, we could use either the switch (as shown) or a logic level from some other portion of the circuit to make the circuit run only when told to do so. In the operate (or *run*) mode, input 2 of the 4093 is held HIGH.

When the output of the 4093 is LOW, all charge in capacitor C1 will drain off to ground. After a period of time equal to approximately $5R_1C_1$ the capacitor will be fully discharged (keep in mind that the LOW output provides a low resistance path to ground).

Recall from our earlier discussion that the Schmitt trigger output will snap HIGH when the voltage on the input reaches a certain positive-going threshold, and the output drops LOW when the input voltage reaches a certain negative-going threshold. Capacitor C1 will charge through R1 when the output is HIGH and discharge (also through R1) when the output is LOW. The output terminals will snap back and forth between HIGH and LOW as the capacitor charges and discharges between these two threshold levels.

The frequency of the output signal will depend upon the R_1C_1 time constant, and the width of the hysteresis band for the 4093. Since the hysteresis band varies according to the supply voltage levels, we cannot offer an easy formula to determine the frequency.

Figure 10-6B shows a somewhat more conventional form of RC CMOS astable multivibrator. This circuit uses cross-coupled inverters and could also use inverter-wired NAND or NOR gates. The linearity of the circuit is improved by the 1 megohm resistor in the input of the first inverter. The frequency of oscillation is given approximately by

$$F = 1/2.2RC$$

Where:

F is the frequency in hertz
R is the resistance in ohms
C is the capacitance in farads

This CMOS oscillator uses the 4049 inverter, so it will be TTL-compatible if the V+ supply is set to +5 volts DC. The use of inverter-wired NAND or NOR gates, however, will not allow TTL-compatibility. In those cases, it might be wise to use either a 4049 inverter or 4050 buffer to make the device TTL-compatible (these must be operated from +5 volts, but the gates could be operated at any CMOS potential).

We could theoretically operate either of the two preceding circuits at frequencies up to several megahertz. The CMOS element would oscillate at those frequencies, but there is a problem in selecting resistor and capacitor values to work in those ranges. The circuits of Fig. 10-6A and Fig. 10-6B should, therefore, be operated in the frequency range of 500 kHz and under.

A piezoelectric crystal resonator element could be used to make an astable operate in frequency ranges above 500 kHz. The circuit in Fig. 10-6C operates from 20 kHz to just over 1 MHz. This circuit uses a piezoelectric element to make a 4001 NOR gate oscillate. The output buffer is a 4050, so TTL compatibility is possible, provided that the 4050 is operated from a +5 volt DC power supply.

The 4001 NOR gate is connected with the "unused" input to a run switch. If switch is open, then the input is HIGH so the device

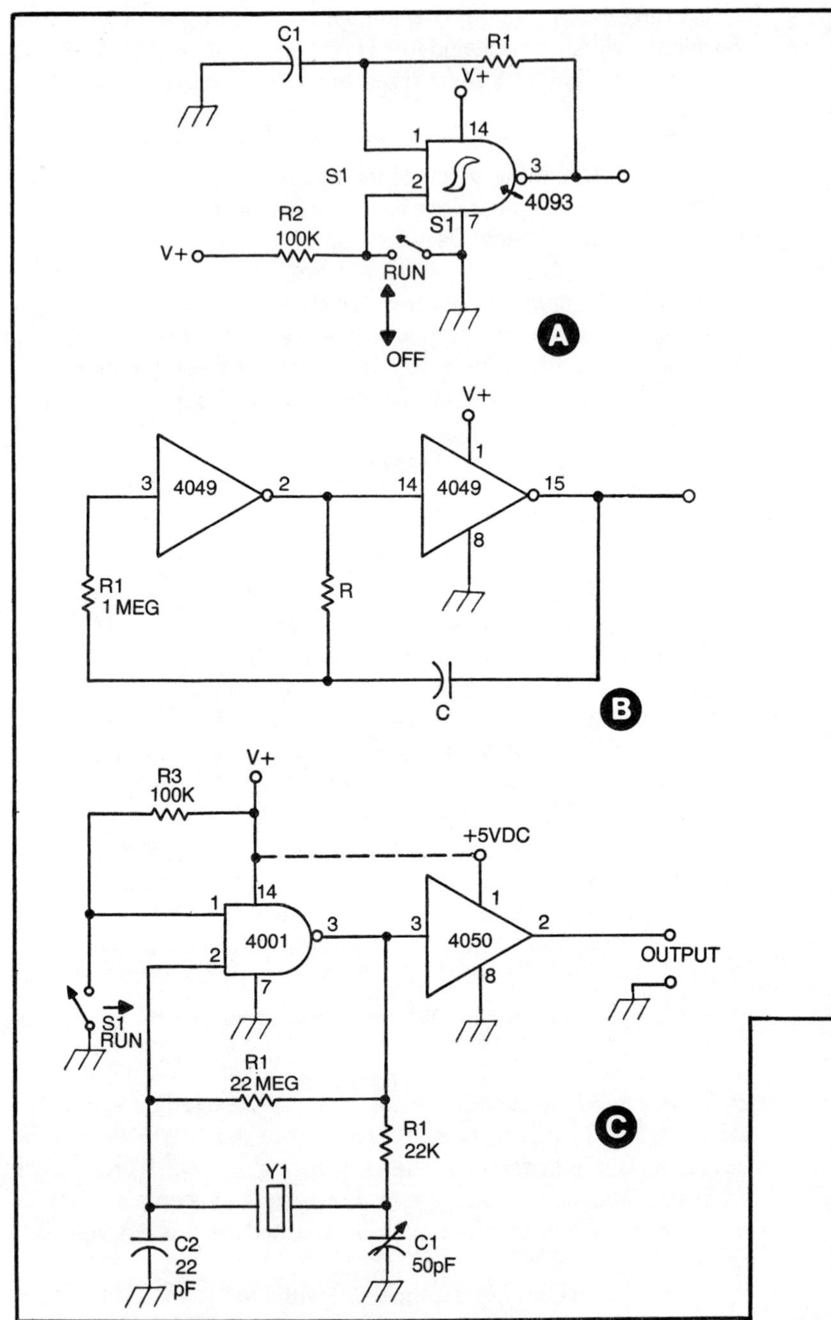

Fig. 10-6. Astable multivibrators from CMOS devices.

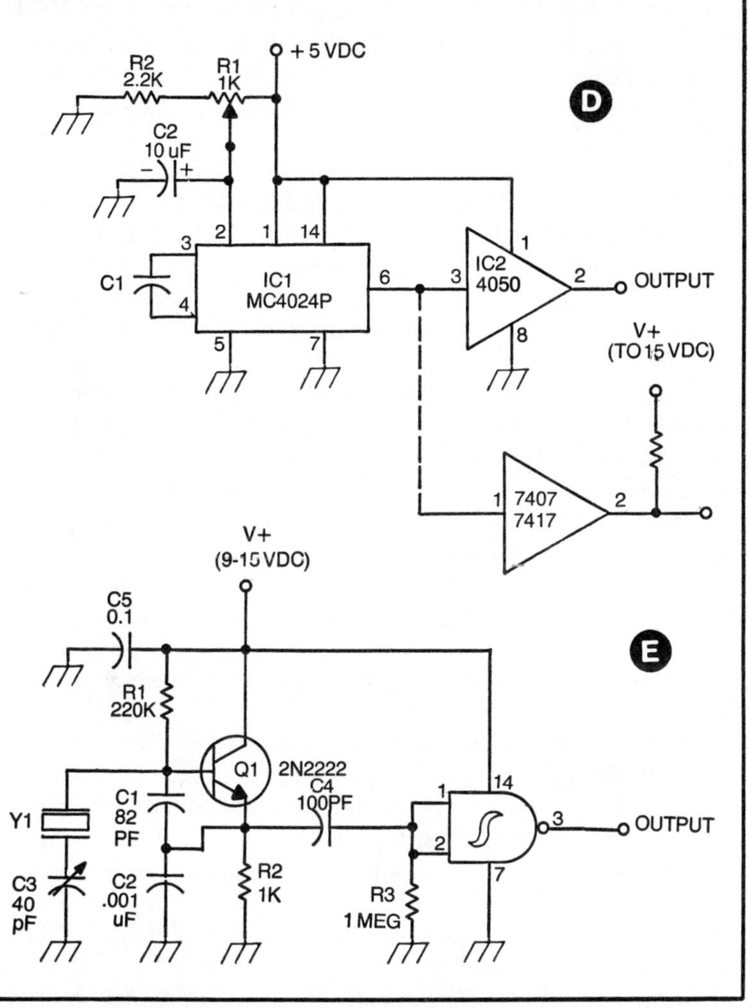

will not operate. If, on the other hand, the switch is closed, the input is grounded (LOW) so the circuit will oscillate. If we do not need manual control of the circuit, then we could use a HIGH/LOW logic level from some other circuit to turn the oscillator off and on.

The variable capacitor in the circuit of Fig. 10-6C is used to trim the oscillating frequency to some precise value. This is a principle advantage of crystal oscillators: precision and stability at reasonable cost.

All oscillators/clocks should be provided with an output buffer circuit. The loading of the output, especially in digital circuits, might change drastically, and this will cause the frequency of oscillation to pull. The high impedance CMOS input of the 4050 allows the oscillator to see a constant high-impedance load. It will, therefore, be less sensitive to changes in external load conditions.

Figure 10-6D shows the use of a pseudo-CMOS astable multivibrator in a CMOS circuit. The actual oscillator is a Motorola MC4024P dual voltage controlled oscillator (VCO). *NOTE:* The MC4024P is NOT a CMOS 4024. The CMOS 4024 is a seven-bit binary counter, while the *MC*4024P is a dual VCO. The MC4024 designation is a Motorola house number, not a standard CMOS number. This unfortunate confusion of type numbers can cause some misunderstanding, so be careful. The MC4024P, incidentally, is a TTL device.

The frequency of oscillation in the circuit of Fig. 10-6D is controlled by two factors: the value of capacitor C1 and the voltage applied to pin 2. The approximate frequency, when the voltage at pin 2 is +5 VDC, is given by $300/C_1$. The voltage can trim this frequency downwards over a range of approximately 3:1 using the voltage at the wiper of potentiometer R1.

If we want to maintain TTL compatibility, then connect the 4050 buffer (or a 4049 if phase reversal is no problem) between the output of the MC4029P and the load. Alternatively, if we want to use this circuit to drive a CMOS device with a voltage higher than +5 volts DC, then connect one of the common TTL open-collector devices as a buffer. In Fig. 10-6D, we have used the 7407 or 7417 devices in this role. A pull-up resistor (10 kohm for 12 volts) is connected from the output of this noninverting buffer to the V+. The V+ is the same supply that powers the CMOS devices—not the +5 volts DC of the TTL MC4024P! If we can tolerate the phase inversion, then we can also use any of the open-collector TTL inverters in this role. Again, a pull-up resistor to a CMOS potential is needed.

Our final astable multivibrator is shown in Fig. 10-6E. In this case, we are using a bipolar NPN transistor in a crystal Colpitts oscillator circuit. This oscillator will work up to 15 MHz, or more, but the frequency of the overall circuit is limited to several megahertz by the CMOS Schmitt trigger (4093) device.

The transistor crystal oscillator will produce sufficient output signal voltage to cross the positive and negative going input thresholds of the Schmitt trigger on each cycle. This results in an output wavetrain of square waves that have the same frequency as the oscillator signal. Since the 4093 is a CMOS device, we may conclude that the output signal is CMOS compatible.

The circuit shown in Fig. 10-6E is representative of many in high precision applications. The designer will specify a *temperature-compensated crystal oscillator* (TCXO) at the required frequency, and then square up the output with a 4093 Schmitt trigger. Some TCXO manufacturers are now offering TCXO building blocks with either TTL or CMOS outputs—they build in the 4093.

Some of the circuits discussed in this section will be used in projects to follow. In any event, the section will serve as a handy guide to CMOS clock circuits.

ADDRESS SEQUENCER

A *binary address sequencer* is used in microcomputer front panels (where it is used to step through memory locations, one step at a time), or in places like *read only memory* (ROM) programmers. The device will start from location 00000000 and will increment one digit at a time to 11111111. Incrementing occurs when the *advance* switch (S1) is closed. The sequencer will return to location 00000000 when the *reset* switch (S2) is opened.

The actual sequencer in Fig. 10-7 is a 4024 seven-bit binary counter. The output word on the 4024 is incremented on digit every time the input (pin 1) is brought LOW. The pulses counted by

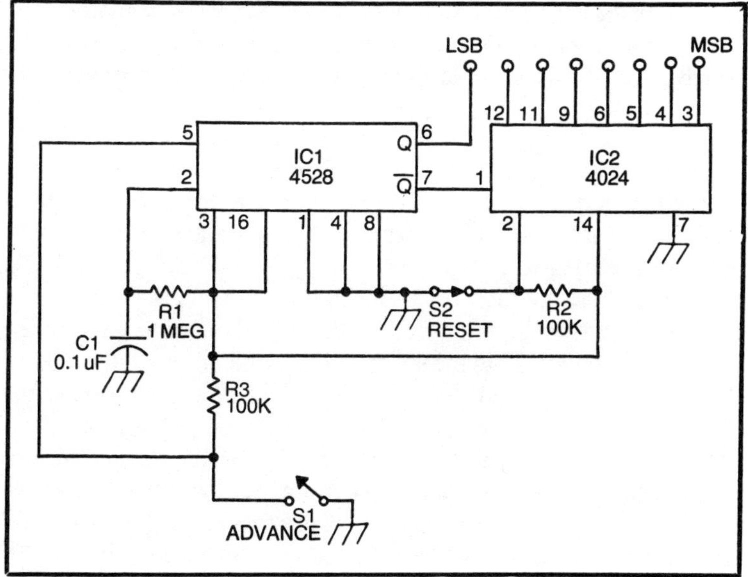

Fig. 10-7. CMOS Binary address sequencer.

the 4024 are supplied by a 4528 one-shot multivibrator. The period of this one-shot is relatively long in order to eliminate contact bounce from switch S1. If there seems to be a problem in this department, then increase the time constant of R1/C1 to make period of the one-shot longer.

The NOT-Q output of the 4528 (pin 7) will drop LOW every time that the advance switch is operated. This is the criterion for the 4024 input, so the NOT-Q output is connected to pin 1 of the 4024 device. The Q output of the 4528 is used to form the least significant bit of the binary address sequencer (there are only seven output bits on the 4024, but most microcomputers require an eight-bit format).

TWO-PHASE CMOS CLOCK

Many digital projects (especially a certain low cost microprocessor) require a two phase clock. Such a clock will have two signal lines that are arranged such that one is LOW when the other is HIGH. In the circuit of Fig. 10-8 we have a configuration that will produce alternate HIGH/LOW conditions with a rest period of one clock pulse between the two conditions (total rest period on each line is two clock periods). This arrangement is wasteful of time, but it serves to allow circuits to *settle* before the next transition comes along.

The circuit of Fig. 10-8 uses three sections of a 4049 CMOS inverter, a 4027 J-K flip-flop, and two sections of a 4001 two-input NOR gate. The outputs of the two NOR gates form the two clock phase lines.

The master clock of this circuit is formed by an RC CMOS astable multivibrator such as seen earlier in this chapter. The two cross-coupled 4049 sections will oscillate at a frequency of $1/2.2R_1C_1$ (up to 500 kHz). An output buffer stage serves to isolate the clock from the flip-flop.

Recall the operation of a J-K flip-flop in the clocked mode: the outputs will change state only on the negative-going edge of the input clock pulse. This means that there will be one output pulse for every two input pulses. Now, let's recall the operation of the NOR gate: if either input is HIGH, then the output will be LOW. The J-K flip-flop works together with the NOR gate sections to synchronize the output clock lines. When the clock pulse applied to the J-K input is HIGH, the outputs of the two clock lines are LOW (rules for NOR gate). When the master clock line of the J-K FF drops LOW, the Q and NOT-Q outputs will change states. Now, we

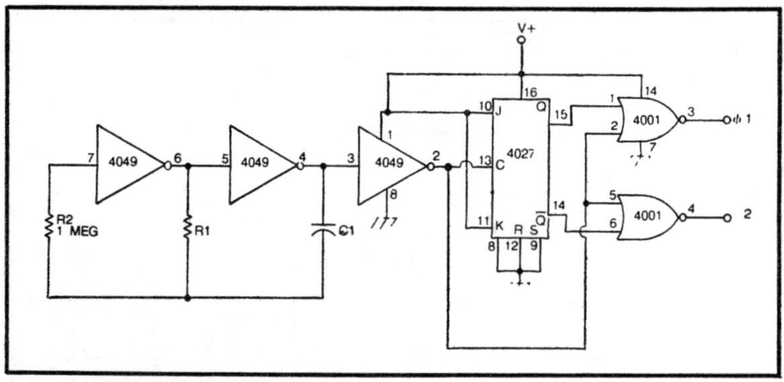

Fig. 10-8. Two-phase clock.

have the common NOR gate inputs LOW and one of the other inputs HIGH. This means that one NOR gate will have a LOW output while the other has a HIGH output. On the next master clock pulse, the situation reverses because the state of the Q and NOT-Q outputs of the FF change. There will, therefore, be only one output clock line HIGH at any one time, and there will be a one master-clock-period offset between the HIGH conditions on the two outputs.

MULTIPHASE CLOCK

There are times when we want to synchronize more than two functions at a time, and do not want any overlap between the functions. This situation requires a multiphase digital clock circuit (Fig. 10-9). This particular circuit operates as an eight-phase clock with active-HIGH outputs. The heart of the circuit is a CMOS 4022 device that is described as a divide-by-8 (i.e., octal) counter with decoded 1-of-8 outputs. The input of the 4022 device is driven by a Schmitt trigger (4093) operated as a buffer, and a Schmitt trigger astable oscillator. We could, incidentally, use any of the astable multivibrator circuits of this chapter, or step through manually using a monostable circuit "bounceless pushbutton." This latter alternative is often used in troubleshooting computers and other large-scale digital projects. We can check each stage at leisure when we manually step through the sequence because the next clock phase will not come along until we mandate it by closing the *advance* switch.

There is no "waiting" period between phases in this circuit (as seen in the two-phase clock earlier): each output snaps HIGH as soon as the previous output snaps LOW.

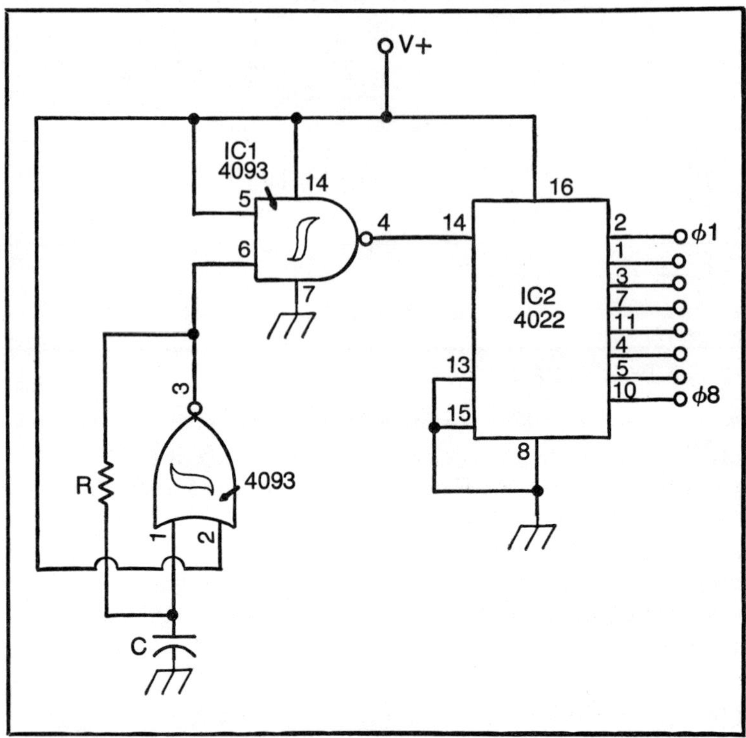

Fig. 10-9. Eight-phase clock.

OSCILLOSCOPE DUAL-TRACE SWITCH

A multi-trace oscilloscope is very useful when it is necessary to compare two waveforms that operate on the same time base. The dual-trace oscilloscope is used in both analog and digital electronics. The circuit in Fig. 10-10 will allow us to make a single-trace oscilloscope into a dual-trace oscilloscope (using two of these devices, one on each channel, can make a two-beam oscilloscope into a four-beamer!). We obtain two traces from one through the use of *switching*.

The switching section consists of a pair of CMOS switches driven out of phase with one another. The CMOS switch is designed to have a low series resistance when the control terminal is LOW and a high series resistance when the control terminal is HIGH. In this project, we are using two sections of the quad switch, type number 4066.

The output amplifier is a 741 operational amplifier. The output of this stage will be a summation of the three input signals. One of

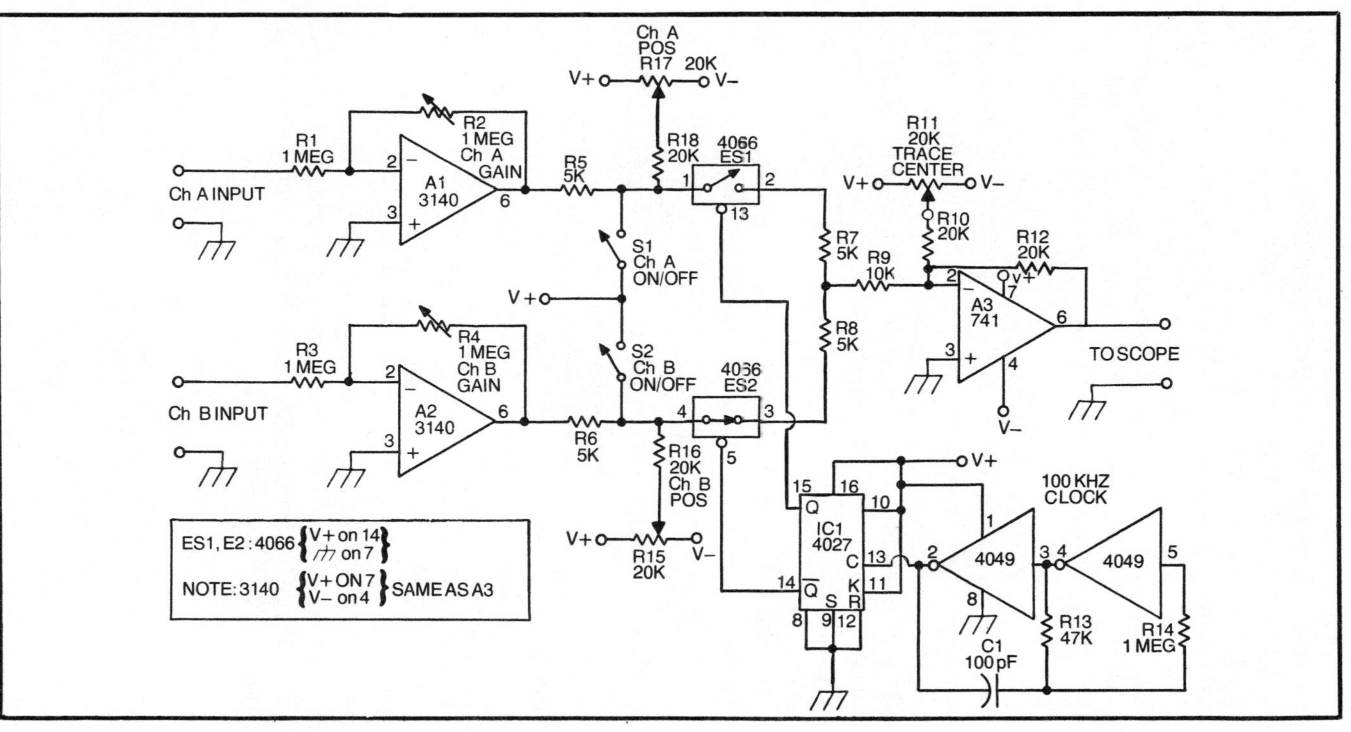

Fig. 10-10. Oscilloscope dual-trace switch.

these inputs is a master position (or *trace center*) signal, while the other signals are the two channels from input amplifiers A1 to A2. The action of the electronic switches is to connect only one input amplifier to the output stage at a time.

The control lines of the two CMOS switches are driven from the complementary outputs of a 4027 CMOS J-K flip-flop. The clock input of the J-K FF is driven from an astable multivibrator, such as seen earlier in this chapter (any of the circuits could be used). In most cases, we will want a frequency above 30 kHz for the clock frequency.

There are two input amplifiers in this circuit, A1 and A2. These should be premium types, such as the RCA CA3140 devices selected. The input impedance of each channel is approximately 1 megohm, shunted by a small capacitance.

The channel gain for each stage is set by the feedback resistors of A1 and A2, respectively. In this case, the feedback resistors are 1 megohm potentiometers, so the gain will vary from zero to unity.

The channel on/off switches (S1 and S2) have the effect of placing a high potential on the A3 input during the periods when the off channel is active. For example, when channel A is active the Q output of the J-K flip-flop is LOW so ES1 is closed. If switch S2 is closed also, then the V+ potential is applied to the input of the A3 amplifier, causing it to throw the trace offscreen.

The clock frequency will set the maximum frequency that we can display on the two channels. The display frequency must be "much less" than the clock frequency. The time base of the oscilloscope must be adjusted to make the "chopped" nature of the two waveforms disappear.

Channel A and channel B position controls are provided by potentiometers R15 and R17. These potentiometers will place a DC current component into the output amplifier (A3) input circuit only when the correct electronic switch is closed. This has the effect of placing the trace for the respective channel at a selected position on the CRT screen.

Chapter 11
TTL Timers

The TTL logic line has been around long enough to be one of the most popular logics used in digital electronics. The TTL line is fast enough to operate well into the megahertz region, yet it will work quite nicely at subhertzian frequencies. The TTL logic draws substantially more current than does CMOS, but this is the cost of high speed in digital logic devices. The most frequent error perpetrated by the logic designer is to build a power supply with too little current available. For most modest TTL projects, count the number of ICs and multiply by 30 mA. This will give you a good approximation of the current requirements (alternatively, you can look up the current requirements of each chip and then add them together). I prefer a 1 ampere power supply for any requirement over three or four chips, but then again, I am overly conservative about a lot of things in design.

TWO-PHASE DIGITAL CLOCK

The project shown in Fig. 11-1 is a two-phase digital clock for use with certain types of digital circuits, especially microcomputers. A two-phase clock must produce clock pulses on two separate lines that are non-coincident with each other. In addition, there is often a need to ensure that the settling time of the circuits being driven is exceeded before a new clock pulse is received. This becomes especially important when interfacing devices such as analog-to-digital converters (ADC) or digital-to-analog converters

R. LANGEVIN

(DAC), or when using ADCs and DACs with sample-and-hold (S&H) devices.

Figure 11-1 shows a two-phase clock circuit that accomplishes the non-coincidental requirement by driving the NOR gates with the output of the J-K flip-flop. We might be tempted to use the flip-flop without the NOR gates. After all, the two outputs of the J-K flip-flop are complementary, so one will be HIGH when the other is LOW. Although this fills our requirement of coincidentally, it does not allow for settling time of the clocked circuits. In order to accomplish the latter goal we must synchronize the system with the NOR gates. Recall from Chapter 4 the basic operating rules for the NOR gate:

- ☐ A HIGH on either input produces a LOW output.
- ☐ Both inputs must be LOW for the output to be HIGH.

With these rules in mind, let's figure out how the circuit of Fig. 11-1 works. Note that the master clock (MC) signal is inverted prior to being applied to the input of the J-K flip-flop. The inverted MC signal ($\overline{MC}$) signal is applied to one input of both NOR gates. The other input of each gate is connected to one output of the J-K flip-flop. We find that the Q output is connected to the gate that controls phase-1 (ϕ_1), while the NOT-Q output of the J-K flip-flop controls the phase-2 (ϕ_2) output of the two-phase clock. The rules of the NOR gate tell us that it is necessary for both inputs of either NOR gate to be LOW before an output pulse can occur.

Let's consider phase-1. The two inputs to the phase-1 NOR gate are the inverted master clock ($\overline{MC}$) and the Q output of the J-K flip-flop. When the first clock pulse arrives at time T1, the $\overline{MC}$ line drops LOW. This makes the Q output of the FF to go HIGH. Since there is a HIGH on one input of the NOR gate, the phase-1 output is LOW. It is not until the second clock pulse arrives that we find a situation in which both inputs of the NOR gate governing phase-1 are LOW. When this pulse comes along, the phase-1 pulse is generated. At that same time, the phase-2 clock pulse is inhibited by a HIGH on one input of the phase-2 NOR gate.

This circuit produces output pulses at a frequency of one-half the clock frequency. The complementary nature of the Q and NOT-Q outputs of the FF ensure that only one pulse is generated at a time, while the NOR gates ensure that a full clock period will exist between each pulse. This will allow circuits being clocked to settle and get ready for the next "happening."

MULTIPHASE DIGITAL CLOCK

The digital clock shown in the previous figure is capable of driving a large number of digital circuits, including several very popular microcomputers. But what do you do if you want to create a multiple-phase clock? The use of gating can create a large array, and that can lead to errors, not to mention the large bill for +5 volt DC current! The circuit in Fig. 11-2 allows us to generate a sixteen-phase digital clock using just two integrated circuits.

The heart of the circuit is a 1-of-16 data distributor integrated circuit. This 24-pin DIP IC examines a four-bit binary input word and decodes the word to a unique output condition (i.e., in which one input is LOW and all others are HIGH). The binary word on the four input lines to the 74155 in Fig. 11-2 will cause only one output at a time to be unique. If switch S1 is open, then the selected output will be HIGH and all others will be LOW. But if S1 is closed, then pin 18 is LOW so the selected output will be LOW and all others are HIGH. We can call switch S1 an output polarity select switch. Table 11-1 gives the input codes and the corresponding phase selection:

There are no "settling" states between phases in this circuit (as in the last circuit). The next phase becomes active immediately after the previous phase goes inactive.

The input word for the 74155 is formed by a four-bit binary counter, the TTL type 7493. This counter will continue to

Table 11-1. Input Codes and Phase Selection for Circuit of Fig. 11-2.

Binary Word	Phase
0000	1
0001	2
0010	3
0011	4
0100	5
0101	6
0110	7
0111	8
1000	9
1001	10
1010	11
1011	12
1100	13
1101	14
1110	15
1111	16

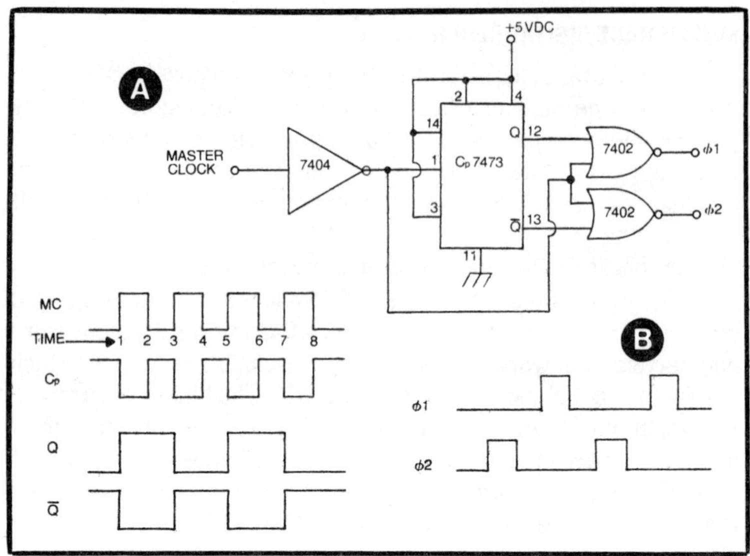

Fig. 11-1. TTL two-phase clock.

increment through its range until power is turned off because it is connected with the reset terminals (pins 2 and 3) to ground.

We may select phase lengths considerably less than sixteen using this same circuit. What is required is some feedback to reset the counter to 0000 when the desired count is reached. The reset pins of a 7493 want to be LOW for operation and HIGH for reset. We could, therefore, select the switch option that gives a HIGH for the desired output and a LOW for all others. We could then connect the terminal output to the reset pins of the 7493. For example, suppose that we needed a twelve-phase clock. We would then connect phase-12 of Fig. 11-2 (i.e., pin 13) back to pins 2 and 3 of the 7493. As long as any other output is selected, the reset pins of the counter remain LOW so the count will proceed in the usual manner. But when the clock has incremented the counter the twelth time, a HIGH is sent to the reset line, and this causes the counter (and the 74155) to go back to state 0000.

If we select the switch option in which the desired output is LOW and all other outputs are HIGH, then it will be necessary to feedback the reset pulse through an inverter stage.

TRANSISTOR-TTL CLOCK

Every now and then we see an application where a transistor is used as a clock in a digital circuit. This poses several problems for the designer. One is the fact that the transistor will produce a sinewave (or nearly so) output waveform. Secondly, the output levels will not be TTL compatible. These problems are overcome by an arrangement such as Fig. 11-3. Transistor Q1 is a Colpitts oscillator that uses a piezoelectric crystal for the resonator. The frequency of oscillation can be trimmed slightly by capacitor C1, in series with crystal element Y1. The feedback for oscillation is formed by the classical Colpitts arrangement, C2 and C3. The transistor is biased from the collector by the current flowing in resistor R1. The transistor used can be almost anything that will oscillate at the Y1 frequency (it isn't too critical in most applications, try a 2N2222). The actual value for resistor R1 may differ slightly from one transistor to another but, in any event, will be between 150 kohm and 390 kohms.

Output signal is coupled to the 7414 TTL device through a capacitor (C4) to the emitter. This emitter-coupled output is

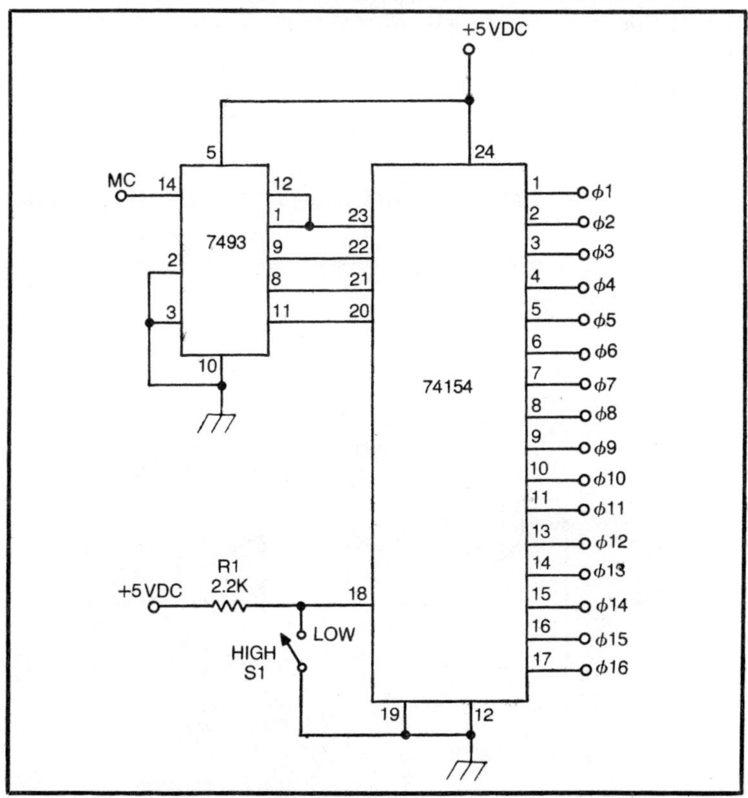

Fig. 11-2. TTL sixteen-phase clock.

relatively standard on simple oscillator circuits. The TTL device used in this circuit is a 7414 Schmitt trigger. A Schmitt trigger is a circuit or device that is used to "square" an input signal, making it suitable for use in digital circuits. The output of the Schmitt trigger used in Fig. 11-3 will snap HIGH when the positive-going input voltage passes 1.7 volts. The output remains HIGH until the input voltage passes its peak and has dropped back down to 0.9 volts. This means that there is a 0.8 volt hysteresis in the device. The output of the 7414 will be TTL compatible, even though some maverick input levels are used.

The 7414 Schmitt trigger is used sometimes to interface high precision, high stability, clocks to digital circuits. Few, if any, TTL circuits that oscillate make stable clocks. When you have a need for a very stable clock, then it might be necessary to use one of the transistor circuits, or buy a temperature compensated crystal oscillator module from a supplier. These modules are small metallic or epoxy building block "function modules" and are quite costly.

TTL CRYSTAL OSCILLATOR

The standard TTL crystal oscillator (Fig. 11-4) uses at least two inverters with a crystal in the feedback loop between them. In most cases, the "inverters" will actually be 7400 NAND gate sections wired as inverters by tying both inputs together. We can

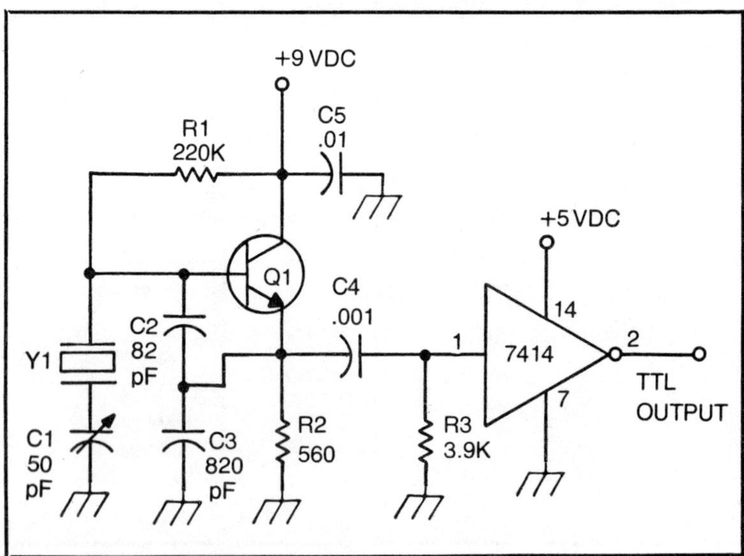

Fig. 11-3. TTL-Crystal controlled clock.

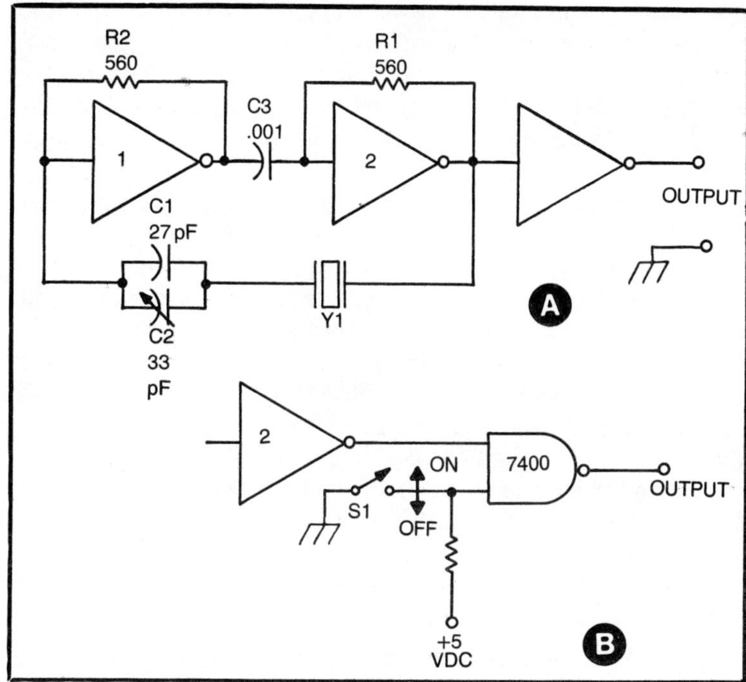

Fig. 11-4. TTL-Crystal controlled oscillator.

build such a circuit with left-over sections of a 7400 device used elsewhere in some circuits, and so may not have to obtain a hex inverter chip.

Resistors R1 and R2 are used to bias the inverter stages to a point where they will oscillate. The circuit is sometimes a little flakey, meaning that it is hard starting. We sometimes have to mess with the values of R1 and R2 to make the circuit work with any specific crystal. The values shown, however, are the most often used values and will work most of the time. The capacitor is used to isolate the two inverter sections from each other.

The frequency of oscillation can be "pulled" a little bit by the combined action of C1 and C2. The trimmer capacitor is used to set the actual operating frequency; a range of several kilohertz is possible. Note that the same technique can be used to partially temperature compensate this circuit. We can select capacitors for C1 that have a temperature coefficient that tends to cancel the temperature coefficient of crystal Y1.

The output inverter is used to buffer the oscillator against the rather sudden and dramatic load variations normal to TTL circuits.

If a NAND gate is used for the output buffer, then we can use it to control the flow of clock pulses to the outside world while simultaneously allowing the oscillator to remain turned on. This will offer both the control required by some applications and the stability provided by leaving the oscillator running. The on-off switch is shown in Fig. 11-4B. One input to the 7400 NAND gate is connected to the output of the oscillator (i.e., the output of inverter 2), while the other input to the NAND gate is connected to the on-off switch. When switch S1 is open, this input is connected permanently HIGH. This means that it is essentially inert, and that the pulses at the other input control the output signal. But when S1 is closed, then there is a LOW on the NAND gate input. By the normal rules of operation for the NAND gate, this condition will force the output permanently HIGH. The output remains HIGH until the switch is again opened. In a practical circuit the switch may be replaced by some logic function from another IC device. In that case, the logic circuit will determine when the clock pulses are transmitted.

MC4024P TTL OSCILLATOR

The Motorola MC4024P is a TTL integrated circuit that contains two independent voltage controlled oscillators (VCO) with TTL-compatible outputs. The device is not to be confused with a device with a similar type number (4024) in the CMOS line of IC devices. These are two different devices, and are not the least bit interchangeable.

There are three grounds and three +5 volt lines on the MC4024P device. One of each are for the package as a whole, while the other two are specific to one or the other oscillator section. In Fig. 11-5 we ground pins 5 and 7; and we connect pins 1 and 14 to the +5 volt line, turning on one of the oscillators. Pin 2 is the voltage control line. We connect it to +5 volts in this application because we do not want voltage control of the oscillating frequency. In this circuit, the operating frequency is given approximately by $F = 300/C$, where C is in picofarads and F is in megahertz. This circuit requires a little tweaking of the C1 value if we require an accurate frequency. But such accuracy is not always wanted in digital circuits. We can often get away with a little sloppiness in the clock frequency in order to obtain a quick and dirty clock circuit (ain't synchronous circuits wonderful?).

I personally prefer using the MC4024P device because it is very easy to tame. Some TTL clock oscillators are somewhat

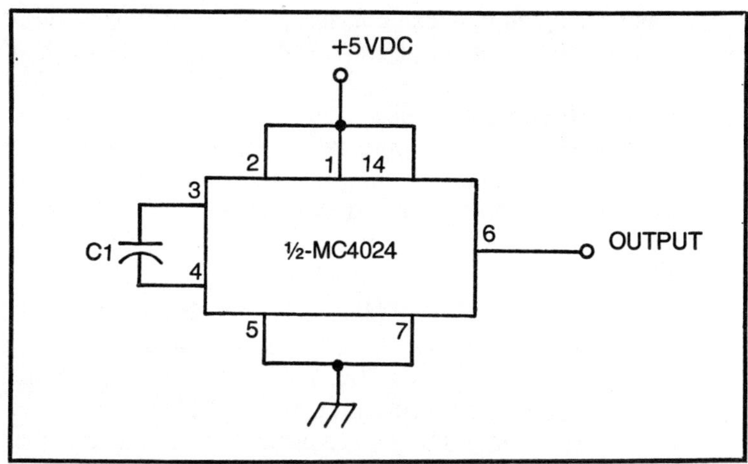

Fig. 11-5. Motorola MC4024P as a clock.

pesky to start. My Digital Group microcomputer, for example, uses a TTL clock. In early production it was found that many people were having trouble making the oscillator start. The result was a white square on the TVT screen; or 1024 of them, one for each memory location: the CPU wasn't running for the lack of a clock signal. The cure recommended by DGI was a 220 pF capacitor to add some more feedback, sort of kicking the crystal in the pants. Using a chip like the MC4024P solves a lot of those problems and works like aspirin on a headache.

The circuit in Fig. 11-5 is competent to work from approximately 1 Hz to over 25 MHz. Some selected MC4024P devices will work nicely to 35 MHz.

MC4024P CRYSTAL OSCILLATOR

The MC4024P device will work nicely as a crystal oscillator at frequencies from (about) 3 MHz to 25 MHz. The circuit shown in Fig. 11-6 will not work reliably at frequencies below 3 MHz unless some additional capacitance is added, and then there is the question of whether the crystal or the capacitor is controlling the oscillation frequency.

The same pinouts are used in this circuit as were used in the previous circuit, with the exception that pin 2 is now connected to a voltage source. The piezoelectric crystal (Y1) is connected across pins 3 and 4, replacing the capacitor as the frequency control element.

We can use the potentiometer (R1) to trim the frequency of oscillation (a very little bit) and to ensure that the crystal starts easily and oscillates in a stable manner. One well-known author published a crystal oscillator circuit in which pin 2 is connected to +5 volts DC. In my experience, this arrangement leads to flakey starting—one thing *I* don't need.

Remember that these circuits are using just one section of the MC4024P. The remaining section is available for use independently of the section being used in this circuit.

MC4024P VCO

The Motorola MC4024P device is intended as a *voltage-controlled* oscillator. The operating frequency is a function of the capacitor value selected for C1 (Fig. 11-7) and the voltage applied to pin 2. This potential must be approximately 2.5 to 5 volts. In this circuit, we have the MC4024P connected as in the previous examples, except for the control input (pin 2).

The control voltage is generated in an RCA CA3140 operational amplifier circuit. The minimum voltage (2.5) is set by adjusting the operational amplifier offset voltage to 2.5 VDC with potentiometer R5.

The control voltage is given by the expression:

$$E_c = \frac{-(R1 + R2)(E)}{R3}$$

The negative sign is needed because this circuit is an inverting follower amplifier. The output polarity will have the opposite sign of the input polarity. Reference voltage source E, therefore,

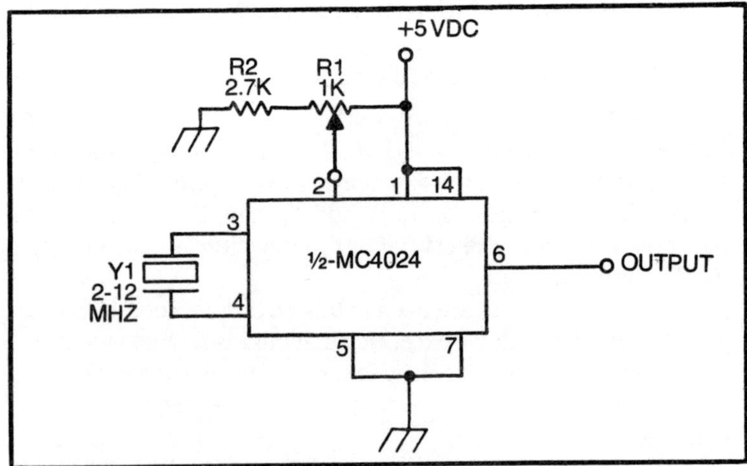

Fig. 11-6. MC4024P as a crystal controlled oscillator.

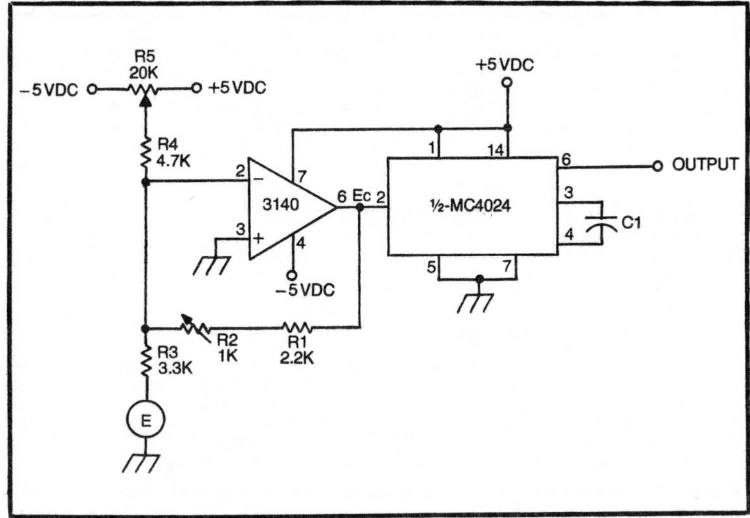

Fig. 11-7. MC4024P voltage controlled oscillator.

should be negative with respect to ground. This polarity will make the output voltage polarity positive, as required by the MC4024P device.

We can adjust the operating frequency of this device over approximately a 3:1 range by using various values of −E at the input. The gain-control potentiometer in the feedback loop of the operational amplifier allows us to calibrate the input voltage to produce a specified frequency at some given input potential.

The operational amplifier selected was the RCA CA3140 BiMOS device. This operational amplifier was selected because it offers the ability to operate well on +5 volt DC power supplies.

60-Hz CLOCK

Some of the circuits to follow will use the 60 Hz AC line as a time base source. This is a reasonably accurate source and is suitable for many digital projects. But sometimes AC is not available, and we will need a circuit to generate the 60 Hz clock. Figure 11-8 is such a circuit. It is our old friend the 555 astable multivibrator, with component values selected for proper operation at 60 Hz. Potentiometer R2 is used to adjust the output frequency to exactly 60 Hz.

The AC power line frequency is usually accurate to within ±1 percent in the short term and much better over the long term (i.e., when the errors of the short term tend to "add out"). This circuit

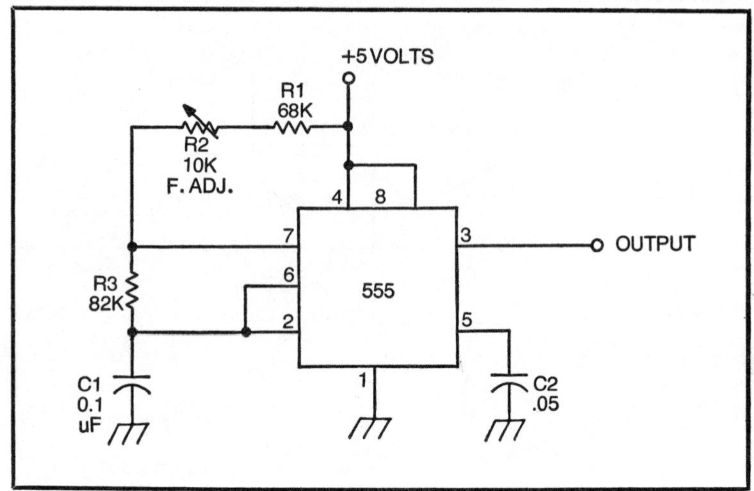

Fig. 11-8. Sixty-Hertz clock.

will not approach that accuracy, but will do in a pinch—like when there is no available AC power.

The output of this circuit is probably TTL compatible. I have found, however, that not all 555s are created equal. Some of them have difficulty driving most TTL gates—and that is especially true when low cost 555s are used. If this proves to be a problem, then follow the 555 output with a device that will translate the level to TTL (see preceding chapters, where several examples are presented).

ONE-SECOND CLOCK

Many of the projects in this book call for a one-second digital clock (i.e., 1 Hz oscillator). We can make a relatively accurate one-second clock by using the 60-Hz AC line frequency as the time base. The short-term accuracy of this form of time base is around 1 percent, while the long-term accuracy is better. The long-term accuracy is superior to the short-term because the errors tend to average out to zero over the long term. The circuit in Fig. 11-9 allows us to scale the 60Hz AC line frequency down to the required 1 Hz. This magic trick is done in a divide-by-60 circuit.

The 60-Hz signal is derived from the secondary to the filament transformer (vacuum tube terminology—oooops) used to derive the rectified DC low voltage. The normal rectifier would be connected to points "X" on the transformer. Diode D1 is an extra diode (not part of the rectifier), and may be any of the 1N4000 series devices. The AC from the transformer secondary is rectified by diode D1. The lack of filtering makes this signal a 60-Hz chain of pulses.

A 7414 TTL Schmitt trigger device is used to convert the 60 Hz pulsating DC signal produced by the rectifier into squarewaves of the same frequency that are TTL compatible. The operation of the 7414 was covered earlier in this chapter.

The output of the 7414 is a series of square wave pulses with a frequency of 60 Hz. These pulses are applied to the input of a 7490 digital counter. The 7490 is billed as a decade (divide-by-10) counter, but in reality it is a biquinary (2 × 5) counter. The signal from the 7414 is applied to the input of the divide-by-2 section of the 7490 (i.e., pin 14). We connect the two sections in cascade by strapping together pins 12 and 1. Pin 11 forms the divide-by-10 output of the 7490. The frequency at this point is 60/10, or 6 Hz.

The 6-Hz signal from the 7490 is fed to the divide-by-6 input of a 7492 counter. The 7492 is normally billed as a divide-by-12 counter, but it (like the 7490) is actually a two-stage counter. The 7492 is a bisextary counter (2X6); we are using only the divide-by-6 section. The signal from the 7490 output is applied to pin 1 of the 7492, and the output is taken from pin 8 of the 7492. The frequency at the output is 1 Hz. Note that these counters are being

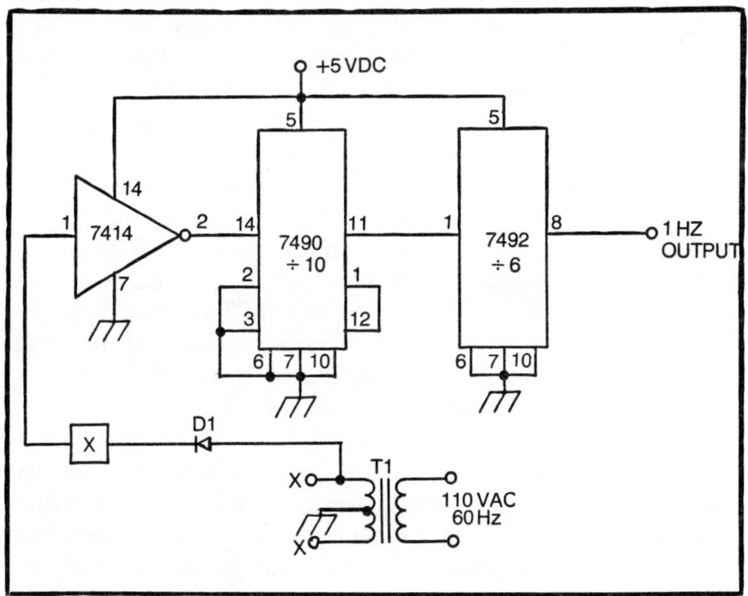

Fig. 11-9. One-second timer.

used as simple frequency dividers. The reset terminals (2,3,6, and 7 on the 7490; 6 and 7 on the 7492) are therefore grounded.

The project shown in Fig. 11-9 can be used for any of the projects to follow that call for a one-second clock, or a 1 Hz clock (the same thing). If you require a portable circuit, then substitute the 60 Hz 555 oscillator shown previously for the 60-Hz transformer. The stability won't be quite as precise because of the RC timing circuits used in the 555.

SIXTY-SECOND TIMER

A 60-second timer will output one pulse every 60 seconds. The circuit in Fig. 11-10 will do this job. The clock for the 60-second timer is the circuit shown previously in Fig. 11-9. For the sake of simplicity we will delete that portion of the circuit. You may refer back to the clock circuit of Fig. 11-9 if you try to build this circuit.

This circuit is a lesson in digital decoding techniques. The heart of the 60-second timer is a cascade divide-by-100 counter consisting of a pair of 7490 decade counters. Since we only want to count to 60, we must conspire somehow to reset the counters to zero after they have counted to 60.

The BCD code for 60 is 0110 0000. We want to generate an output pulse and reset the counters when this code appears on the output of the counters (i.e.,0000 on IC1 and 0110 on IC2).

The 7430 device is an eight-input TTL NAND gate. It will produce a HIGH output if any of the eight input lines are LOW. The output of the 7430 will drop LOW only when all eight input lines are HIGH. Lines 6 and 7 (i.e. pins 6 and 11 of the 7430) will normally be HIGH when the counter is at the desired count of 0110, so we connect these lines directly to the corresponding counter outputs on IC2. All of the other counter outputs will be LOW when the desired state is reached. These output lines, then, must be inverted prior to being applied to the 7430 inputs. With the arrangement shown in Fig. 11-10, the only possible code that will generate an output pulse (i.e., LOW on the output of the 7430) is 0110 0000, which produces the required 1111 1111 code.

When the correct count is reached, the 7430 output will drop LOW. An inverter stage is used to reset the counters with the output pulse. The output level is normally HIGH, so the inverted output which is applied to the reset terminals of the counters will be LOW. The normal operation of the 7490 counters is LOW, so the count will proceed.

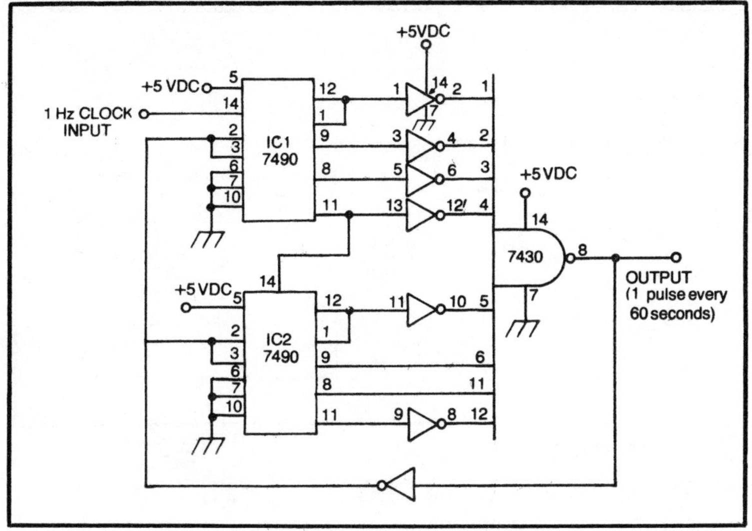

Fig. 11-10. Sixty-second timer.

When the counter "times out," at 60 seconds, however, the 7430 output drops LOW, and this is reflected through the inverter as a HIGH condition. The HIGH on the reset lines will reset the counters to 0000 0000, allowing the process to begin over again.

THREE-MINUTE TIMER

We can take the 1 Hz clock (Fig. 11-9) and the 60-second timer (Fig. 11-10) and make a three-minute timer (Fig. 11-11). We output an active-LOW from the 60-second timer. This signal is inverted by a 7400 NAND gate connected as an inverter (i.e., both inputs tied together). The output of the 7400 is applied to the input of a 7490 decade counter.

Three minutes is 180 seconds, and the clock pulse feeding the 7490 in Fig. 11-11 has a 60-second repetition time. This means that we will want to output a pulse, and reset the 7490 every third pulse. The code for 3 in BCD terms is 0011. Pins 12-1 form the "1" output of the 7490, while pin 9 is the "2" weighted output. We connect these pins to alternate inputs of a 7400 NAND gate. The operation of the circuit after that is pretty much the same as in the circuit of Fig. 11-10. As long as either input of the 7400 NAND gate is LOW, then the output is HIGH. We need to see both inputs HIGH in order to generate a LOW output pulse. This condition is met only when the count is 0011 (i.e., 3_{10}).

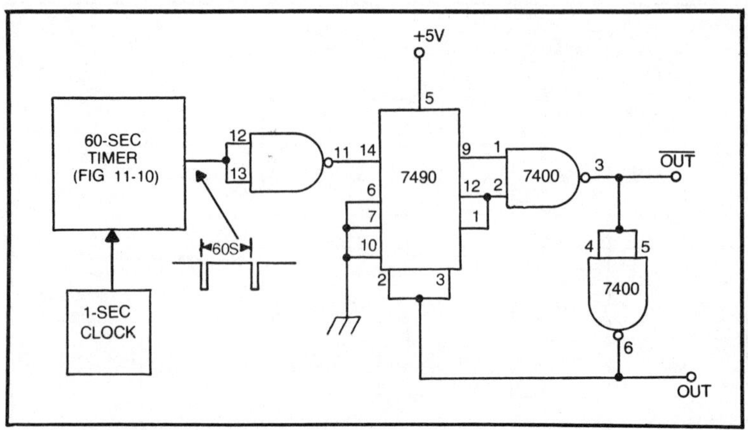

Fig. 11-11. Three-minute timer.

The reset lines of the 7490 are normally kept LOW by inverting the HIGH at the output line. But when the counter "times out," the output line drops LOW. This LOW is reflected through the inverter as a HIGH that resets the counter to 0000.

The main output of both this and the previous circuit can be designated as an $\overline{\text{OUT}}$ line, indicating the active-LOW status. If we select the reset line as an alternate output, it will be active-HIGH, so can be designated OUT.

SHIFT-REGISTER ONE-SHOT

The 74164 is a TTL shift register. OK, what the hell is a shift register? If you don't know, then go back to Chapter 4 and read that section. The 74164 is a serial-in-parallel-out (SIPO) shift register with eight stages. The data applied to the serial inputs (pins 1 and 2) will be shifted one stage to the right for each clock pulse that is received. There are a total of eight stages, each with its own output line. The contents of the register are cleared (i.e., outputs go to zero) by bringing the clear input (pin 9) briefly LOW.

In Fig. 11-12 we use the clear input of the 74164 as a trigger. All this stage actually does is to clear the shift register to 00000000, so that the count will begin on the next clock pulse. The serial inputs are wired HIGH, so the data that shifts will be HIGH. We could make the data LOW (thereby making the device active-LOW output) by making *either* serial input LOW. In this case, however, we have tied the serial inputs permanently HIGH through a current limiting pull-up resistor.

Clocking the 74164 causes the HIGH at the serial input to be shifted one position to the right for each clock pulse. If the period of the clock signal is T, then we can select periods of 1T to 8T by selecting the correct output line.

AUTORESETTING SHIFT REGISTER TIMER

The circuit of Fig. 11-12 will simply time out, and then become dormant. The circuit operates, therefore, as a one-shot. We can make the circuit operate more like an astable multivibrator by using the output line to reset the 74164.

The operation of the circuit in Fig. 11-13 is identical to that of Fig. 11-12, except for the addition of the gate (G9). This circuit, like several to follow, were used in an "evoked potentials" stimulator that I designed for brainwave research in a university medical engineering laboratory. This accounts for the odd numbering of the components and for the spurious "bullets" marked X and Y on the diagram. These refer to the printed circuit card edge connectors and may be safely deleted by you.

Gate G9 is a NOR gate that is connected as an inverter stage. We could also use a regular hex inverter section or a NAND gate that has been wircd as an inverter. The CLR input of the 74164 is held HIGH in normal operation. The output of the 74164 is LOW, so it will reflect a HIGH through the inverter. But when the counter times out, and the particular output selected goes HIGH, then the gate causes the CLR input of the 74164 to drop LOW long enough to reset the IC. The result is a brief pulse output at the *time-out* of the circuit.

Sensory evoked potentials is a method of removing the apparent randomness of the EEG signal by an averaging technique. A computer is used to digitize the EEG analog signal, and then

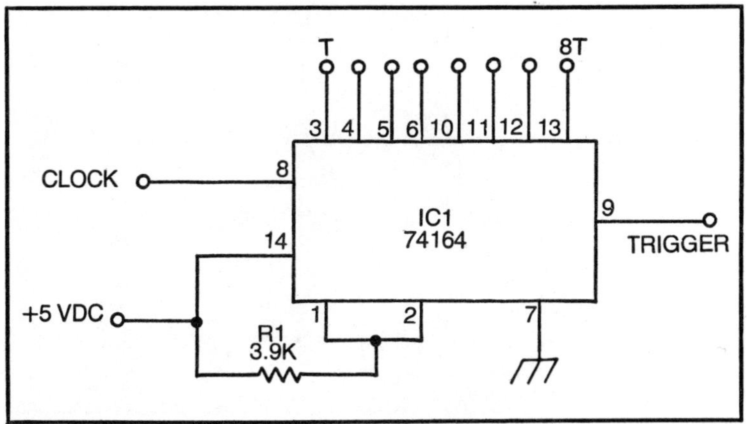

Fig. 11-12. Programmable timer.

average the signal in a coherent manner over a large number of iterations of the same stimulus. If, for example, you were to look at the waveform from the brain while flashing a light at the "victim," it would be impossible to tell which components of the seemingly random process seen on the oscilloscope is due to the flash of light. But if you average the waveform after each of (say) 100 stimulii, then the randomness would average out to near zero leaving only the component that is due to the flash of light. This technique gives physicians and life-sciences investigators another tool for looking at the human brain. The project in Fig. 11-13 was used as part of a television stimulator. The idea was to flash a stimulus pattern on the screen of the TV at the proper moment, and then time its duration before shutting it off (i.e., reverting back to the program). It is necessary that the pattern switching be done between vertical frames, so as to avoid a spurious stimulus caused by on-screen switching. We use this timer to count from one to eight vertical intervals (16.7 to 133.3 milliseconds), and to ensure off-screen switching (more in a moment).

The evoked potential in the human brain will ring for about one full second before the phenomenon dies out. We do not wish to restimulate the victim until all ringing has died out. As a result, there must be an interstimulus timer in the circuit.

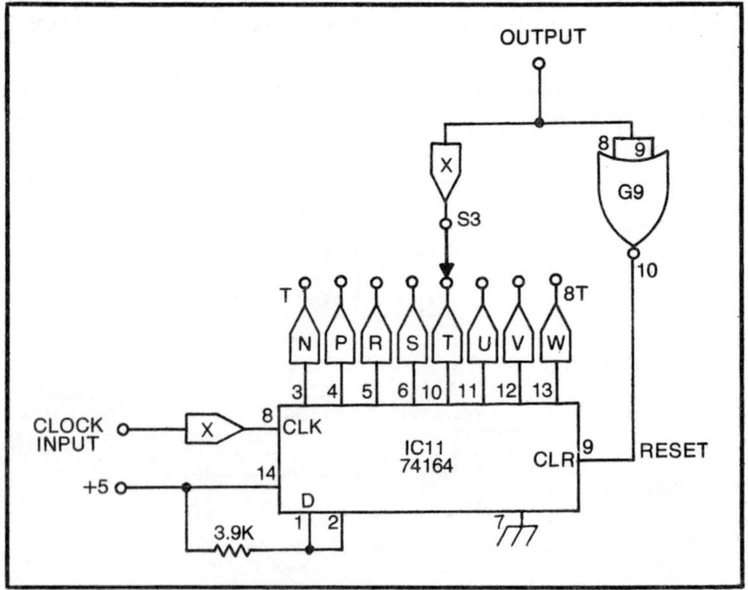

Fig. 11-13. Resettable programmable timer.

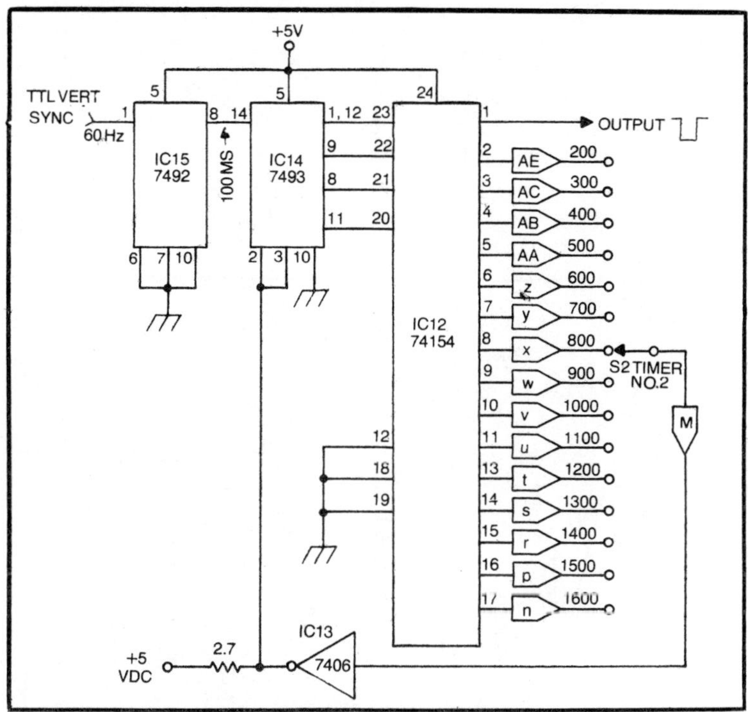

Fig. 11-14. Duration timer.

INTERSTIMULUS TIMER FOR EVOKED POTENTIALS STUDIES

The circuit in Fig. 11-14 delivers interstimulus durations of 200 to 1600 milliseconds, in 100 ms increments. The vertical synchronization signal from the television is used as the clock. IC15 works as a divide-by-6 stage to produce a 10 Hz (i.e., 100 ms) output pulse for the main counter.

The main counter is a 7493 connected as a divide-by-16 (hexadecimal) stage. The four-bit binary output lines are connected to the four-bit input of a 74155 data selector. This 24-pin IC will produce a unique LOW condition on an output dictated by the binary word applied to the inputs. Output 1 (input code 0000) is selected as the master output for the circuit. The time between stimulii is set by resetting the main counter (IC14) to 0000 at the specific time after the last reset command. This is done by setting switch S2 to one of the other 74155 output lines.

When the time is up, the selected output drops LOW. This LOW is reflected as a HIGH to the reset terminals of the main counter because of the action of inverter IC13. I had to use a 7406

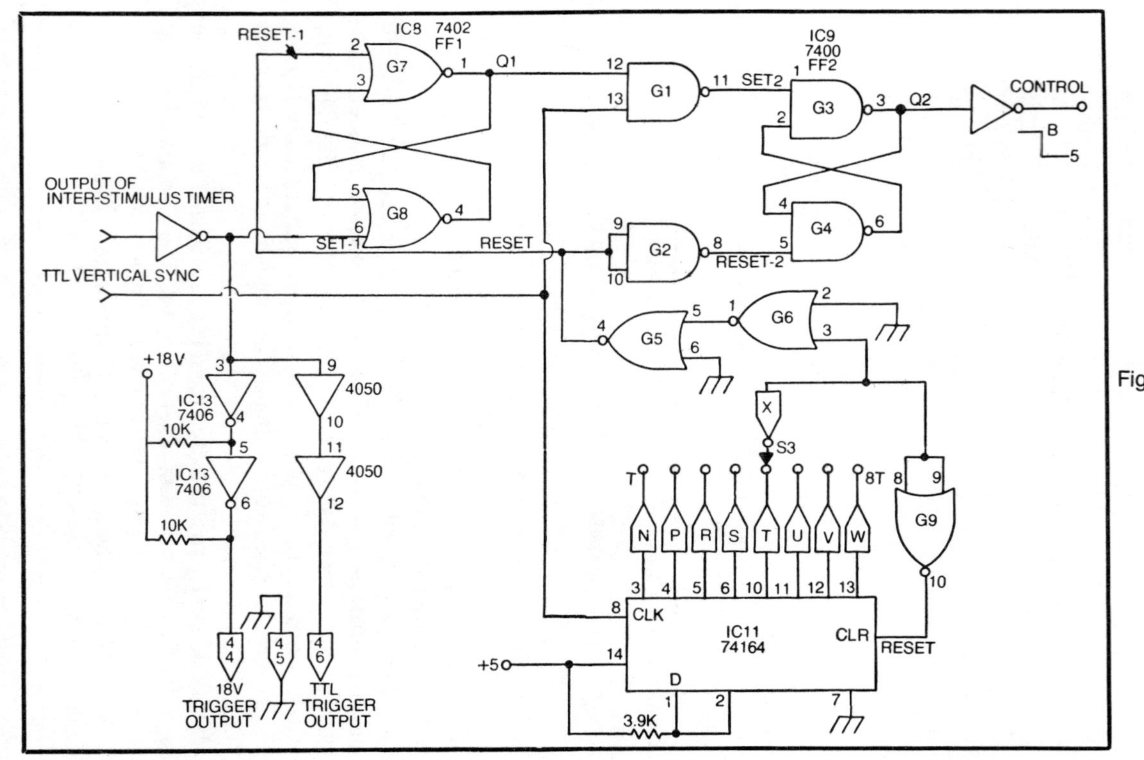

Fig. 11-15. Timer controller.

inverter because this chip was needed elsewhere in the circuit and had one section left over. You could use any of the TTL inverters (which don't require a pull-up resistor to +5 volts DC) or connect a NAND or NOR gate as an inverter.

MAIN CONTROL UNIT

This section could easily be called "putting it all together." We take the 74164 circuit of Fig. 11-13 and the output of the interstimulus timer of Fig. 11-14, and use them to control the operation of the stimulator. In my case, the stimulator was a color-bar-dot generator connected to the antenna terminals of the television receiver.

A TTL-compatible rendition of the vertical sync signal is available from the Heathkit color-bar generator that I selected. This signal is used as the synchronization clock in the circuit of Fig. 11-15. Flip-flop FF1 is a NOR-input RS flip-flop. The output of the interstimulus timer is inverted and then used to trigger the set input of FF1. The Q_1 output of FF1 is normally LOW, so that will lock up NAND gate G1. But when the interstimulus timer sets the FF1 output, then a HIGH is applied to this input of G1. The other input of G1 is connected to the vertical sync line. When the vertical pulse goes HIGH, gate G1 will pass a signal that will set flip-flop FF2. This event is the trigger that turns on the stimulator by causing the control BS line (no jokes, please) to go LOW.

The vertical signal is also used as the clock for the 74164 shift register timer. We select the duration desired for the stimulus by selecting 74164 outputs T to 8T for a stimulus duration of 17 to 134 ms. When the selected time has expired, the corresponding output of the 74164 goes HIGH causing a signal to be propagated through gates G5 and G6 to the main reset line. This signal will cause both RS flip-flops (FF1 and FF2) to go back to their reset, or cleared, condition. The output of the 74164 is also used to reset the 74164 in the manner described previously.

The operation of this circuit allows the selection of one to eight vertical frames for the stimulus length and allows a new stimulus to occur only after the old stimulus has finished completely. Depending upon the experiment, and to some extent the subject, the duration of the interstimulus interval will be 200 to 1600 milliseconds. In most experimental situations, a duration of 1300 ms is deemed satisfactory.

Chapter 12

Retriggerable Timers

A monostable multivibrator is a binary circuit with only one stable state. There are two *permissable* states, but only one of these is stable. The monostable multivibrator, also called the one-shot, is designed to remain in whichever of the two states is the dormant condition until a trigger pulse causes it to shift to the transient state. But the transient state is unstable. The circuit will remain in that state only for a specified, predetermined period of time. After the period expires, the circuit reverts to the dormant state. It will then remain dormant until another trigger pulse is received. In most circuits, the duration of the transient time is set by an RC time constant.

Many multivibrator circuits also have a brief *refractory period* following the transient period in which no further trigger pulses are honored. This is usually the period required for the capacitor in the RC timing network to recharge (or discharge) back to its dormant condition.

Regardless of whether or not there is a refractory period, most monostable multivibrator circuits cannot be retriggered until the transient period has expired. Those that have refractory periods require the transient period *plus* the refractory period before any additional trigger pulses will be honored. This behavior is not a circuit defect; it is absolutely essential in some applications. For "bounceless pushbuttons," for example, the whole purpose of the one-shot is defeated if the circuit responds to additional trigger pulses. In that application, the first "bounce" of the switch contacts will trigger the one-shot, which proceeds to trigger some additional circuitry while ignoring the multiple secondary bounces of the switch contacts.

But there are applications where it would be nice to retrigger a one-shot. Such applications include any circuit where the one-shot is triggered every time some event occurs. If the event occurs again before the circuit is ready to accept another trigger pulse, then some function of the circuit will be lost.

OPERATION

Figure 12-1 shows the operation of the retriggerable monostable multivibrator. The period of the one-shot output pulse is T. At time t_1, a trigger pulse is received, so the output pulse goes HIGH for time T. No additional pulses are received on the trigger line, so the output drops LOW again as soon as its period (T) has expired (i.e., t_2). This is normal operation for a one-shot circuit and is the same as for the non-retriggerable circuits. At time t_3, another trigger pulse is received. The output of the one-shot again goes HIGH and would remain HIGH until time t_5 (i.e., period T). But at time t_4 another trigger pulse occurs before the expiration of period T. This causes the circuit to retrigger; the output remains HIGH for an additional time equal to period T. The total duration of the HIGH condition is not T, or 2T, but is equal to T plus the expired portion of the first trigger pulse (i.e., duration $t_4 - t_3$). The total time, therefore, is

$$T^1 = T + (t_4 - T_3)$$

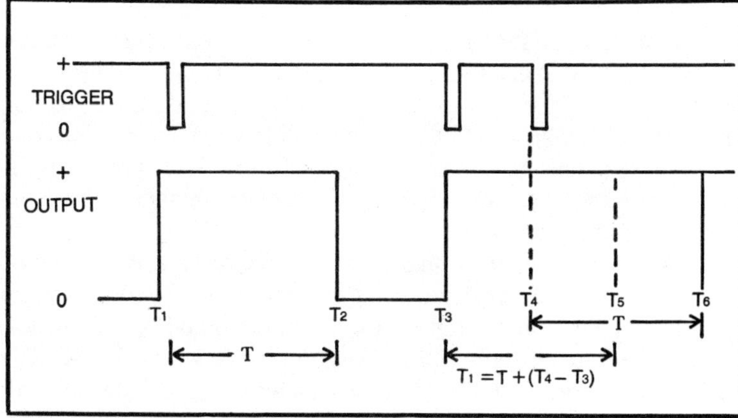

Fig. 12-1. Waveform for a retriggerable one-shot.

This is the behavior expected of a retriggerable monostable multivibrator. There are several different types of retriggerable circuits available. Some are special-purpose retriggerable multivibrator chips (i.e., 74122), while others are clever circuits involving normally non-retriggerable circuits.

Figure 12-2 shows two methods for retriggering the 555 timer IC. This circuit can be operated as a non-retriggerable monostable multivibrator (Chapter 6) with an output duration that is a function of the RC time constant R_1C_1. Figure 12-2A shows a method for making the 555 retriggerable from a negative-going pulse. In ordinary operation, the trigger input of the 555 (pin 2) is held HIGH (i.e., greater than ⅓V+, usually very close to V+). A negative-going trigger pulse must pull the trigger input to a LOW of less than ⅓V+. If this is done, then an internal comparator sets an RS flip-flop and the output of this flip-flop is (inverted to form) the chip output.

The circuit is not retriggerable because the charge on the capacitor is not dumped by the trigger pulse. As long as the capacitor continues to charge, and is less than ⅔V+, then the output remains HIGH and the trigger input is not affected. The trick is to reset the internal RS flip-flop by discharging capacitor C1 using the trigger pulse.

In Fig. 12-2A we connect a transistor switch across timing capacitor C1. Transistor Q1 is a PNP type, so it wants to see the base more negative than the emitter in order to be forward biased. The emitter of Q1 is connected to the "hot" end of capacitor C1, so Q1 is forward biased if the trigger line is HIGH. The negative-going trigger pulse serves two purposes: It will cause the trigger line to be actuated (i.e., pin 2 is brought to less than ⅓V+), and it will momentarily forward-bias transistor Q1. With the transistor forward biased there is a low-impedance short circuit across the capacitor, so the capacitor discharges. The duration of the retriggering pulse must be great enough to allow the discharge of C1 to take place. This is not too much of a problem in most cases, unless the amplitude of the trigger pulse is insufficient to drive the transistor into saturation.

Another method is shown in Fig. 12-2B, and this method allows the use of a positive-going trigger pules. Two transistors are used: One is connected across capacitor C1 and will cause it to be discharged whenever the transistor is forward biased. The other transistor is connected to the trigger terminal. The collector of transistor Q2 is also connected to V+ through a pull-up resistor.

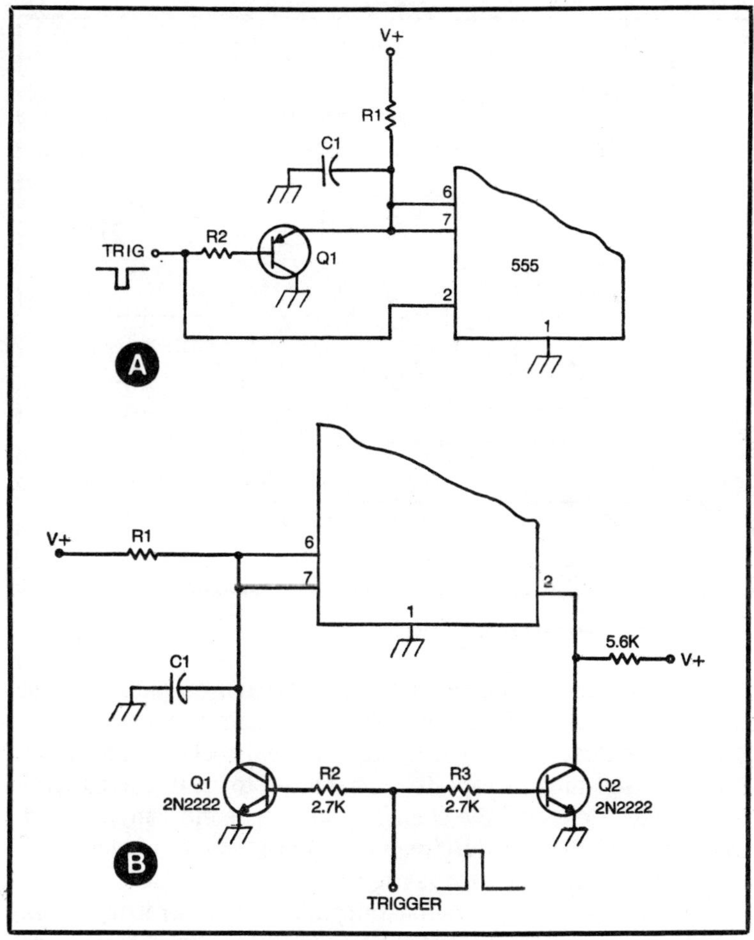

Fig. 12-2. Methods for making a 555 retriggerable.

This arrangement ensures that the trigger terminal is kept HIGH when no trigger pulse is present. But when a positive-going trigger pulse is applied to the bases of the two transistors, the 555 will be retriggered. When transistor Q1 is forward biased, the charge (if any) in capacitor C1 is dumped. Also, when transistor Q2 is forward biased it grounds pin 2 of the 555 for the duration of the trigger pulse.

A similar tactic is shown in Fig. 12-3, this time using the XR-2240 IC timer. This will also work on the second-source XR-2240 (i.e., the Intersil 8240) and on the 8250/8260 BCD versions.

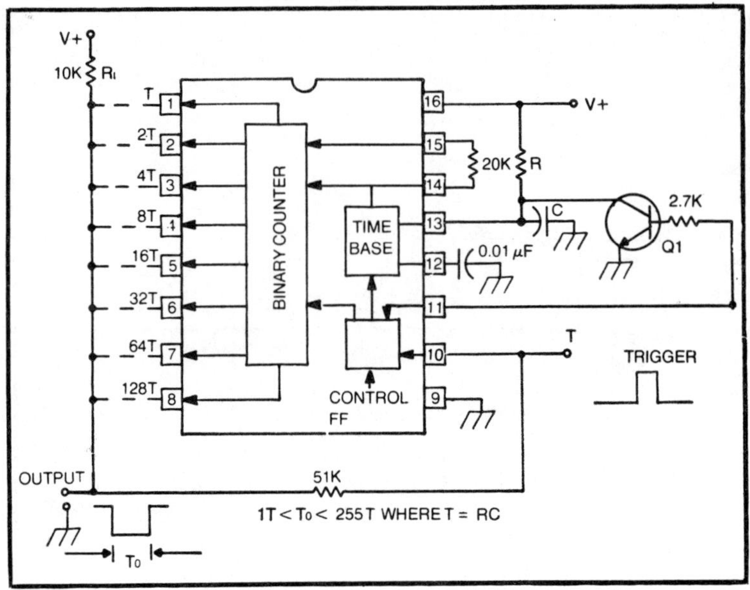

Fig. 12-3. Retriggerable XR-2240 circuit.

The XR-2240 timer requires a positive trigger pulse applied to pin 11 of the chip. But we have the same problems on the XR-2240 as existed on the 555. The timer will not honor additional trigger pulses until the capacitor has charged to the threshold level required to "time-out" the chip. The solution is the same: we add transistor switch Q1 to the circuit in order to "dump" the charge on capacitor C1 in response to the trigger or retrigger pulses.

An example of an operational amplifier used as a retriggerable one-shot is shown in Fig. 12-4A. The circuit uses the op-amp in its capacity as a voltage comparator. The noninverting input is biased through voltage divider R2/R3 to a value of:

$$\beta V = \frac{R3(V+)}{R2 + R3}$$

The timing network consists of resistor R1 and capacitor C1. There is a JFET (Q1) in parallel with C1. In the dormant condition, however, JFET Q1 is pinched off by the V− power supply (through the 470 kohm resistor). This means that the capacitor will charge to very nearly V+. Since the voltage across C1 is applied to the inverting input, and is greater than βV+, the operational amplifier sees essentially a positive potential applied to the inverting input. By the ordinary rules of operational amplifiers, therefore, the output will be LOW- or HIGH-negative, depending upon whether monopolar or bipolar power supplies are used.

When a positive-going pulse is applied to the gate of the JFET, the channel resistance of the JFET goes from damn high to very low, thereby discharging the capacitor. The input of the operational amplifier now sees a negative voltage because $\beta V+$ is greater than zero. The op-amp output snaps HIGH-positive, and, after the trigger pulse vanishes, capacitor C1 begins to recharge. The output of the operational amplifier remains HIGH for the period required to recharge to $\beta V+$. At that time, the operational amplifier output drops LOW again. We can retrigger the circuit at any time because the trigger pulse drops the charge on C1 to zero, and the process has to begin anew.

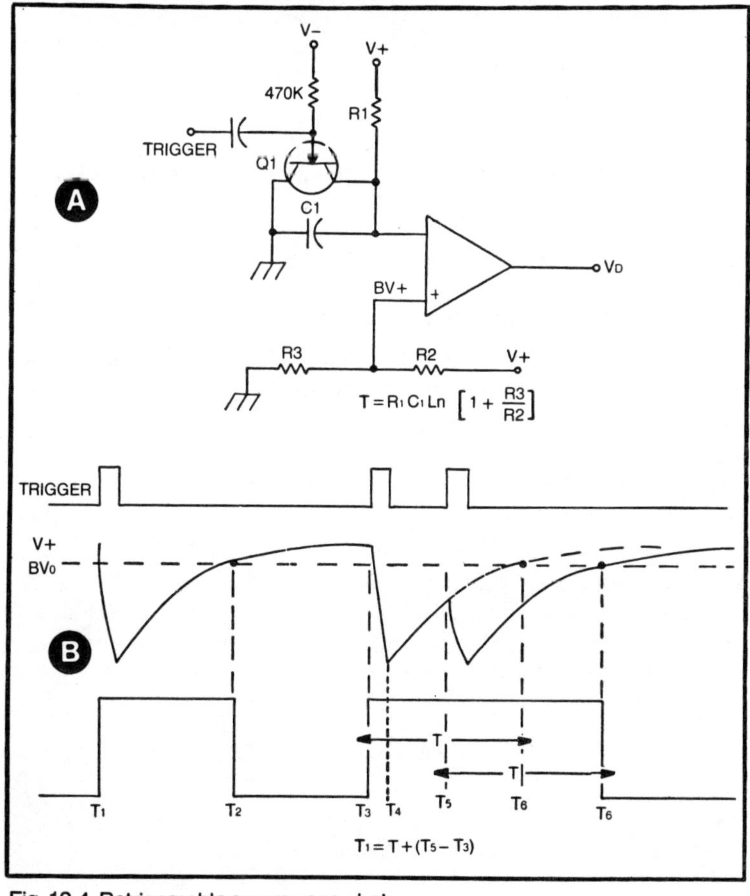

Fig. 12-4. Retriggerable op-amp one-shot.

The timing waveforms for this circuit are shown in Fig. 12-4B. The normal monostable operation of the circuit is shown at time t_1. When the positive-going trigger pulse is received, it rapidly discharges C1 from V+ to zero. As soon as the voltage across C1 drops lower than βV+, the output of the op-amp snaps HIGH. When the capacitor voltage rises to βV+, then the comparator sees a zero input condition, so the output voltage drops to LOW.

The retriggerable operation of the circuit is shown at t_3 to t_6. A trigger pulse is received at time t_3, and the same operation takes place. But, at time t_5, another trigger pulse is received. This pulse turns on the JFET, thereby forcing the charge on capacitor C1 back to zero. The output states does not change because the op-amp input is relatively the same—at least in polarity.

The duration of the output pulse in Fig. 12-4A is given by the expression

$$T = R_1 C_1 \mathrm{Ln}\,[1 + \frac{R3}{R2}]$$

The total time that the monostable will be in the output HIGH state in the example in Fig. 12-4B is:

$$T^1 = T + (t_5 - t_3)$$

TTL MONOSTABLE MULTIVIBRATORS

In Chapter 3, we introduced you to some TTL monostable multivibrator chips. Some were retriggerable and some were not. Some were dual one-shot; others were single one-shot devices. A general circuit for all of these devices is shown in Fig. 12-5. This particular version uses the 74122 device, which is a single, retriggerable monostable multivibrator from the TTL family. The device has complementary outputs, so it will produce both Q and NOT-Q levels. The trigger is a negative-going pulse that makes the transition from +5 volts to zero when triggering the one-shot. Resistor R1 and capacitor C1 form the timing for the one-shot, and obey the expression:

$$T = 0.37\, R_1 C_1\, [1 + \frac{0.7}{R1}]$$

Where:

R_1 is the resistance of R1 in kohms
C_1 is the capacitance of C1 in picofarads

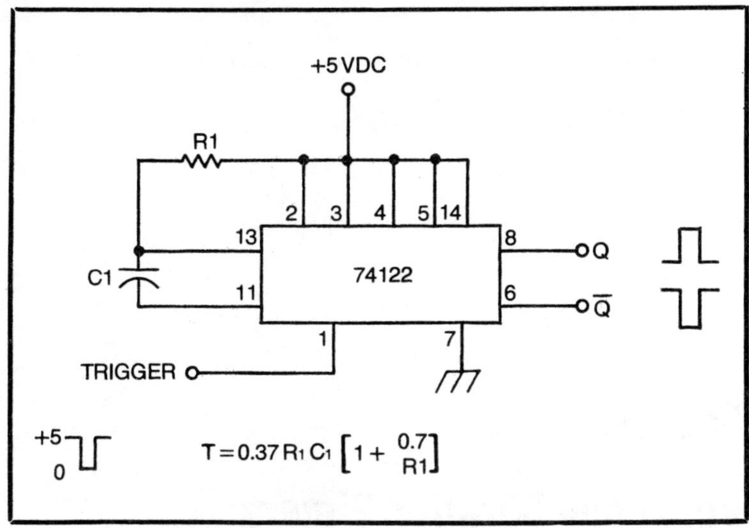

Fig. 12-5. Retriggerable TTL one-shot.

T is thc timc in nanoseconds

A certain minimum time, a refractory period of sorts, must expire before the 74122 can be retriggered:

$$T_{RT} = 0.22C_1$$

Where:

T_{RT} is the time to retriggerability in nanoseconds

C_1 is the capacitance of C1 in picofarads

The 7412X series monostable multivibrators are usually not the best choice in practical circuits unless the very fast pulses they can produce are needed. If you must have a pulse in, say, the millisecond range, then it would be better to use the 555, XR-2240 or an operational amplifier circuit. The 7412X devices are highly susceptible to noise triggering.

Chapter 13

Very Long Duration Timers

The long-duration timer presents a special problem to the electronic designer. The drift and other errors of some technologies tend to accumulate, and the total error over long periods becomes large. In fact, whenever we have errors that are functions of time, then the long duration timers will suffer markedly. Any long-duration RC network, for example, will suffer from several time related problems. We must find some way to overcome these errors if we are to use an electronic circuit for long durations.

A viable alternative, incidentally, is to use electromechanical timer elements in long-duration applications. We can buy timer elements that can be either manually or electrically set to whatever time we want up 24 hours. The timer will close (or open) an electrical switch when the set duration has expired. But this is not an elegant approach and is not satisfying to the electronic designer. How do the various circuits perform?

The unijunction transistor is a form of non-IC timer. It depends upon the voltage across a capacitor (a function of the RC network time constant), and the applied voltage to the UJT, to set the time. Unfortunately, most UJT circuits are not terribly good at short durations! For long-duration applications they are abominable! Don't use them, unless you like problems.

The 555 device is useful only up to 10 or 100 seconds. The values of the capacitors needed in the timing network become too large above those ranges. We find that large-value capacitors are typically aluminum or tantalum electrolytics, and these have undependable values. We find that most electrolytic capacitors are rated −20 to +100 percent of the label value. In addition, there is a dependence upon the applied voltage and a time dependence: the capacitance is rarely what you think it is. Another problem with the electrolytic is the most large-value capacitors have a lot of shunt resistance, and this resistance will foul up RC time constant determinations.

The timebase section of the XR-2240 IC timer suffers from the same problems as does the 555—the RC values needed for long durations. But, and it is a BIG BUT, the internal binary counter can make the XR-2240 into long-duration timer with RC component values that are easier to tame. Recall that the duration of the output pulse, when the wired-OR output configuration is used, can be anything from 1RC to 255RC. If we set the RC time constant to produce a duration of 1 second, then we can make a 255-second timer with the stability of a 1-second circuit. We can use RC time constants as long as 10 seconds safely in the XR-2240 circuit. This means that a single XR-2240 can be used in timers up to 2550 seconds (i.e., 42.5 minutes). Greater times, however, require that we either use some other device or cascade two or more XR-2240 devices.

Figure 13-1 shows the use of two XR-2240 timers in cascade to produce a very long-duration timer. Unit 1 is used as the

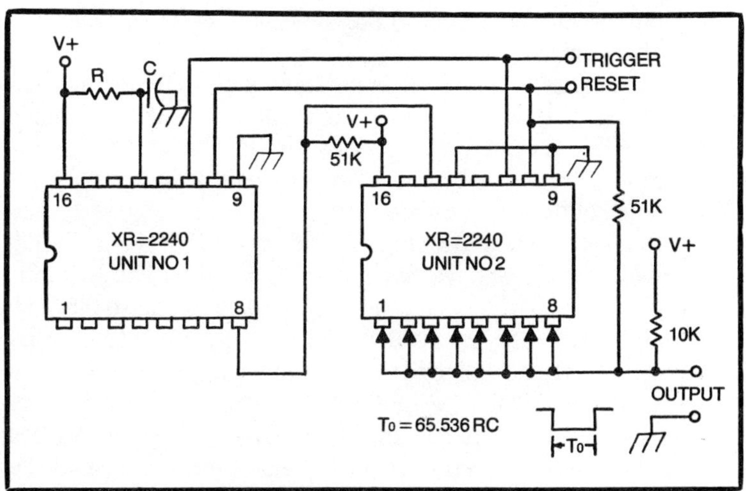

Fig. 13-1. XR-2240 long duration timer.

timebase, and has a frequency that is set by the usual equations. We use only the eighth bit of the binary counter. The output of the first unit is used as the timebase pulse for the second unit. All of the outputs of the second unit are wired together, so the output time is $(256)^2$T, or 65,536 T., where T is the period of the RC time constant. The output period T_0, when the RC time constant (T) is 1 second, therefore, would be 65,536 seconds (18.2 hours!).

The circuit in Fig. 13-1 is probably the cheapest and most viable method to obtain a long-duration timer. We can, however, also press certain CMOS ICs into service. In Fig. 13-2, for example, we see the use of a CMOS 4060 device (the 4020 will also work). This timer is a fourteen-stage ripple counter, so it can count to division ratios up to 16,384. If we apply a 0.5 Hz clock signal to the input of this counter IC, then the 2^{14} output will go HIGH once every (2 seconds) (16,384), or 32,768 seconds. This period is 9.1 hours! Since we can use clocks to 0.1 Hz (10 second period) made by RC techniques, then it becomes possible to build a 163,840 second timer (45.5 hours) using relatively easy to tame RC components!

The 4060 contains its own clock, but in this circuit we are not using the internal clock. The clock terminals are wired here as an input to a Schmitt trigger. This connection means that we must supply a noise-free clock with a good strong amplitude. There are no outputs for 2, 4, 8, or 2048 counts, but we can select any other output up to 16,384. Using the outputs directly limits us to durations that are powers of two, but by using them in combination, with gate circuitry external to the 4060, allows us to select any period that is an integer multiple of the input period.

The reset terminal (RST) is normally kept LOW for continuous counting. If we want to reset all outputs to zero, then we would bring the RST terminal momentarily HIGH. But for the type of application represented by Fig. 13-2, we would want to ground the RST permanently.

Be careful when using the 4060 device. It is a CMOS integrated circuit, and as such it will be sensitive to static electricity. Static charges on your body or tools can cause the 4060 to pop. In some cases, the 4060 will be zapped immediately, while in others the damage will tend to get worse with time, and you will see a "spontaneous" failure later on. The 4060A (or 4060) is more sensitive to this problem than the 4060B. B-series CMOS devices use internal gate protection zener diodes to protect against the high-voltage static problem.

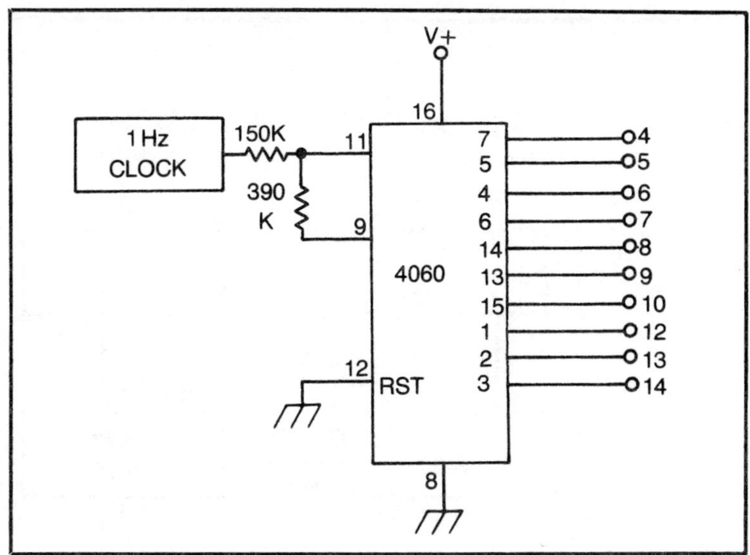

Fig. 13-2. CMOS long duration timer.

TTL LONG-DURATION TIMER

Figure 13-3 shows the circuit for a rather complex TTL long-duration timer. If a one-second clock signal is applied to the input terminal of this circuit, then we can generate time periods of 0 to 9999 seconds. Of course, changing the period of the clock signal will also change the total time. This circuit is designed to count to a preset number up to 9999, so we can calculate the duration of the signal from the preset number entered into the counter and the period of the clock.

The counters used in this circuit are TTL 7490 devices. These counters are decade counters, meaning that they offer a division ratio of ten. The output of the 7490 device is in binary coded decimal (BCD), so it will have a four-bit format. The counter stages are cascaded such that the D output of on counter drives the input of the next counter in the chain. A total of four decades are used, so we can have a count up to 9999.

The heart of this timer is the 7485 magnitude comparator. These devices are designed to compare two four-bit binary words and issue outputs that tell us whether word-A is less than word-B, greater than B or equal to B. In this circuit, we apply the BCD outputs of the 7490 decade counters to the *word-A* inputs of the 7485, and the outputs of BCD thumbwheel switches to the *word-B* inputs.

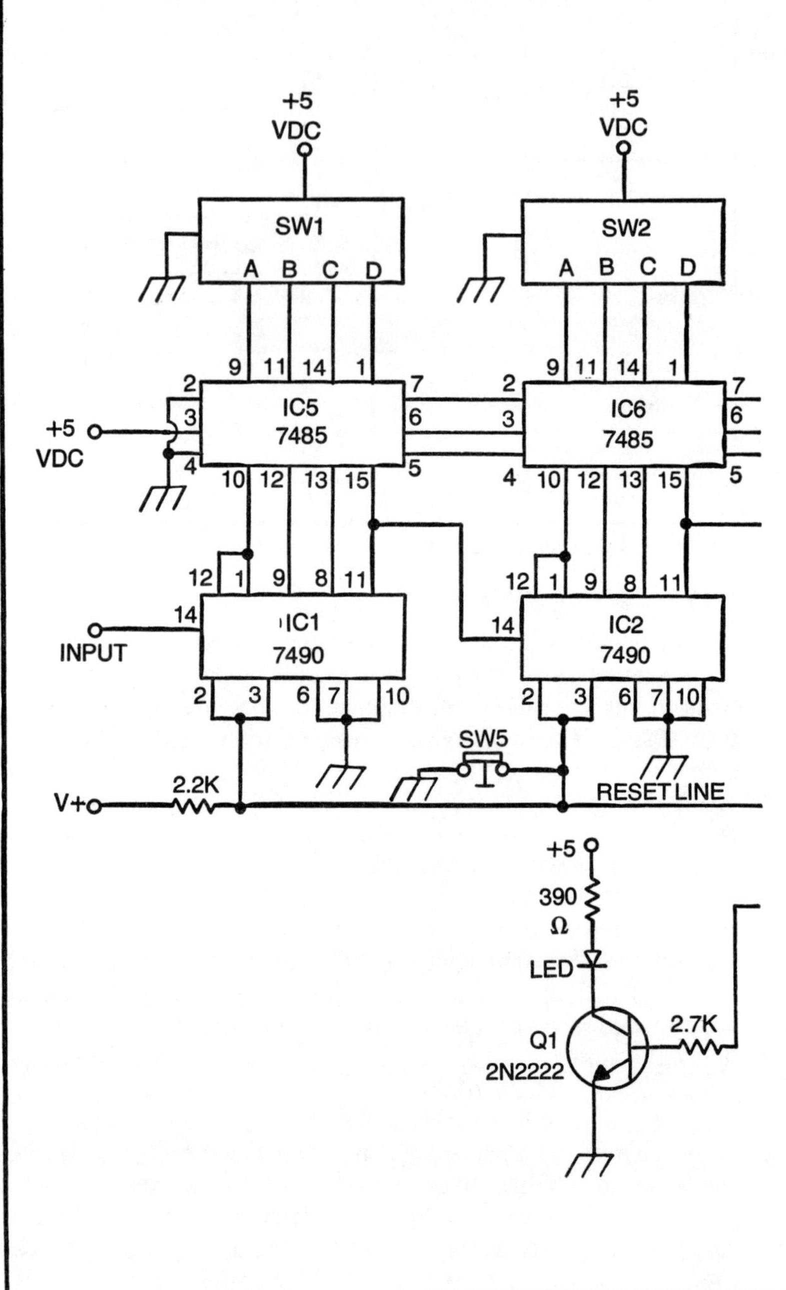

Fig. 13-3. TTL long duration timer.

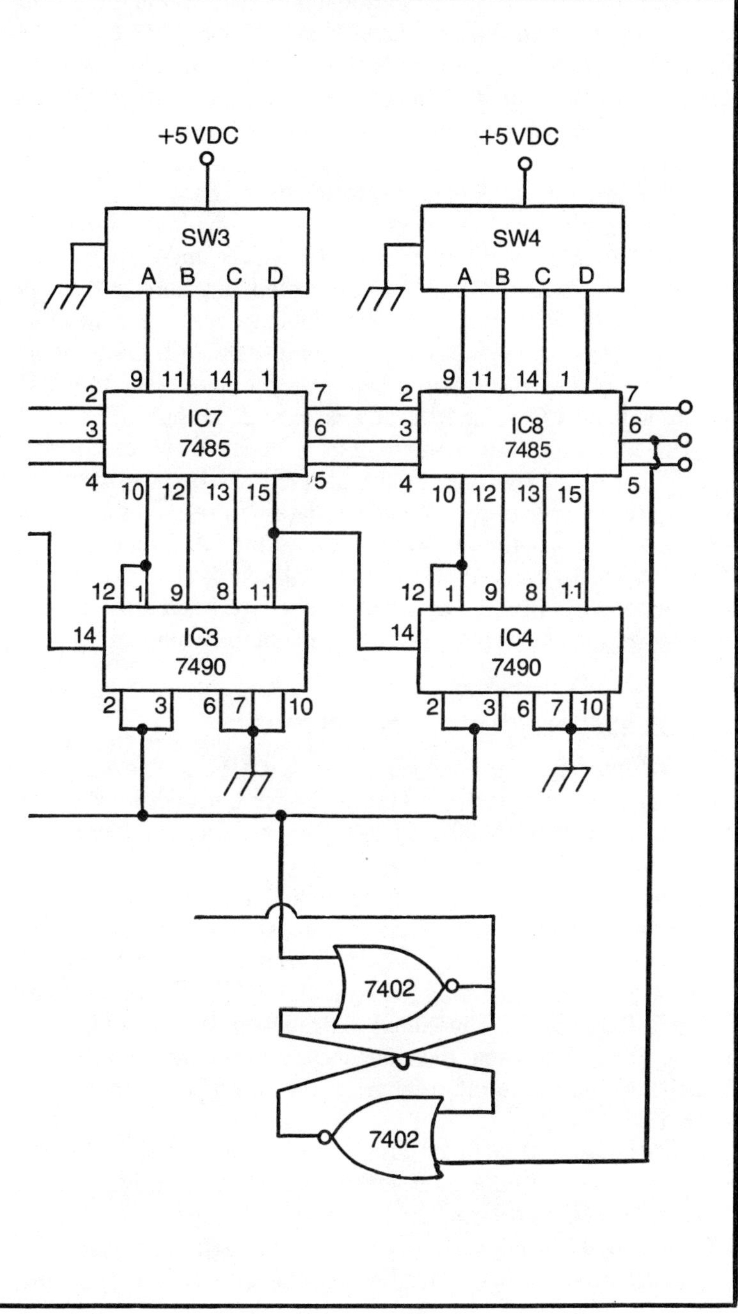
+5 VDC
+5 VDC
SW3
A B C D
SW4
A B C D
9 11 14 1
9 11 14 1
2
3
4
IC7
7485
7
6
5
2
3
4
IC8
7485
7
6
5
10 12 13 15
10 12 13 15
12 1 9 8 11
12 1 9 8 11
14
IC3
7490
14
IC4
7490
2 3 6 7 10
2 3 6 7 10
7402
7402

The 7485 devices have cascade inputs that examine the state of the outputs from the preceding stage. If these are connected properly, then the last outputs in the cascade chain (i.e., those of IC8 in this example) will reflect the condition of the chain as a whole. This means that the A=B output (IC8, pin 6) will not go HIGH until both words match on IC5, IC6, IC7, and IC8. Only when this condition is met will the A=B output go HIGH.

The user will set the desired period by dialing in a number from 0000 to 9999 on the thumbwheel switches and then pressing the reset button (SW5). This latter step guarantees that the counter will begin incrementing from 0000, rather than some other value that may or may not have anything to do with the previous count. As long as the count is less than the set count, the A=B output will be LOW, and the *A less than B* output will remain HIGH. But when the counter output is equal to the comparator input, then the *A less than B* output goes HIGH causing the RS flip-flop (consisting of two 7402 NOR gates) to set. This condition will turn on transistor Q1 and illuminate the LED indicator. An alternative plan, incidentally, is to have the A=B output turn on an independent RS flip-flop and simultaneously reset the counter. Otherwise, the clock continues to increment the counters.

INTERSIL 8250/8260 LONG-DURATION TIMER

The Intersil people second-source the XR-2240 device under their own type number 8240. They also offer two additional timers in the same line: the 8250 and 8260. These are almost identical to the 8240/2240 device, except that coding of the internal counter is different. The 8260 device is optimized for time delays programmed in seconds and minutes. The maximum count of the 8260 is 59, with the 60th pulse creating a carry output to the next counter in the sequence. The coding is two-digital BCD. The 8250 is also coded in BCD, but it is intended to produce counts from 0 to 99, instead of 59. The same pins are used in these counters, except that the weighting is different from the 2240/8240 device. See Table 13-1.

The long-duration timer circuit shown in Fig. 13-4 uses the 60 Hz AC power line as the clock source. A half-wave bridge rectifier receives the AC signal from the line and rectifies it to drive the first 8260 device. This stage is used to divide the 60 Hz AC signal by 60, so that a 1-second clock is available to drive the following stages. The seconds and minutes stages are two additional 8260 devices,

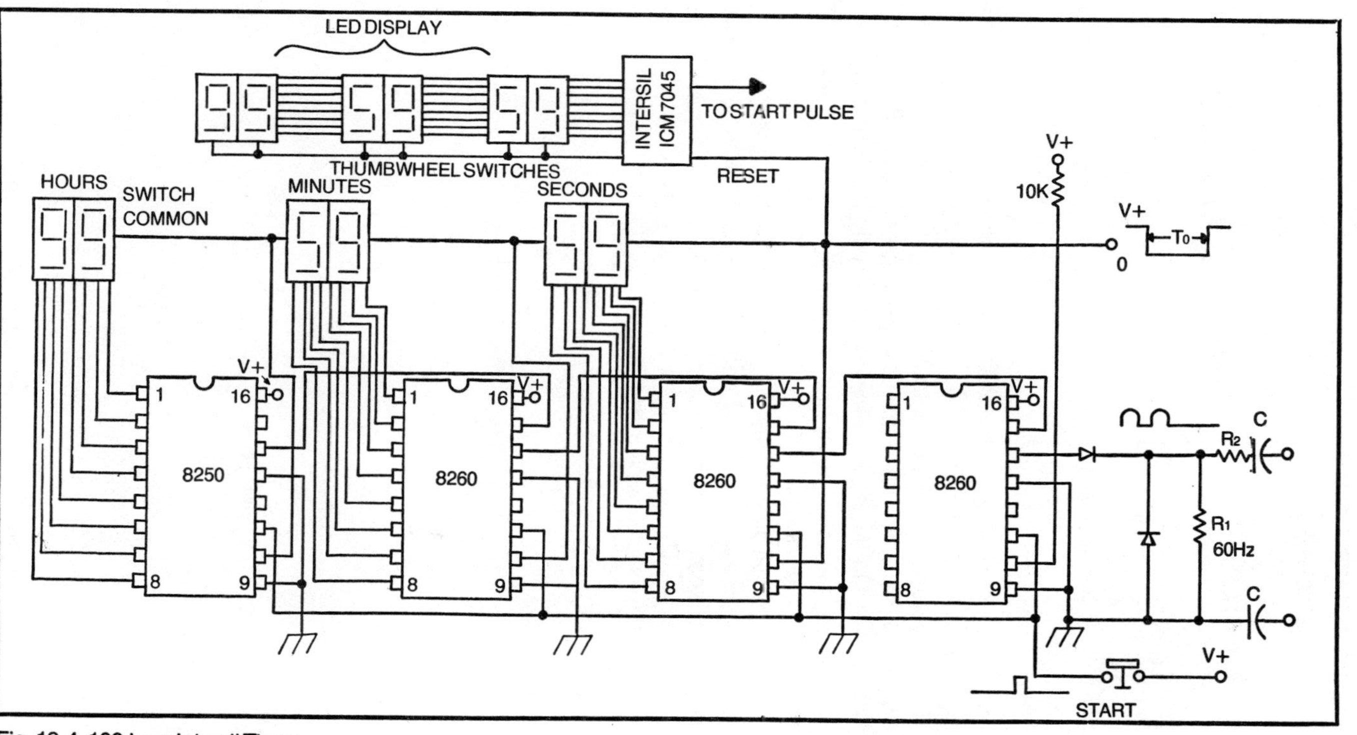

Fig. 13-4. 100-hour Intersil Timer.

Pin	8250	8260
1	1	1
2	2	2
3	4	4
4	8	8
5	10	10
6	20	20
7	40	40
8	80	(not used)

Table 13-1. Pin Coding.

while the hours stage is an 8250. This arrangement allows us to count a total time of 99 hours, 59 minutes, 59 seconds. The desired time is entered with the thumbwheel switches, which have the effect of wired-OR with respect to the counter outputs. The output will remain LOW until all selected outputs are again HIGH, which occurs only when the total count equals the set count entered on the thumbwheel switches.

An optional digital counter (Intersil ICM7045) is connected to the output lines. This device will begin to increment as a period counter as soon as a start pulse is received and forces the timer output LOW. This period counter will allow you to monitor the output of the counter as it increments.

Chapter 14
Op-Amp Timers

Almost any linear or digital IC element can be pressed into service to make timer circuits. Some of them are not terribly well suited to this application, but will "make do" if the circuit is designed by a clever person. The operational amplifier, on the other hand, is disgustingly simple to make work in most cases. The op-amp follows simple rules, and it is both obedient and consistent. We can make use of the well-defined properties of the operational amplifier to make some rather nice timer circuits. If you are not familiar with the theory of operation for this form of linear integrated circuit, then reread Chapter 1.

The operational amplifier is especially suited to those applications where most of the elements in the circuit are operational amplifiers or other linear ICs. If the predominant type of device is digital, then some of the special-purpose timers may be indicated.

OP-AMP AS COMPARATOR

The comparator is the operational amplifier configuration that is most commonly used in timer applications. The comparator is merely an amplifier with *excessive gain*. The functions of the comparator are to issue a *zero* volts output when the two input voltages are *equal* and distinct high outputs when the two input voltages are not equal. There are two possible situations in this case, given two voltages V1 and V2: either V1 is greater than V2, or V1 is less than V2. The ideal comparator will issue unique outputs for all three possible conditions:

- ☐ V1 = V2
- ☐ V1 less than V2
- ☐ V1 greater than V2

Figure 14-1A shows how an operational amplifier can be used as a voltage comparator. We obtain the "excessive gain" by deleting the feedback resistor between the output terminal and inverting input of the operational amplifier. The gain for the comparator, then, is the raw open-loop voltage gain (A_{vol}) of the operational amplifier device. Depending upon the grade, quality, and type number, this figure might be from 10,000 to 1,000,000. Let's consider the implication of having a gain of, say, 10,000 in an amplifier. Assume that the maximum output voltage is 12 volts (set by the power supply potential and the drop required to the supply rail). In this case, we find that 12 V/10^4 is 1.2 mV. The implication is that 1.2 mV will saturate the amplifier! If we use this operational amplifier as the comparator in our timer circuit, then we will find that a few dozen microvolts around zero difference results in a zero output, while at all other times, the output is either very high positive or very high negative (see Fig. 14-1B).

The diodes in Fig. 14-1A are to limit the output amplitude to some value short of the maximum permissable output voltage. Some operational amplifiers will latch up if allowed to saturate totally. Besides, the "knee" is sharper for the zener diodes than for the maximum op-amp output voltage (see Fig. 14-1B).

The graph in Fig. 14-1B shows the transfer curve for both inverting and noninverting operational amplifier comparators. The ideal comparator will have a vertical slope between states, but any real device will have a slight slope for low differential input voltages that are not sufficient to saturate the amplifier. Also, we would find on a time scale that the slew rate of the operational amplifier would tend to cause the vertical increase to be less than zero time.

In an ideal operational amplifier comparator, the zero crossing point for both curves would be the (0,0) point on the graph. Real operational amplifiers, however, have a certain amount of offset (V_{off}) that tends to shift the zero crossover either positive or negative some small amount. This offset is the smallest differential input voltage that will cause the output transition. In some-low-grade special-purpose comparator ICs, it has been found that 10 mV differential input was needed in order to affect the change. In some other causes, premium comparators and high-grade opera-

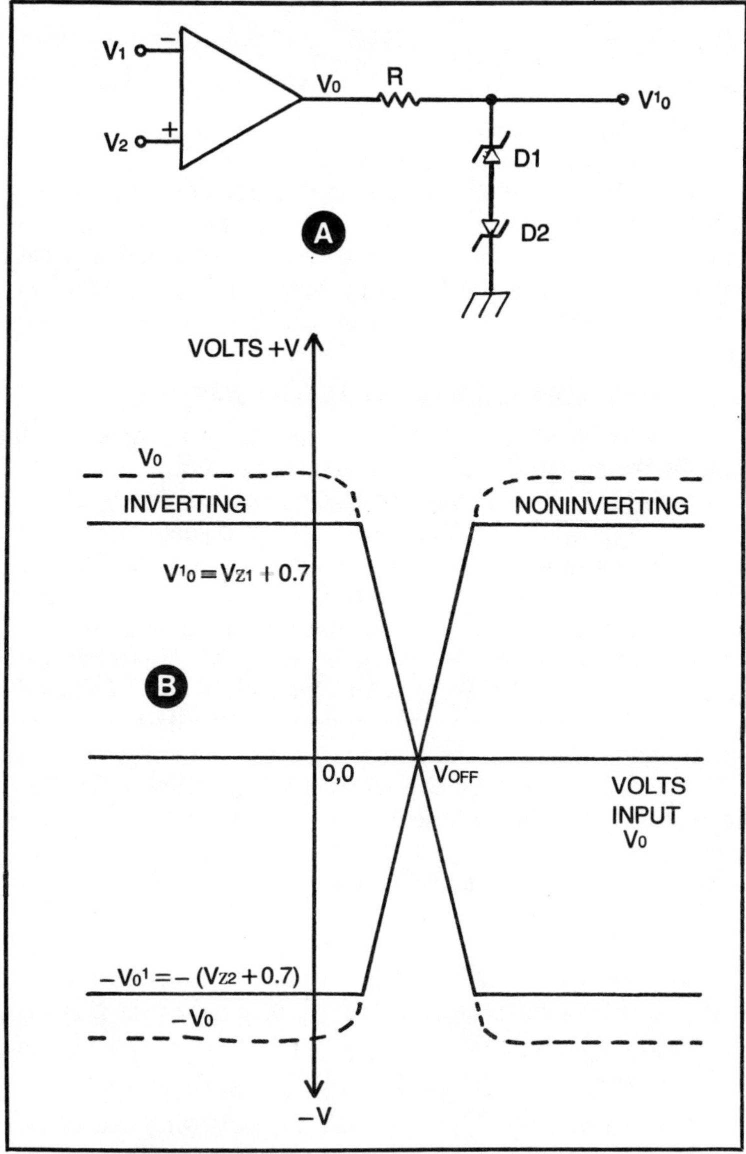

Fig. 14-1. Op-amp comparator.

tional amplifiers used as comparators had offset potentials of less than 1 mV.

The output voltage will be 700 mV (or so) higher than the zener voltage selected for D1 and D2. If identical zener diodes are used for these positions, then the output voltage for the stack would be:

$$V^1_0 = V_z + 0.7$$

Where:

V^1_0 is the output voltage

V_z is the zener potential of the diodes

The 0.7 volt factor allows for the junction potential of the other diode in the circuit. When the operational amplifier output is positive, for example, we find that diode D1 is reverse biased and will go into avalanche at the zener potential. The other diode, D2, is forward biased and will have the normal PN silicon voltage drop of 0.7 volts.

OPERATIONAL AMPLIFIER MONOSTABLE MULTIVIBRATOR

The circuit of Fig. 14-2A shows the use of the operational amplifier comparator as a monostable multivibrator, or one-shot. The one-shot will have only one stable state, which I label the *dormant* state. In this case, the dormant state keeps the output HIGH. When a one-shot is triggered, the output state changes to the alternate state. The output remains in the alternate state only for a predetermined period of time, after which the circuit returns to dormancy. In the one-shot shown in Fig. 14-2A, the output will remain HIGH when the circuit is dormant and will drop LOW for a predetermined period of time once triggered. After the time period expires, it goes HIGH again.

The circuit in Fig. 14-2A is based on the operational amplifier comparator discussed previously in this chapter. The diodes are used for exactly the same purpose, so are therefore unremarkable in this circuit. Like the similar resistor in Fig. 14-1A, resistor R4 is used to limit the current to the zener diodes, preventing damage to them.

The noninverting input of the comparator examines a sample of the output potential (i.e., $V_f = \beta V^1_0$) where the term β is the feedback factor:

$$\beta = \frac{R3}{R2 + R3}$$

Feedback voltage V_f is applied to the noninverting input of the operational amplifier. The voltage at the inverting input of the operational amplifier is the capacitor voltage V_{C1}. Diode D4 is used to clamp V_{C1} to 0.7 volts positive. When the output of the operational amplifier is HIGH on the positive side, then the value of V_{C1} when the capacitor is fully charged will be 0.7 volts. In the case where the operational amplifier is HIGH negative, however,

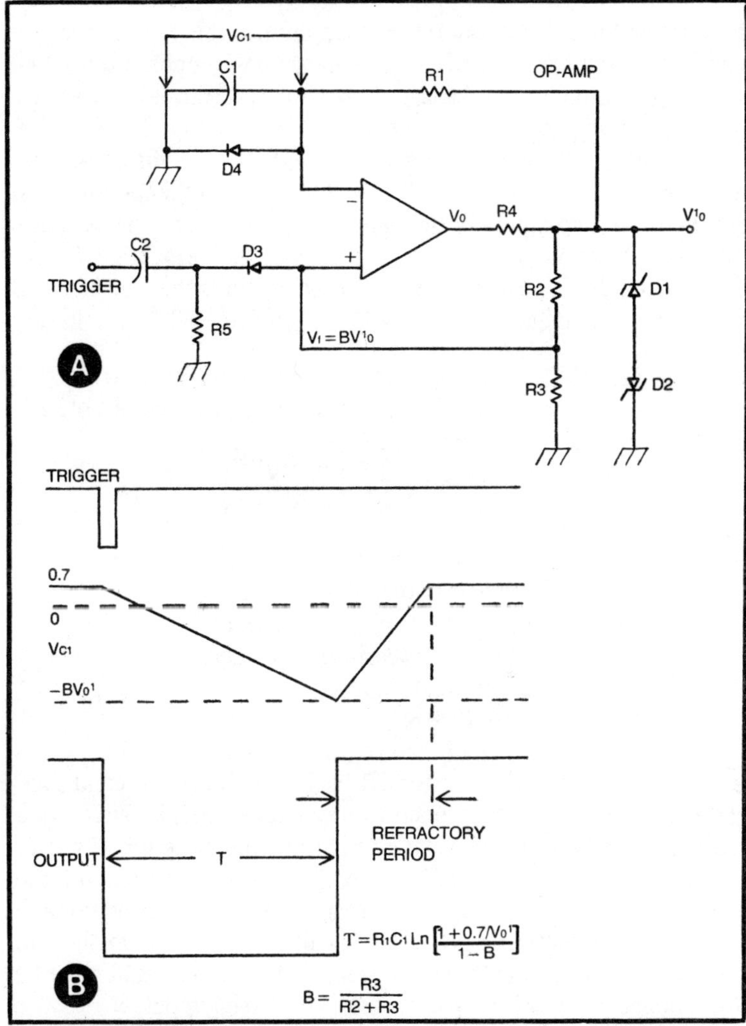

Fig. 14-2. Op-amp one-shot.

diode D4 is reverse biased, so it will not affect the operation of the circuit.

Let's assume that the one-shot of Fig. 14-2A is in its dormant state. In this case, the output is HIGH positive, the capacitor is charged to 0.7 volts, and there is a slight positive bias on the noninverting input. Now, let's assume that a negative-going trigger pulse is applied to the trigger input of this circuit. This will bring the noninverting input negative for a brief instant. But, because of the rules of operational amplifiers, this is sufficient to cause the output to snap LOW. The operational amplifier output is now negative and will discharge capacitor C1 from +0.7 volts to some negative potential (see the second trace in Fig. 14-2B). The output will remain negative (i.e., LOW) until the capacitor takes on a negative charge that is equal to $-\beta V^1_0$. When this potential is reached, the two inputs of the operational amplifier see the same voltage, so the output state reverts back to its dormant condition (i.e., snaps HIGH positive again). The capacitor is then discharged from the high negative value and charged to +0.7, its initial condition.

The period of the monostable multivibrator is the time that the output stays in the transient state. In the case of the circuit of Fig. 14-2A, the period T is given by:

$$T = R_1 C_1 \, \text{Ln} \left[\frac{1 + 0.7/V^1_0}{1 - \beta} \right]$$

Where:

T is the time, in seconds, that the output is LOW
R_1 is the resistance, in ohms, of resistor R1
C_1 is the capacitance, in farads, of capacitor C1
Ln indicates that natural logarithms are used
V^1_0 is the output potential, in volts
β is the factor R3/(R2 + R3)

The *refractory period* shown in Fig. 14-2B is the period required to recharge capacitor C1 back to its dormant condition. The one-shot will not respond to any trigger pulses during this period. No trigger pulses will be honored, after the initial pulse, until the refractory period has expired. The end-point of the refractory period might be the instant that the capacitor voltage reaches 0.7 volts, but more likely it will be some voltage short of that level. The idea is that the differential input potential must be low enough to allow it to be overcome by the trigger pulse.

OPERATIONAL AMPLIFIER ASTABLE MULTIVIBRATORS

The word "astable" means *no stable states*. The astable multivibrator, therefore, has no stable states but oscillates back and forth between the two permissable states, producing a square wave or other form of waveform. In this section we are going to consider two forms of operational amplifier astable multivibrator circuit: *square wave* and *triangle*. The squarewave circuit can be used as a clock circuit in timers and also as a timer in its own right because the HIGH and LOW durations are fixed. The triangle circuit is not terribly useful as a timer, by itself, but can be used in combination with other components to produce a timer. The sloping waveform of the triangle makes it unusable as a digital clock. But, if we know the slope and the amplitude of the triangle, we can use the circuit with a comparator biased to some reference level as a timer. In that case, we would need to calculate the time required for the output to increase to the reference level.

The circuit shown in Fig. 14-3A is the circuit for an operational amplifier square wave generator. This astable mul-

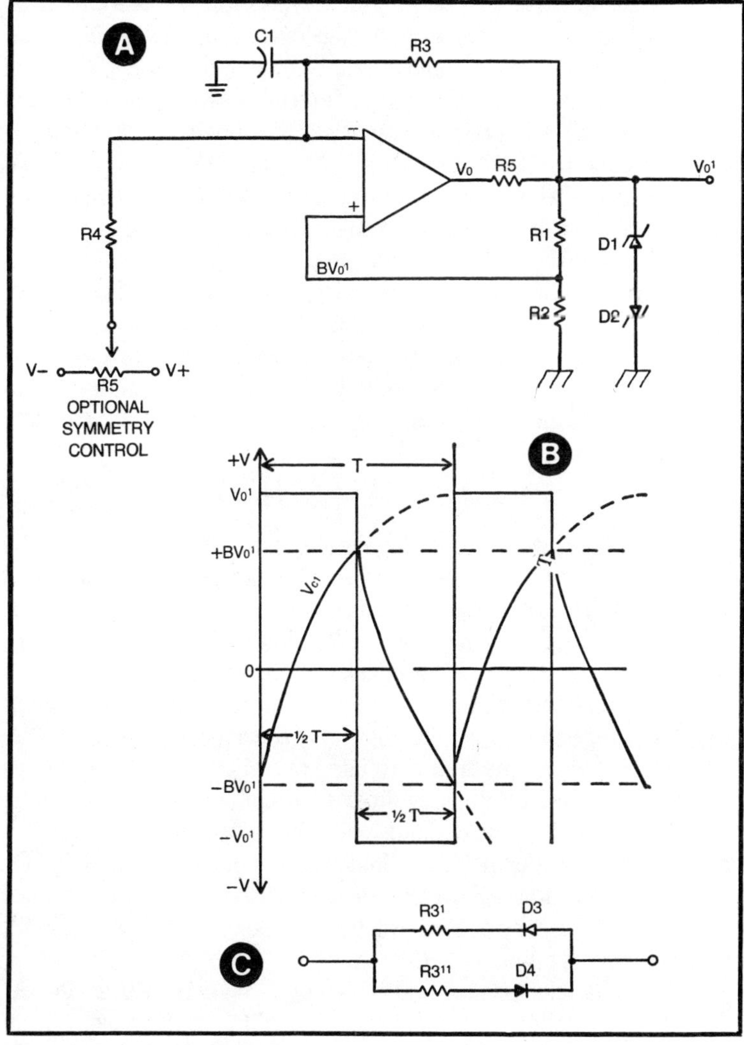

Fig. 14-3. Op-amp astable.

tivibrator will produce constant-amplitude square waves at a fixed frequency. In general, the circuit will operate from 1 Hz to 15 kHz using commonly available operational amplifiers. The amplitude symmetry of the square wave depends upon the zener potentials of D1 and D2. If both zener potentials are the same, then the output waveform will have baseline, or amplitude, symmetry.

The potential applied to the noninverting input is, like the previous case, a sample of the output amplitude, expressed by:

$$\beta = R2/(R1 + R2)$$

The voltage at the inverting input is the potential across capacitor C1. This potential depends upon the output amplitude from the operational amplifier (V^1_0) and the time constant R3C1.

For the sake of our discussion, let's assume that the oscillator is in its condition where the output is HIGH positive. The voltage across capacitor C1 will be $-\beta V^1_0$. The output of the operational amplifier is HIGH positive, so the capacitor will begin to discharge toward the positive direction, continuing on to become positively charged at $+\beta V^1_0$. When this potential is reached the output of the operational amplifier reverses polarity, so it drops HIGH negative. The charging current of capacitor C1 (i.e., the feedback current through R3) is now reversed, so the capacitor voltage begins to reduce as the capacitor is charged in the negative direction. This cycle will continue indefinitely, with the capacitor voltage oscillating back and forth between $-\beta V^1_0$ and $+\beta V^1_0$ with a period of:

$$T = R_3 C_1 \text{ Ln } 1 + \left[\frac{2R_1}{R_2}\right]$$

There are two methods for varying the time symmetry of the output waveform from the operational amplifier multivibrator of Fig. 14-3A. We can use the potentiometer shown in Fig. 14-3A or the diode method shown in Fig. 14-3C. In the first method, we are biasing the capacitor with a small voltage, positive or negative, that is chosen to alter the period that the output is either positive or negative (depending upon the polarity of the biasing potential).

The alternate method is to use two different resistor values for R3 and isolate them using diodes D3 and D4 (Fig. 14-3C). The capacitor charging time depends upon $R3^1$ on positive excursions of the output waveform (i.e., diode D3 is forward biased). On negative excursions of the output waveform diode D4 is forward biased, so the capacitor charging time is set by $R3^{11}$. If $R3^1 \neq R3^{11}$, then the time symmetry is affected.

The triangle astable multivibrator is shown in Fig. 14-4A. This circuit consists of an operational amplifier comparator and an operational amplifier than is connected as a Miller integrator (A2).

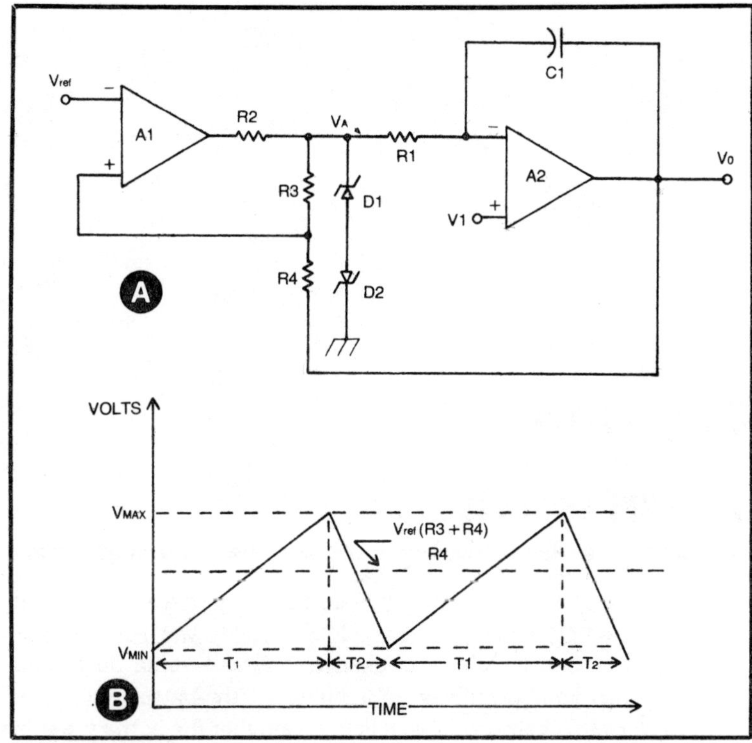

Fig. 14-4. Triangle generator.

The Miller integrator produces an output signal that is the *time integral*, or *time average* if you prefer, of the input signal voltage. Since the input end of resistor R1 is connected to the output of the comparator, which is constant, the output slope of the integrator is a constantly rising slope (for a description of how the integrator works, TAB Book No. 1012, *How to Design and Build Electronic Instrumentation*).

The comparator is biased at the inverting input with reference potential (V_{ref}) and at the noninverting input by the difference between potential V_a and output potential V^1_0. The frequency of oscillation is:

$$F = \frac{R_3}{4\,R_4\,(R_1\,C_1)}$$

The integrator capacitor charges and discharges with slopes that depend upon the output amplitude polarity.

Chapter 15

Alarm Projects

An alarm is a device that is designed to warn you of some untoward event. A low-light alarm, for example, is used to warn of low light conditions. A burglar alarm is used to warn of some ding-dong breaking into your premises or vehicle. A low-battery alarm is used to warn that battery voltage is low, and that the battery needs charging. There are a number of different types of alarm, but only a few different types of circuit. We can adapt some circuits from one type of alarm to another with few modifications. I will leave some of those applications to your vivid imagination.

LOW-BATTERY ALARM

The automobile battery is a nominal 12 volt DC lead-acid unit. That "12-volt" figure, incidentally, is *nominal* only and exists only when the engine of the vehicle is not running. The voltage across the battery when the engine is running will be somewhat higher. Most U.S. autos will produce from 13.2 to 14.4 volts when the engine speed is 1500 RPM. The voltage will be higher at higher RPM, but it is limited by the voltage regulator. If the voltage drops below 10 volts DC, the engine won't even crank over, and this means that we cannot obtain the engine speed needed to charge the battery (sigh).

Automotive applications are not the only use for the lead-acid battery. For almost any non-portable alternate energy scheme we will want to be warned of low voltage. We might see the lead-acid battery forming the energy storage system (they are, after all, called *lead-acid storage batteries*), while the input energy might be from a windmill, solar panel, or some heat recovery system attached to the house, furnace, or fireplace. The time to do something about low battery voltage in such a system is before the battery is totally depleted. In that case, we might want to redesign the circuit in Fig. 15-1 to trigger at 12.6 volts (i.e., change the zener diode to a 12.6 volt unit).

The circuit is shown in Fig. 15-1, and consists of an operational amplifier (741 unit) that acts as a voltage comparator and a 555 astable multivibrator. The comparator is used to monitor the voltage level of the battery, while the 555 is used as the alarm sound source.

The battery voltage will be 13.2 to 14.4 volts when the battery is fully charged. This potential is applied to the inverting input of the comparator (i.e., pin 2 of the 741 operational amplifier). The noninverting input of the comparator sees a reference potential, in this case 10 volts DC. The reference potential must be lower than the nominal battery voltage and is selected to be the trip point at which the system operates. As long as the battery voltage is greater than the reference voltage, the comparator thinks it sees a positive potential applied to the inverting input. This means that the output potential (i.e., on pin 6) will be LOW. The comparator output potential is used to bias a gating transistor (Q1) that turns on and off the astable multivibrator. When the output of the comparator is LOW, then Q1 is unbiased. This places the collector at a HIGH level, thereby lifting the ground of the 555 astable multivibrator.

When the battery voltage drops below 10 volts (by only a few millivolts), then the comparator thinks that it sees a negative potential at the inverting input (i.e., V1 is less than Vz). This causes the comparator to produce a HIGH output at pin 6. A HIGH on pin 6 will cause transistor Q1 to be forward biased, thereby grounding the 555 astable multivibrator and turning on the alarm.

The 555 will begin oscillating as soon as the battery voltage drops below the reference potential. The loudspeaker (LS1) is a low-cost 8 ohm unit, and is connected to the 555 through capacitor C3. The Resistor (R4) is optional and is used to reduce the volume of the alarm, if desired. The value of this resistor will be 0 to 50 ohms, depending upon the desired volume, the nature of the loudspeaker, and the V+ terminal voltage (in other than 12 volt systems); the value of R4 is found experimentally.

AUTOMATIC BATTERY CHARGER

The circuit of Fig. 15-1 is sometimes modified to work as a battery monitor and automatic charger. This is not always wise, or even possible. A popular version (seen in some other "timer" books by other publishers!) is designed to work with Gel cells (lead-calcium batteries). A colleague of mine used that circuit to monitor the battery in a portable (cadaver) kidney transport unit. The unit was designed to keep an excised human kidney "alive" by perfusing it with a physiologically compatible solution during the time it was being transported from the cadaver donor to the recipient (a transplant patient who had experienced renal failure). If the battery failed and the kidney was lost, then the patient would have to go back on the hemodialysis machine until another valid kidney could be found. It was deemed *no good* to lose kidneys! But, unfortunately, a few were lost because the Gel cell in the transport unit kept going dead. Several different charger circuits were tried and none worked (including the popular circuit based on the 555). A better solution, offered by a manufacturer of Gel cells, was to buy an ordinary OEM power supply (i.e., a *Powermate* model) that had both adjustable output voltage and adjustable output current limiting. The voltage control is set to the desired terminal voltage of the battery, while the current limiting control was set to provide a short-circuit current equal to the maximum allowable charging rate of the selected battery (i.e., in this case 500 mA).

The circuit of Fig. 15-1 can be used in non-critical designs by changing IC2 into a monostable multivibrator circuit. The output (pin 3) is then used to trigger a relay that turns on a battery charger circuit. The comparator output is connected in a manner that will trigger the 555 one-shot. Similar circuits are shown in this book as other applications, and can be easily adapted to your needs.

POWER FAILURE ALARM

There are times when it becomes imperative to know when the AC mains power has failed. We might, for example, be using the AC mains to power an electrical heater in a greenhouse or some other area that must be kept warm. If the mains power fails, many plants could be lost. We can use the circuit of Fig. 15-2 to warn of a failure.

The circuit of Fig. 15-2 is a 555 astable multivibrator connected so that it turns on only when gating transistor Q1 is forward biased. The bias on the base-emitter junction of transistor Q1 is the potential that appears across capacitor C4. This capacitor

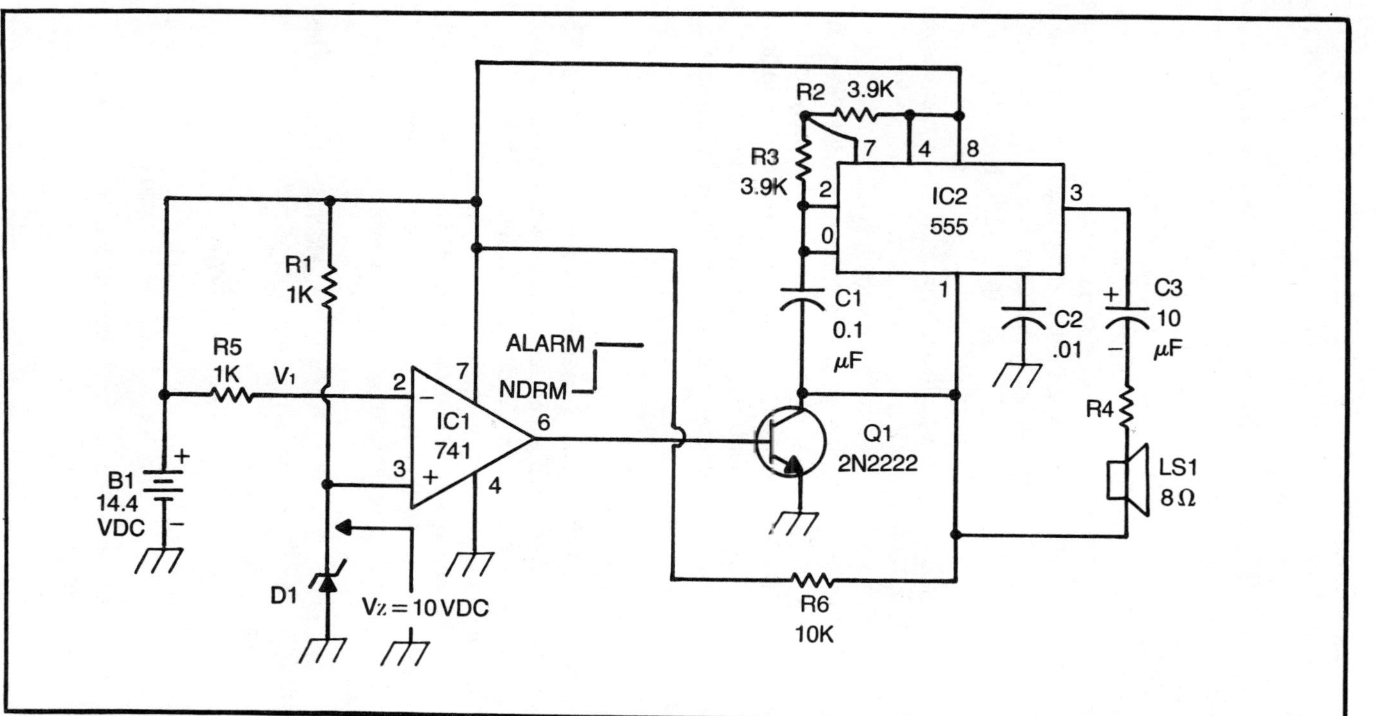

Fig. 15-1. Low battery alarm.

is charged from two sources: the current through R5 and the current through R6. The current through R5 is derived from the V+ power supply (9-volt battery B1), while the current through R6 is derived from a negative pulsating DC power supply that is operated from the 115 volt AC mains. If the AC mains voltage is present, then the potential across C4 will be close to zero or slightly negative. This situation keeps transistor Q1 reverse biased and the astable multivibrator is silent. But, if the mains voltage fails, the only source of current to charge C4 is through resistor R5 from the V+ power supply. The capacitor will soon discharge from the negative value and take on a positive polarity. When this polarity reaches 0.7 volts (or a little less, depending upon the individual transistor selected for Q1), the transistor is turned on, grounding the astable multivibrator. The astable begins to oscillate at 1000 Hz and this tone is heard in the loudspeaker.

We can make this circuit do something besides make a noise by replacing the astable circuit with a one-shot circuit. We could then use the one-shot to trigger an SCR or latching relay, causing some action. For example, in the case of our greenhouse application, the one-shot/SCR combination could turn on a gasoline-engine AC generator system or ignite a propane gas heater.

LOW LIGHT MONITOR

The circuit in Fig. 15-3 operates as a low light alarm. This type of alarm could be used in cases where it is imperative that a light be left turned on, or to turn on the lights of your home at sundown (an anti-burglary tactic that sometimes works). The circuit is made from a 555 IC timer similar in configuration to the one-shot circuit. In this case, however, we want the output to remain HIGH as long as the trigger terminal voltage is less than ⅓ V+.

The timing input (pins 6 and 7) are held at a potential of less than ⅔ V+ by the action of resistor voltage divider R1/R2. The actual bias on this point is:

$$V_{6,7} = \frac{(6.8\,k)(12\text{ volts})}{(15+6.8)k}$$

$$V_{6,7} = (0.31)(12\,V) = 3.72\text{ volts}$$

This potential keeps the 555 from thinking that it has "timed-out" before we even get started!

The bias on the trigger terminal must be greater than ⅓ V+ during daylight hours, dropping to less than ⅓ V+ at night.

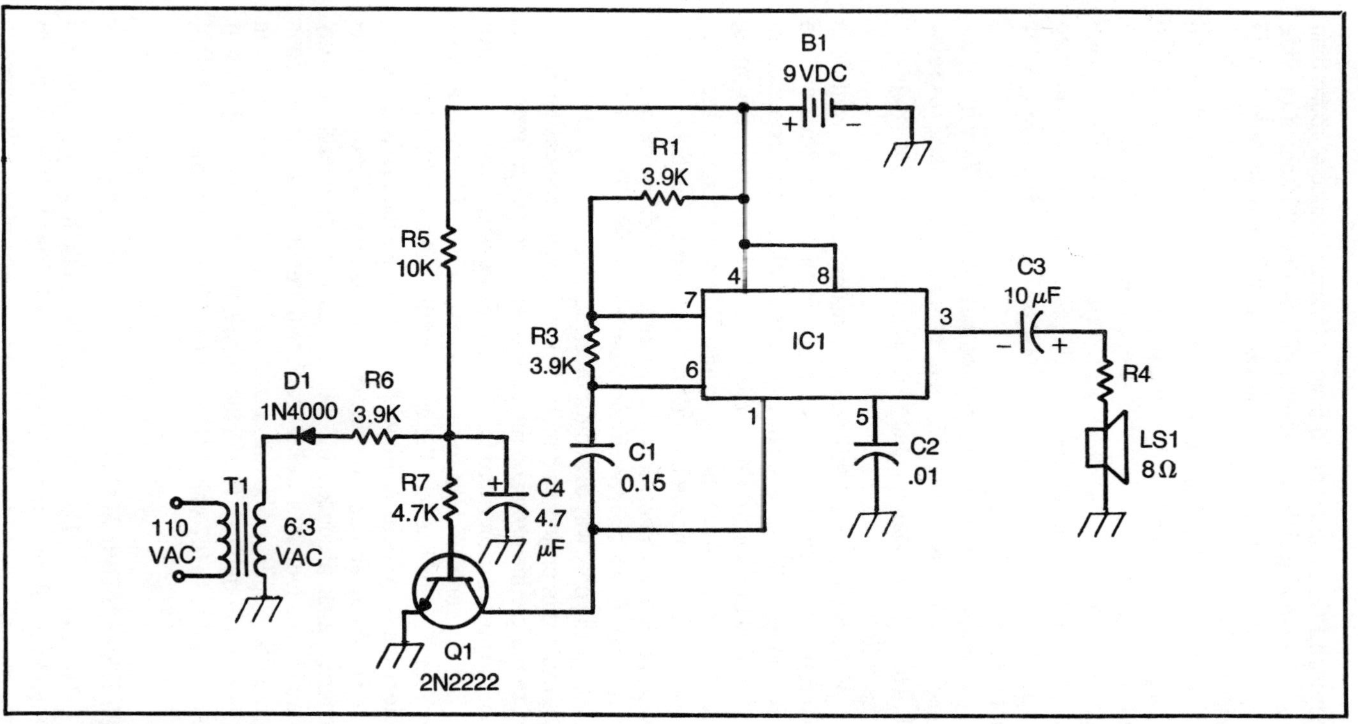

Fig. 15-2. Power failure alarm.

Photocell PC1 is selected to have a dark resistance of more than 150 kohms and a light resistance of less than 20 kohms (a wide range of low cost photocells will fill this bill, and the selection is noncritical).

During the daytime, when the resistance of PC1 is low, the voltage at pin 2 of the 555 will be greater than ⅓ V+, so the device will not trigger. But at night, the resistance of the photocell rises to a point where the voltage on pin 2 is less than ⅓ V+. This causes the photocell to trigger the 555. The output of the 555 will snap HIGH and will remain HIGH as long as the darkness keeps the photocell resistance high.

A HIGH condition on the output of the 555 will cause transistor Q1 to be forward biased, which in turn energizes relay K1. This relay is connected in a manner that turns on the AC power at the outlet.

WARNING! The 115 volt AC mains can KILL you! It is imperative that you wire this circuit correctly, or there will be great danger. The large pin on a standard 115 volt duplex outlet (in the USA) is the *neutral* line of the AC mains. The small, flat prong is to the hot line from the AC mains. You must connect the relay contacts to interrupt the hot side of the AC line. The light will still go on and off if the neutral is interrupted, instead, but there will be a potential shock hazard present on the outlet. The wiring, as shown, is the only way to avoid a shock hazard.

Build this project in a metal box and ground the casing of the box to the ground prong on the outlet. Also ground the green wire from the power mains cord (always use three-wire cords; two-wire cords are a contradiction to safety!) to the case. If there is a short between the hot line and the case, it will cause the fuse to blow. Otherwise, you could be left with a hot chassis, and that's a killer.

The diode across the relay coil (K1) is needed because of the inductive kick produced by the coil as it is de-energized. Any inductor (including a relay coil) will generate a tremendous high-voltage spike when the current is interrupted. This spike can blow the 555 device! The diode serves to dampen the spike and limit its value to 0.7 volts. Use the 1N4007 or any other 1000 PIV diode rectifier.

LOW TEMPERATURE ALARM

We can use the same circuit for both low light and low temperature alarms. If the photoresistor in Fig. 15-3 is changed to a thermistor, then we have a temperature alarm. We want a

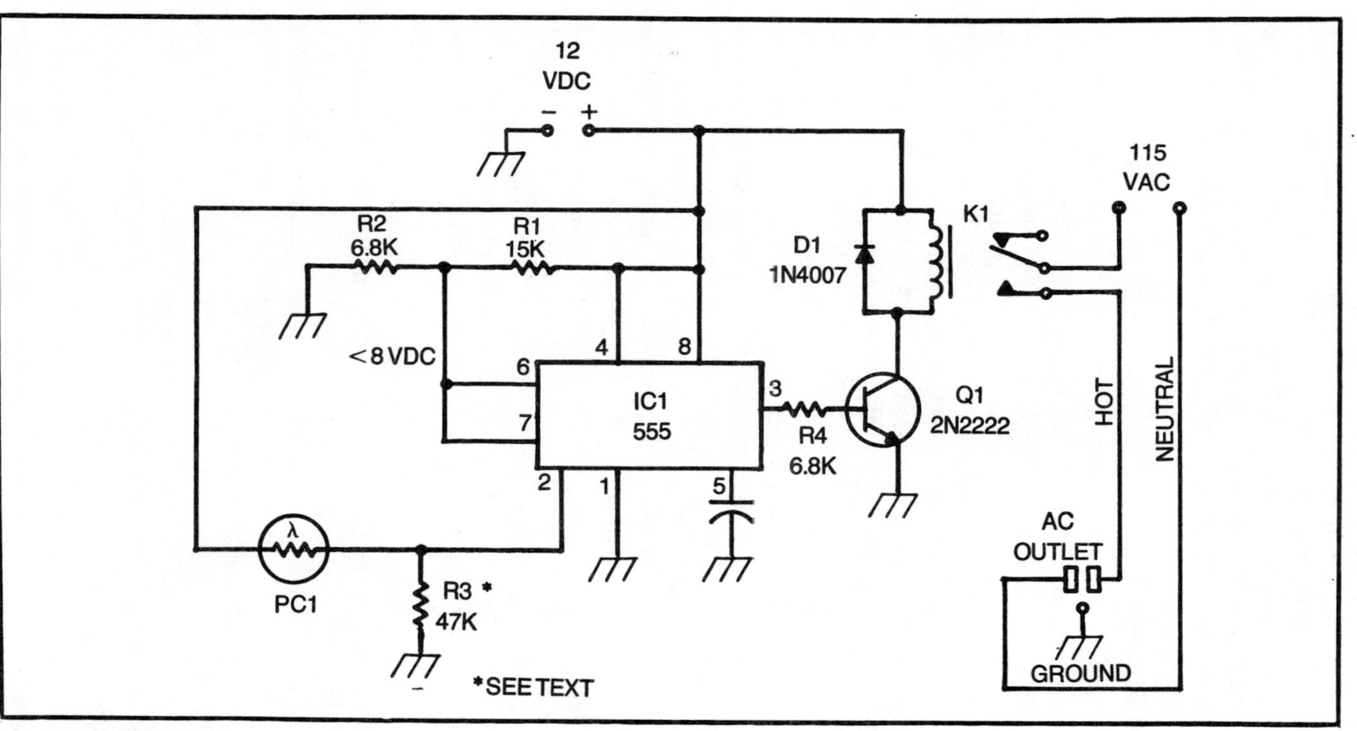

Fig. 15-3. Low light alarm.

thermistor that reduces its resistance with increases in temperature, i.e., a negative temperature-coefficient model. The value of R3 will have to be adjusted for the resistance of the thermistor at the desired trip-point temperature. Note that this resistor is used to adjust the trip-point in both low light and low temperature alarms.

IDIOT LAMP MONITOR

Automobiles and some heavy equipment use idiot lamps to monitor and warn when certain critical parameters are exceeded. In the case shown in Fig. 15-4, an automobile oil lamp, there is a pressure-sensitive switch mounted in the engine block of the car so that will monitor the pressure in the crankcase. If the pressure falls below a safe level, the switch closes and turns on the light (hopefully!!!) Most idiot lamps operate by grounding the cold end of the lamp; the hot end remains connected to the vehicle power supply. The circuit in Fig. 15-4 will monitor the idiot lamp and issue a warning when it is dead. The 555 is operated as a one-shot multivibrator that will turn on SCR D1 when the lamp is out.

In normal operation, when the lamp is good, the potential at the trigger input V1 is approximately ⅔ V+ due to the action of resistor voltage divider R2/R4. This keeps the one-shot from triggering. But when the lamp is bad (i.e., open), the voltage across the pressure-sensitive switch is removed, and this also removes the potential from the lamp end of R2. The value of V1 now drops to approximately ¼ V+ due to the action of voltage divider R3/R4. This potential will trigger the 555 multivibrator.

Once the 555 is triggered, its output terminal (pin 3) snaps HIGH. This action passes a differentiated pulse to the SCR through RC network C3/R7. The pulse turns on the gate of the SCR, causing its anode-cathode resistance to drop from near infinity to a few ohms. The LED will turn on, indicating that the idiot lamp is open.

One potential problem with this circuit is in the hold current of the SCR. The SCR should be a 500 mA or 1 A unit. If the hold current is greater than the LED current, then either replace the LED (and R8) with a lamp, or place a low-value resistor from the anode of the SCR to the V+ power supply. The resistor supplies the rest of the current needed to keep the SCR turned on.

BURGLAR ALARM (I)

Intrusion alarms are designed to warn of intruders with a minimum of false alarms. There are two major faults that cause

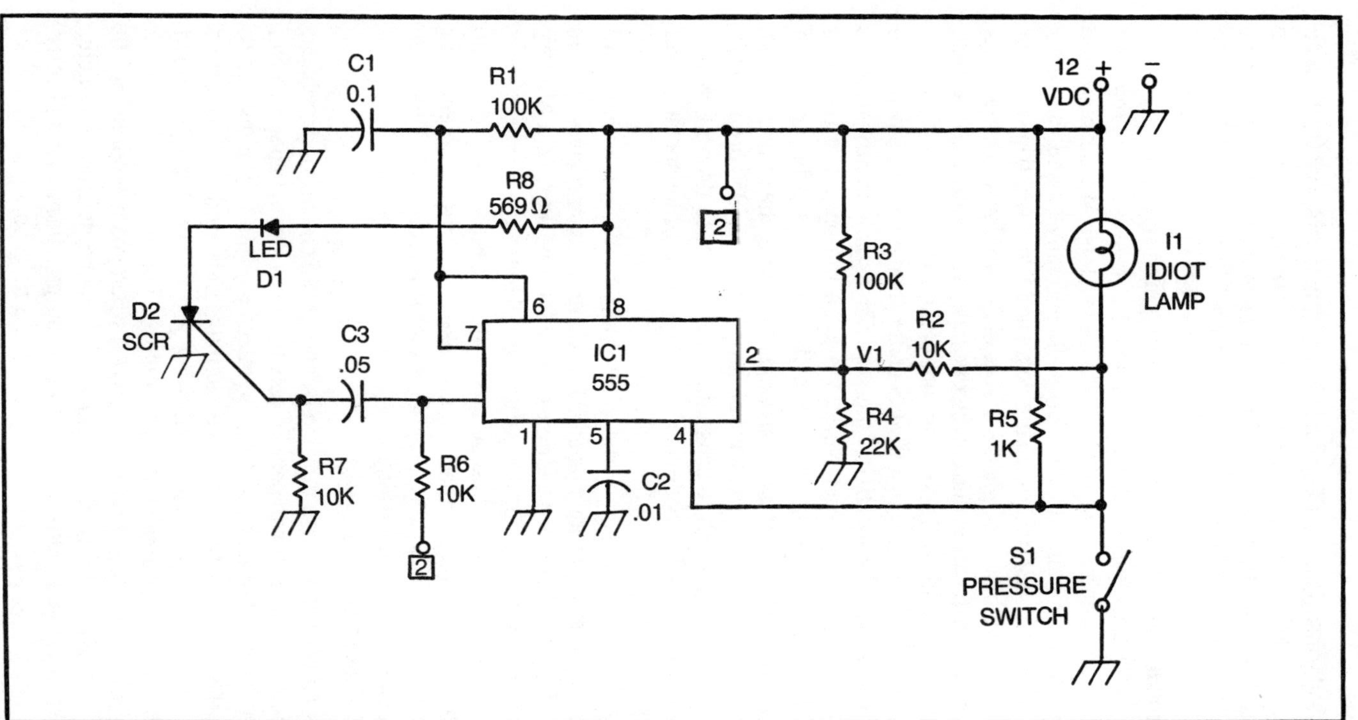

Fig. 15-4. Idiot lamp monitor.

false alarms: poor quality or defective components, and (in the case of electronic systems) triggering on noise pulses. In the alarm circuit of Fig. 15-5, we can limit the false alarm incidence from noise triggering by use of a V+ potential in the 9 to 15 volt DC range. This makes the circuit essentially "high threshold logic," so it would take a helluva noise pulse to trigger the circuit.

The principle on which this alarm is based is the *protection loop*. This requires a continuous circuit in a loop. We can make the continuous circuit in the form of normally-closed magnetic switches (S1, S2) located on the doors and windows, trip wires, or conductive tape on the glass surfaces of the windows. All elements (switches, trip wires, and conductive tape) are connected in series, so that the circuit will open if any one of them opens.

The 4050 device is a CMOS buffer (noninverting). When the output of the 4050 is HIGH, then the 555 astable multivibrator is ungrounded and will not work. We keep the output of the 4050 HIGH by applying a logical HIGH to its input. This is the function of the protection loop. If the loop opens, indicating the presense of an intruder, however, the voltage at the input of the 4050 drops to 1/5 V+, which is enough to cause the 4050 output to drop LOW (the transition in a CMOS device operated from a monopolar power supply is ½ V+).

When the output of the 4050 drops LOW, the astable will begin to operate producing an 800 Hz output squarewave. We can use this pulse output to either drive a loudspeaker, or a Mallory *Sonalert* (as shown), to produce an unusual tone.

This alarm must be operated with a V+ switch that can be turned on and off so that the premises are either protected or unprotected. We do not, after all, want the alert to sound when we legally enter the house (sigh).

BURGLAR ALARM (II)

Our burglar alarm Version I suffers from one major drawback: it isn't very loud. It will serve nicely to alert a sleeping person of an intruder without also alerting the intruder. This will allow you to call the police or take whatever action is indicated. It also serves the very useful purpose of allowing you to avoid neighborhood embarrassment for false alarms—only *you* know!

But that doesn't do you any good when you are not home. In the case of a "nobody home" house or an unattended motor vehicle, we want to use a loud—no, very loud—burglar alarm that will scare hob out of the intruder and wake up the neighbors—so that they will call the fuzz.

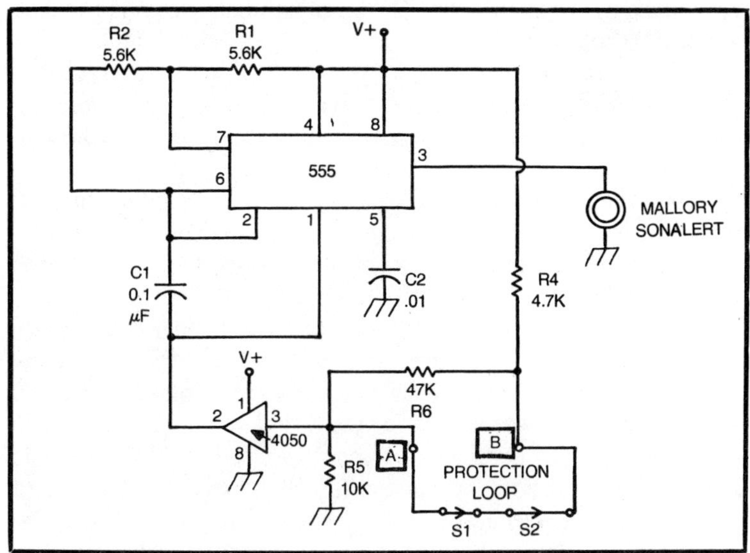

Fig. 15-5. Protection-loop intruder alarm.

Appropriate outdoor alarm devices include bells, Klaxon horns, and sirens. Stores that sell burglar alarm supplies can offer a selection of appropriate devices, and the circuits of Figs. 15-6 and 15-7 will operate them.

The Version II burglar alarm is shown in Fig. 15-6. In this circuit, we are going to project a light beam across a protected path (i.e., a door entrance or hallway), and cause an alarm if the beam is interrupted. The circuit of Fig. 15-6 is basically a 555 one-shot designed to fire SCR D1, turning on the siren or bell. The trigger terminal of the 555 is connected to a resistor voltage divider, of which the high-end member is the photocell. Select a photocell that has a dark resistance of 150 kohms, or more, and a lighted resistance of less than 15 kohms.

As long as the light from lamp I1 falls on photoresistor R2, the resistance of R2 will be low. This means that the voltage across R3, and at the trigger input of the 555, will be high enough to forbid triggering of the 555. But, when someone passes between the lamp and the photoresistor, the resistance of the photoresistor climbs to a value that is sufficient to drop the trigger input potential below the ⅓ V+ needed to cause triggering.

Once triggered, the 555 output will snap HIGH causing a current to flow through resistor R4 to the gate of the SCR (D1). This current will gate-on the SCR causing its anode-cathode resistance to drop markedly. The siren now has one end grounded and will therefore turn on.

The SCR selected can be any 50 volt PIV (or greater) type that will pass 25 amperes, or more. If you use one with a current rating greater than 25 A, however, it may be necessary to modify the gate resistor value. Find out the minimum current needed to gate the SCR on. The 555 will supply up to 200 mA (which is sufficient), but this value is limited by the 100 ohm resistor in series with the gate. Change R4 to a value that is compatible with the selected SCR (i.e., the value is $R4 = (12\ VDC)/(I_{gate})$.

We may not want to use a 12 volt siren or bell in the alarm circuit. This is especially true if it is to protect a premises that has 115 volts AC available. We would not want to make a 25 or 30 ampere 12 volt power supply for the siren, when 115 volt sirens and bells are easily obtained. We could then operate the electronic part of the alarm from a low current 12 volt power supply (or even a battery) and use 115 volts for the siren or bell. In that case, we would want to replace the siren in Fig. 15-6 with the relay circuit of Fig. 15-6B; we find that the relay will turn on when the SCR is gated on. The contacts of the relay can then be connected such that it will energize a 115 volt AC bell or siren.

BURGLAR ALARM (III)

Version III of the burglar alarm (shown in Fig. 15-7) is similar to Version II, except that we are using the protection switches to detect the presence of an intruder. In this example we are using normally-open magnetic switches mounted on doors and windows of a home, or mechanical switches on the doors, hood, and trunk lid of a vehicle.

We can connect any number of protection switches to the circuit, provided that they are all normally-open types and are connected in parallel with each other. If any one of them closes, then the trigger input of the 555 is grounded, and this causes the output of the 555 to snap HIGH.

Once again, a HIGH on the output of the 555 will cause the SCR (D1) to gate on, thereby turning on the siren. Also, if the device is to be used in the home, where 115 volts AC is normally available, then the circuit previously shown in Fig. 15-6B may be used.

BURGLAR ALARM (IV)

A clever burglar can overcome the protected pathway type of alarm (Version II) by shining a light into the photoresistor housing.

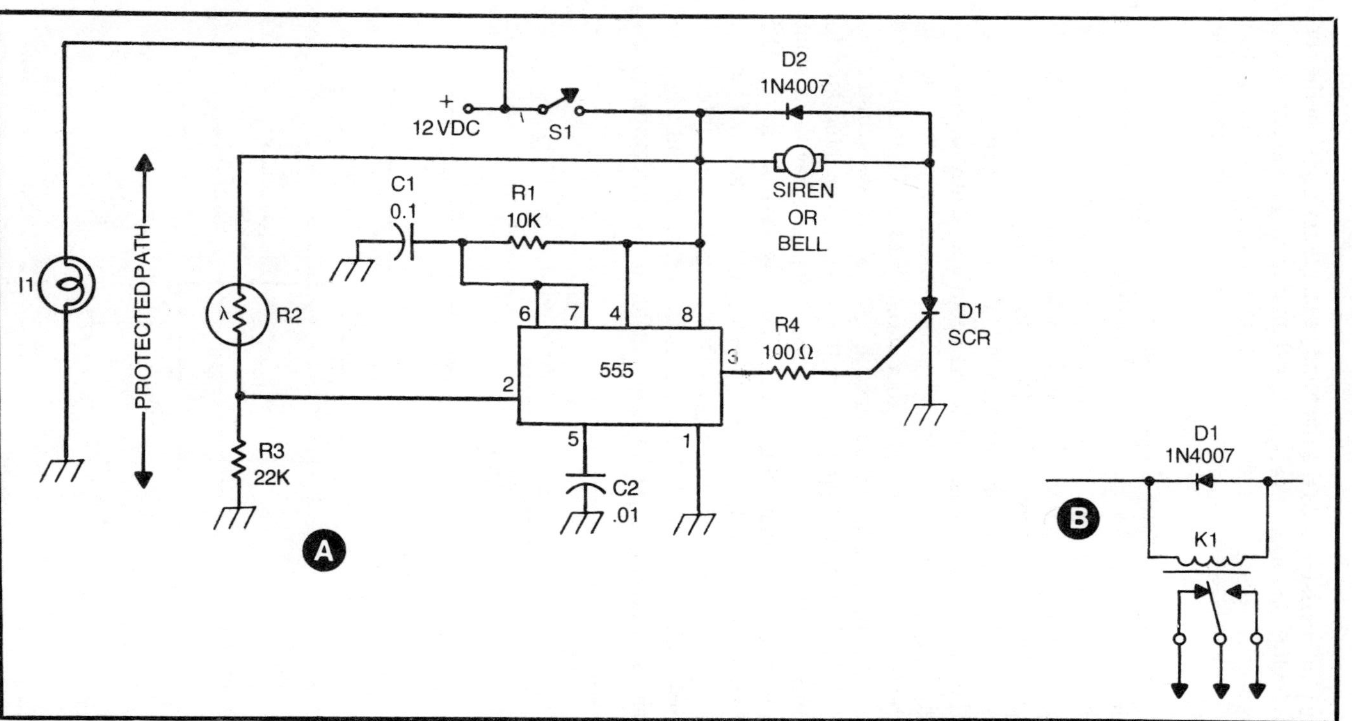

Fig. 15-6. Protected-zone alarm.

The photoresistor sees the light and cannot determine whether or not the light is *correct*. As far as the dumb circuit is concerned, the pathway is free of intruders because it sees a light (sigh). All the burglar needs is a flashlight to shine into the housing containing the photocell.

We can overcome this overcomer by using a pulsating light source, and a receiver that responds only to variations in the light source. Figure 15-8 shows the block diagram of such a project. The lamp is turned on and off by a 10 Hz astable multivibrator (a 555 circuit will do nicely). The transistor (Q1) is a 2N3055, or equivalent, to handle the high current of the lamp needed. The light in the pathway is a pulsating 10 Hz beam that shines on the base region of a phototransistor Q1. The amplified pulses at the collector of Q2 can be fed either directly to the 1 millisecond one-shot, or be squared-up first in the voltage comparator (an LM-311). The output pulses from the one-shot are integrated to form a DC potential. Note that the one-shot pulses have a constant duration and constant amplitude. The only variable in the DC output voltage, therefore, is the frequency of the pulses. As long as the light beam is uninterrupted, the output of the integrator remains high keeping PNP transistor Q3 turned off. But when the beam is interrupted, the capacitor inside of the integrator (don't

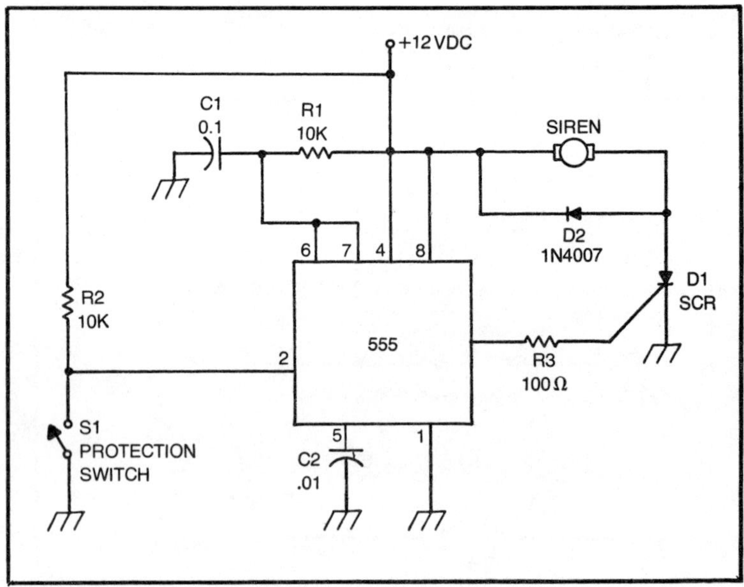

Fig. 15-7. Intruder alarm (parallel form).

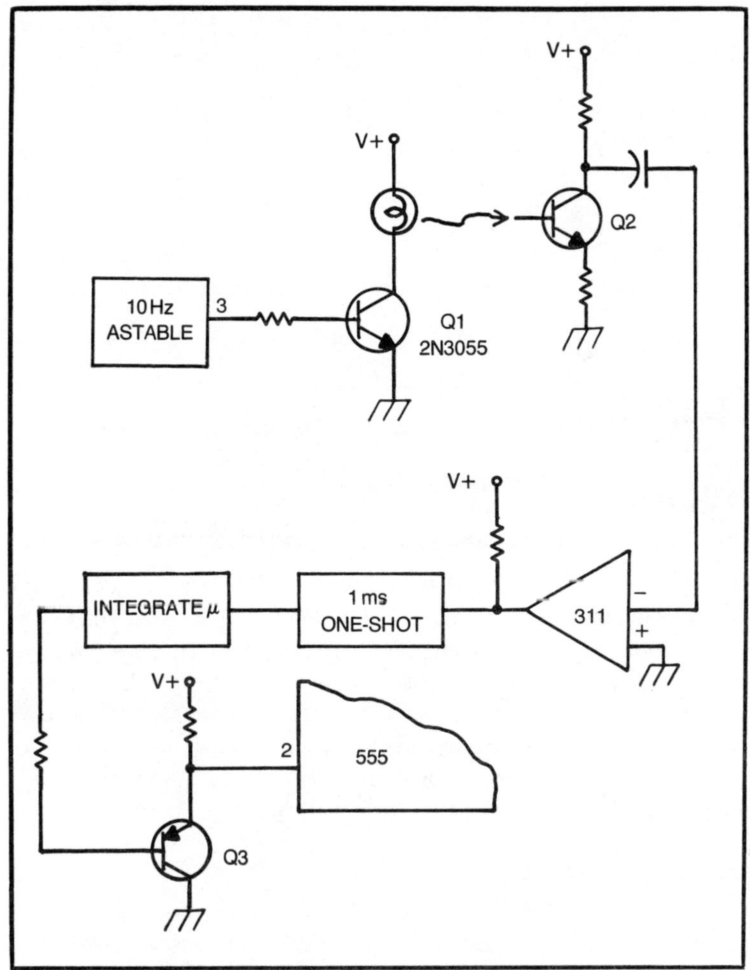

Fig. 15-8. Pulsed light-beam alarm.

look for it, it's not shown) will discharge, and this reduces the voltage at the output enough to forward bias transistor Q3. When Q3 is forward biased, the voltage on the trigger input of the 555 (pin 2) drops below ⅓ V+, so the 555 is triggered. The rest of the 555 circuit could be like either Fig. 15-6 or 15-7.

No burglar alarm is 100-percent effective: there is always some means to overcome the alarm. The use of these circuits, then, must be coupled with a bit of intelligent planning—design the mechanics of your project to discourage tampering!

Chapter 16

Some More Projects

This chapter is dedicated to some miscellaneous projects. Although you may build these as is, it is recommended that you learn to do some design on your own. Modify these projects, where appropriate, to fit your own needs. While it is flattery to build some project exactly as the author specifies, it is a higher form of flattery to build it just like—except. That proves that it was worth the effort to present the theory of operation in preceding chapters, and that you learned something from the material.

DC-TO-DC CONVERTER

Figure 16-1 shows the circuit for a small negative power supply that can be used to supply small amounts of power to operational amplifiers, CMOS devices, and other devices that require a negative power supply. The principle need for this type of power supply is in circuits where the vast majority of devices are monopolar (e.g., transistors, TTL digital logic, or 555 timers), and we want to accommodate a few bipolar devices.

The power supply shown here operates from a positive power supply in the 9- to 15-volt range and consists of a 555 astable multivibrator. The resistors and capacitor of the timing network of the astable are selected to operate at approximately 10 hertz.

The output of the 555 astable multivibrator is connected to a half-wave voltage doubler circuit consisting of diodes D1 and D2, and capacitors C3 and C4. The diodes are connected to produce a

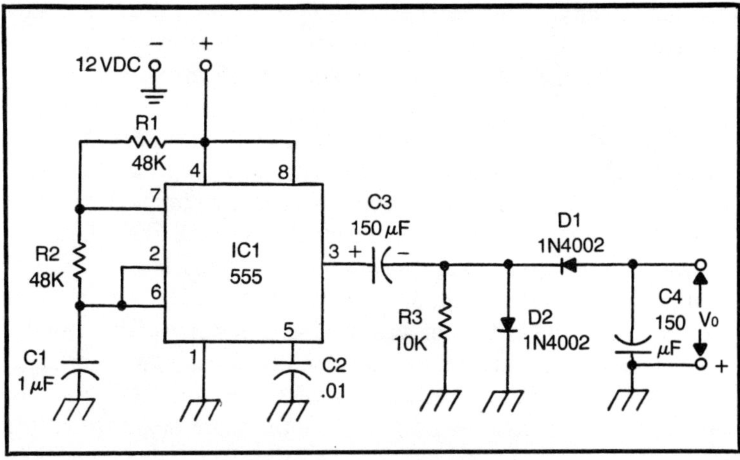

Fig. 16-1. Negative voltage power supply from positive source.

negative output voltage. This voltage will not be exactly double the supply voltage, even though the circuit is a voltage doubler, because of the duty factor of the output signal. The total negative voltage will be nearer the positive supply voltage for these reasons. If this circuit is followed with a negative regulator (7912, LM-320-12, etc.) then the voltage will be more stable. The output voltage of Fig. 16-1 is unregulated.

DICE GAME

Figure 16-2 shows a simplified schematic of a dice game. The totalizer is a digital counter with no gate, i.e., the input pulses are

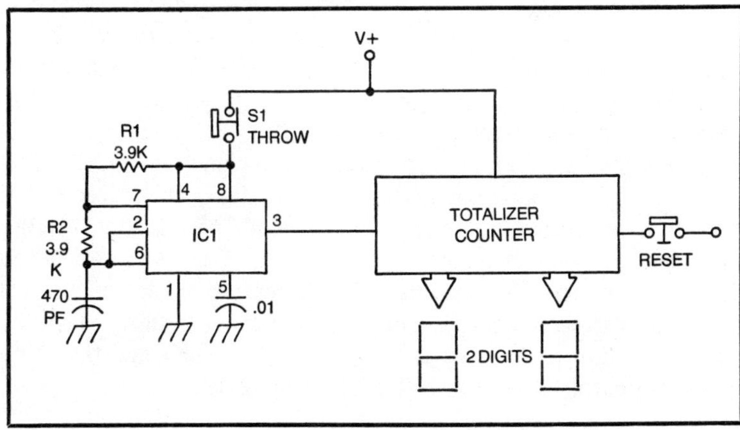

Fig. 16-2. Dice game.

applied directly to the decimal counting assembly. (See Chapter 9 of TAB Book No. 1194, *How To Troubleshoot & Repair Amateur Radio Equipment*). We reset the counter to 00 after each throw of the dice.

The roller is a 555 astable multivibrator operating at approximately 260 kHz. When the *throw* switch is depressed, the 555 is energized and will begin to oscillate. The counter will overflow over 100,000 times per second during the period that the button is pressed. When the button is released, the oscillations will die off rapidly, and the digital output of the counter will indicate the state existing when the last count was received. Because of the speed of the oscillation and the fact that only the two least significant digits are registered by the two-decade totalizer, it is unlikely that any given player will continuously produce the same result or pattern of results.

TEMPERATURE MONITOR

A thermistor is a resistor that is designed to produce a resistance value that is either proportional, or inversely proportional, to the temperature of the resistor. Most thermistors are not terribly linear, and hence will be inaccurate unless placed in a linearization network or operated only over narrow portions of their *resistance-vs-temperature* curve. Most of the devices advertised as thermistors for high accuracy measurements are actually combination networks of thermistors and other resistors to make the circuit look more linear over a wider range of applied temperatures. A prominent supplier of thermistor temperature probes is the *Yellow Springs Instrument Co.*, Yellow Springs, Ohio.

Some thermistors have a *positive temperature coefficient*, (PTC) while others have a *negative temperature coefficient*. (NTC) The PTC thermistors produce an increase in resistance for increases in temperature, while NTC units produce a decrease in resistance for increases in temperature.

Figure 16-3 shows a circuit using a thermistor in the timing network of a 555 monostable multivibrator. This circuit will produce an output pulse that has a width proportional to the temperature. The thermistor (R1) must be a PTC unit, so that the pulse width increases with increases in temperature. Resistor R2 is used to scale the output width to some convenient scale factor, and its value depends upon the parameters of R1 and the scale factor selected (milliseconds per degree kelvin).

We use a second 555 multivibrator, in this case an astable circuit, to trigger the one-shot circuit. The period of the astable is

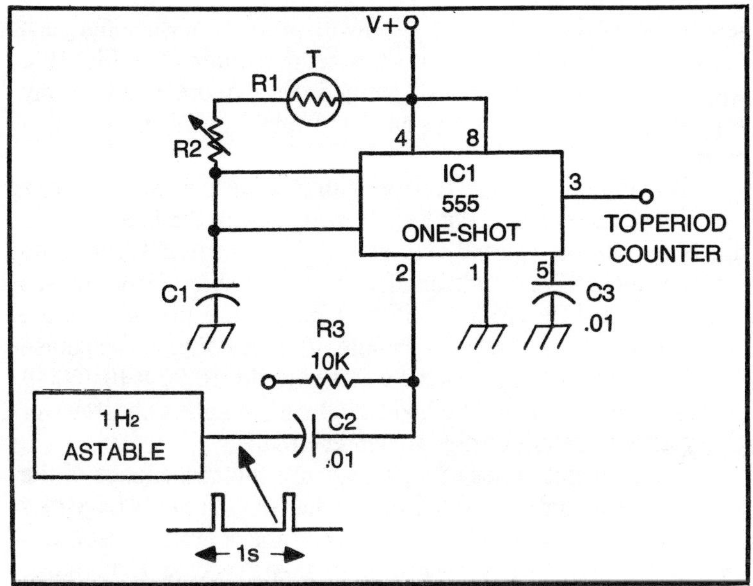

Fig. 16-3. Temperature monitor.

set to be longer than the normal duration of the one-shot pulse (1 second will do).

The width of the one-shot output pulse is measured on a *digital period counter.* Interpretation of the results depends upon the scale factor selected.

DIGITAL CAPACITANCE METER

The period of a 555 monostable multivibrator is a function of both resistance and capacitance (i.e., T = 1.1 RC). This implies that we can measure the capacitance value if we keep the resistor standard and constant, and then measure the period of the monostable multivibrator. This notion has not been lost on test equipment designers, who now offer a wide variety of digital capacitance meters.

Not long ago, an engineering student had to produce a senior design project (a fact of life in most U.S. engineering schools). This young lady had been told by the professor that the *only* way to do the job was to place the capacitor in one arm of an AC Wheatstone (or similar) bridge, and then build an AC voltmeter to measure the voltage at the output when the unknown capacitor was in the circuit, and compare this with the voltage present either with the bridge in the null condition or with a standard test capacitor. One wonders how the U.S.A. gets its technological lead over the rest of the world (well, most of it?) with professors who offer such advice. He claimed that this was the standard method. In Fig. 16-4 we see the block diagram of a method that *is* more standard; my engineering student colleague used it to obtain an "A" in the design project.

The digital counter can be either a period counter or a totalizer. In the example of Fig. 16-4 we have a totalizer counter that is operated as a period counter because of the NAND gate at its input. The NAND gate will admit a certain number of clock pulses from the 100 kHz clock (oscillator) during the period that the one-shot is active. If, for example, the one-shot has a pulse duration of 0.05 seconds, then the gate will admit 0.05×10^5 (5000) pulses to the totalizer. The book cited earlier in this chapter will give you more details on digital counter circuits.

The principle timer used to measure the capacitance of the unknown capacitor C_x is IC1. This 555 one-shot circuit is triggered by the other one-shot, IC2. In this circuit, the operator must press a *measure* button (switch S1) in order to take a reading. If this stage were made into an astable multivibrator, then automatic updating would occur. Make the period of the astable long enough to accommodate any possible durations of the ICI one-shot.

The period of the principle measurement one-shot is given by the expression:

$$T = 1.1\,(R1 + R2)\,C_x$$

Where:

T is the time in seconds

R1 is the resistance of R1 in ohms

R2 is the resistance of R2 in ohms

C_x is the capacitance of the unknown capacitor in farads

We must select a suitable scale factor, and then select values for R1 and R2 that will make the circuit work with the given range of capacitor values. We can select a scale factor of one count per picofarad (1 C/pF), which translates to one-shot durations of 10^{-5} seconds per picofarad. Let's now calculate the value of the two resistors in combination:

$$R1 + R2 \;=\; T/1.1C_x$$

We have rearranged the first equation to solve for the series combination resistance (R1 + R2). Now, let's select a trial value for C_x and then make the calculation. Select 100 pF. This translates to a one-shot period of 100×10^{-5} seconds, or 0.001 seconds. Plugging these values into the equation gives

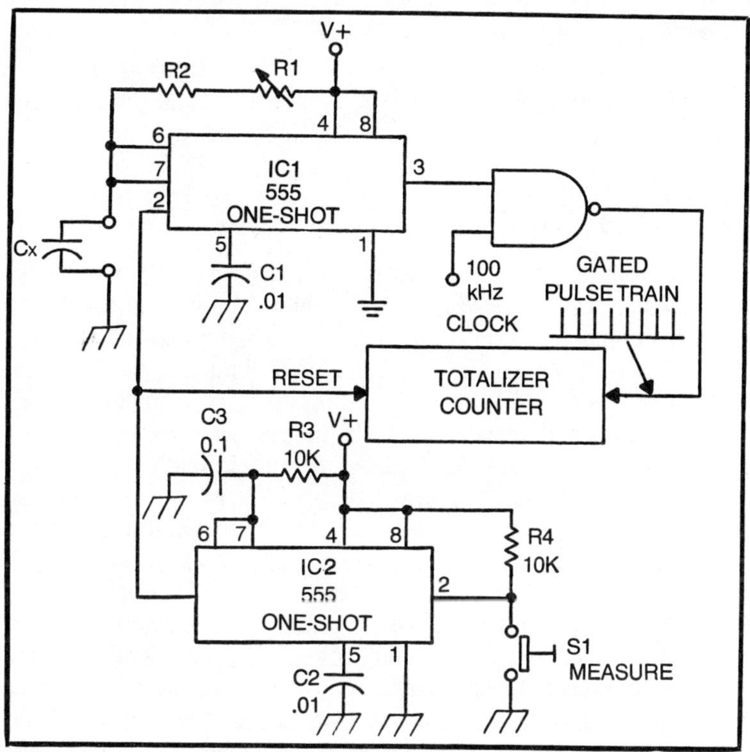

Fig. 16-4. Capacitance meter.

$$R1 + R2 = \frac{0.001}{(1.1)(10^{-10}F)}$$

$$= 9.091 \text{ megohms}$$

The potentiometer is used to trim the value of the resistor network to the correct value. In most cases, we will want to use a value of approximately 10 percent of the total for R1. This means that we will use a 1 megohm potentiometer for this resistor and then make up the rest of the 8.091 mehohms with one or more series connected fixed resistors.

The resistors selected for the timing network must be precision types. Why? Because ordinary carbon resistors will have too great a temperature coefficient for use in a precision instrument. It is necessary to use the precision types, even though a potentiometer is used to "trim" the value, because precision resistors are usually low temperature coefficient models. Also, the potentiometer should be a ten-turn or more trimmer potentiometer. This will allow very good resolution in adjusting the circuit to the correct operating point.

Accuracy and precision will also suffer if the 100 kHz clock circuit is not properly designed. This circuit must be a crystal oscillator that is temperature compensated against thermal changes. The crystal oscillator frequency must be adjusted to exactly 100.00 kHz for best operation.

CONTINUITY TESTER (VISUAL)

Continuity testers are instruments that will tell you whether or not a circuit is *continuous*. An ohmmeter can be used as a continuity tester, as can a combination of a battery and a light bulb. The circuit in Fig. 16-5 is a special type of continuity tester in that it will latch the reading and hold it until the *reset* button is pressed. This circuit will help immensely in the case where a single technician must "dope out" a large circuit. In one case, I had to dope out a large public address system. There was an electrical panel in the control room where all the wires from the rest of the church were terminated, but we had no way of knowing which wires went to which rooms. Fortunately, all of the wires were paired. It took two technicians, working with walkie-talkies, to dope that circuit out. The circuit shown in Fig. 16-5 will allow one person to check out such a circuit. The probes can be connected at the remote room, and then a trial "short circuit" placed across the suspected pair. If the light comes on, then the circuit was continuous—and the light will *latch on* so that the technician can come back and check it.

IC1 in Fig. 16-5 is a 555 wired as a monostable multivibrator. The trigger terminal is permanently held HIGH by resistor R2, connected to the +9 volt DC line. But, when transistor Q1 is turned on, the voltage at the trigger terminal drops LOW thereby triggering the one-shot. Transistor Q1 is turned on when there is a short circuit across the probes. This short circuit will allow a forward bias to be applied to Q1, triggering the one-shot.

The output terminal of the 555 (pin 3) will go HIGH for a period of approximately one millisecond (0.001 s). This pulse is applied to the gate of a silicon controlled rectifier (SCR). The SCR turns on, lighting the LED. If the current of the LED is too small to sustain the *on* condition of the SCR, then add a parallel resistor (R5) to add current to the SCR circuit.

The *reset* button (S1) will interrupt the current applied to the SCR anode, so will extinquish the SCR. The circuit will then be ready to receive another pulse from IC1.

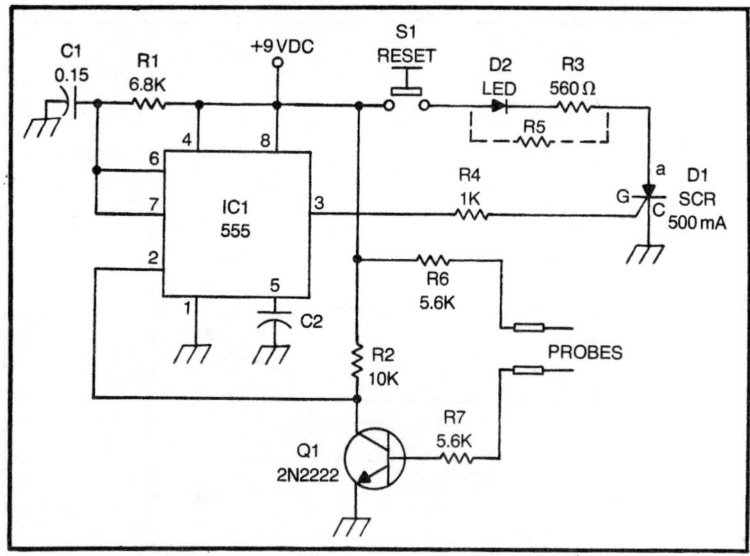

Fig. 16-5. Continuity tester (visual).

AURAL CONTINUITY TESTER (I)

Figure 16-6 shows the circuit of a special type of continuity tester: it allows the operator to *hear* when the circuit is continuous! The use of a visual continuity tester or an ohmmeter requires us to *look* at the tester in order to see if the circuit is continuous. But this

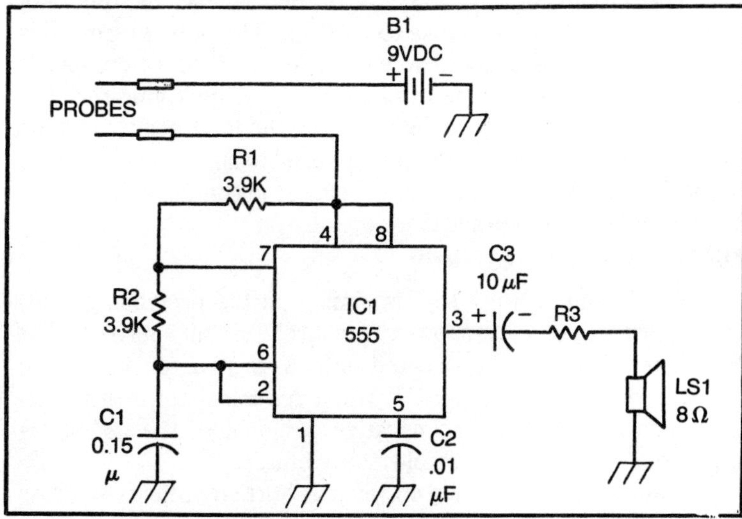

Fig. 16-6. Continuity tester (aural).

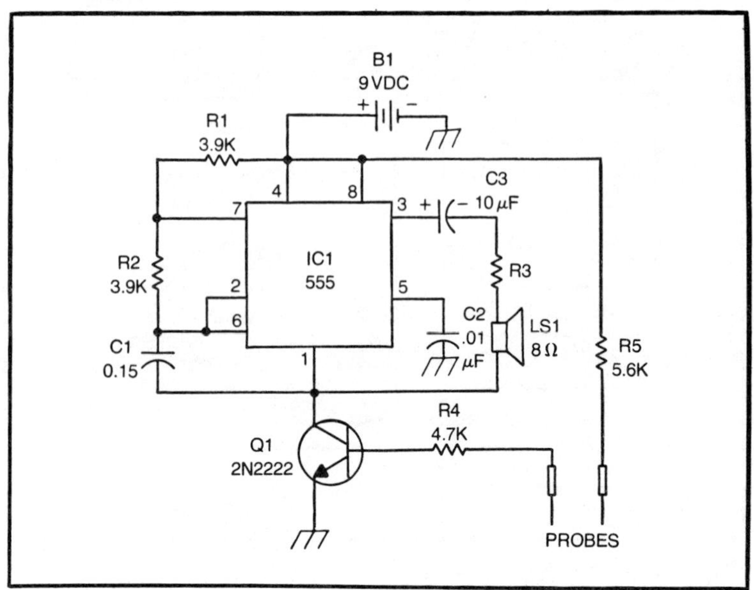

Fig. 16-7. Aural continuity tester.

tester allows you to listen for the indication of continuity. This type of indication is especially useful when working on a large switch box, chock full of wires and thousands of little terminals. Take your eye off the work for just a second, and you lose your place.

The circuit of Fig. 16-6 is a 555 astable multivibrator operating at a frequency close to 1000 Hz. The output terminal is connected to a loudspeaker. Note that resistor R3 is optional and is used to reduce the volume of the 555 astable multivibrator. The value will depend upon the impedance of the loudspeaker and the desired volume level. The probes are used to apply the V+ supply to the 555 astable. If the circuit under test is continuous, then the astable will receive V+ and therefore turn on.

AURAL CONTINUITY TESTER (II)

The circuit in Fig. 16-7 is similar to the previous circuit, except that we are using a low-current probe circuit approach. The astable multivibrator is connected with its negative terminal to the collector of gating transistor Q1. When this transistor is turned on (i.e., forward biased), the negative terminal of the astable is grounded, turning on the astable multivibrator.

Transistor Q1 is turned on when a short circuit exists across the probes. This condition will apply a current through R4 & R5 to the base of the transistor.

Chapter 17

And Still More Projects

This is the "miscellaneous" chapter that all electronics books should have. It allows the author and the reader to indulge in material that may or may not fit the general format of the book. In this chapter we will present several different projects which should be of interest to many readers.

The amateur radio operator seems to have been overlooked throughout most of this book. Since I am a "ham" (K4IPV), this situation cannot be allowed to continue! Hams rise up! Support your right to keep and bear timers! Or, is that "Support your right to time bears"?

Figure 17-1 shows a 100-kHz calibrator based on the 555 integrated circuit timer. The frequency of the astable multivibrator is determined by a piezoelectric crystal. In this case, we have selected a 100-kHz crystal. This frequency allows us to place harmonic markers at frequencies throughout the HF bands at 100 kHz intervals. We can use the markers to calibrate the dial of the communications receiver, and then the receiver is used to make rough measurements of received frequency. The dial will be relatively accurate at frequencies close to the 100-kHz marker (± 10 kHz) with lessened accuracy at frequencies farther away (to ±50 kHz).

The output waveform will be a squarewave, rich in harmonics. The harmonic content can be improved somewhat by placing a transistor-output amplifier between the output of the 555 and the load. The amplifier is designed to have a diode rectifier in the collector load circuit.

The crystal calibrator circuit of Fig. 17-1 will work throughout the HF region but is limited because the markers are available only at 100 kHz intervals. If we had markers at closer intervals, then we could measure frequencies to a greater accuracy. Figure 17-2 shows a wide-range amateur radio crystal calibrator circuit.

The oscillator circuit in the crystal calibrator of Fig. 17-2A is a Motorola MC4024P voltage controlled oscillator. The timing capacitor of the VCO is replaced with a 1 megahertz crystal. This IC will not always oscillate with a crystal at frequencies less than 2 MHz, so some capacitance is placed in parallel with the crystal. These capacitors also allow us to set the oscillator frequency to the exact frequency (1000.0 kHz).

Alternatively, we could operate the crystal oscillator at 4.0 MHz. The 1.0 MHz signal could then be derived from a pair of cascade J-K flip-flops operated in the clocked mode. Recall from Chapter 4 that the J-K FF will divide the input frequency by two because it allows output changes only on negative-going input transitions. A pair of J-K flip-flops operating in cascade, therefore, will divide the oscillator frequency by four, so the 4 MHz signal becomes a 1 MHz signal.

An alternative oscillator circuit is shown in Fig. 17-2B. This circuit uses TTL inverters or NAND gates wired to act as inverters. The crystal sets the frequency of oscillation, while C1 allows us to trim the frequency to a close tolerance.

Some designers prefer to bias the TTL gate inputs with the resistors shown. In most cases, these resistors will have a value between 500 ohms and 1500 ohms.

The output stage is another TTL gate (also inverter wired), and this IC serves as a buffer to the outside world. The gate presents a constant load change, so it will not allow frequency changes due to load variations.

An alternative output scheme, also shown in Fig. 17-2B, allows us to turn the oscillator signal on and off without turning off the oscillator. This tactic allows the oscillator to stabilize in its thermal environment and thereby attain better stability.

The main use of this crystal calibrator is to supply lower frequency interval markers throughout the high frequency spectrum. The marker intervals are produced by the 7490 decade dividers in Fig. 17-2A. Each 7490 is a decade counter, wired as a decade (divide-by-10) frequency divider. No decoding is allowed

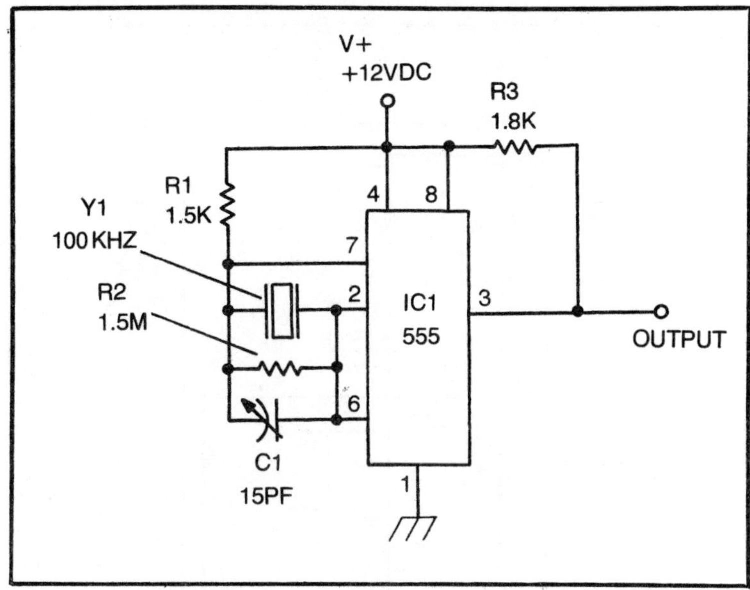

Fig. 17-1. 100-kHz crystal marker oscillator.

on this circuit because its main use is to provide markers in a communications receiver.

A switch selects the output from the marker according to the frequency needed. The output of the crystal oscillator is 1000 kHz, the first divider 100 kHz, second divider 10 kHz, and the third divider is 1 kHz. We could also place a J-K flip-flop in the line from the switch to the output buffer amplifier and obtain frequencies that are one-half of the frequencies stated above; i.e., 500 kHz, 50 kHz, 5 kHz, and 500 Hz.

How do we use a crystal marker to measure frequency? Well, for that let me cop out (this is a timer book) and refer you to TAB Book No. 1194 p. 320, *How to Troubleshoot and Repair Amateur Radio Equipment*.

SIDETONE OSCILLATOR

A CW (morse code) operator needs some tone feedback from the transmitter in order to send the code characters correctly—he must *hear* what is being sent. If the transmitter or transceiver does not provide a sidetone oscillator, then it becomes necessary to provide one yourself. An example is shown in Fig. 17-3.

The tone is formed in a 555 astable multivibrator operating at a frequency in the neighborhood of 500 Hz (which many CW

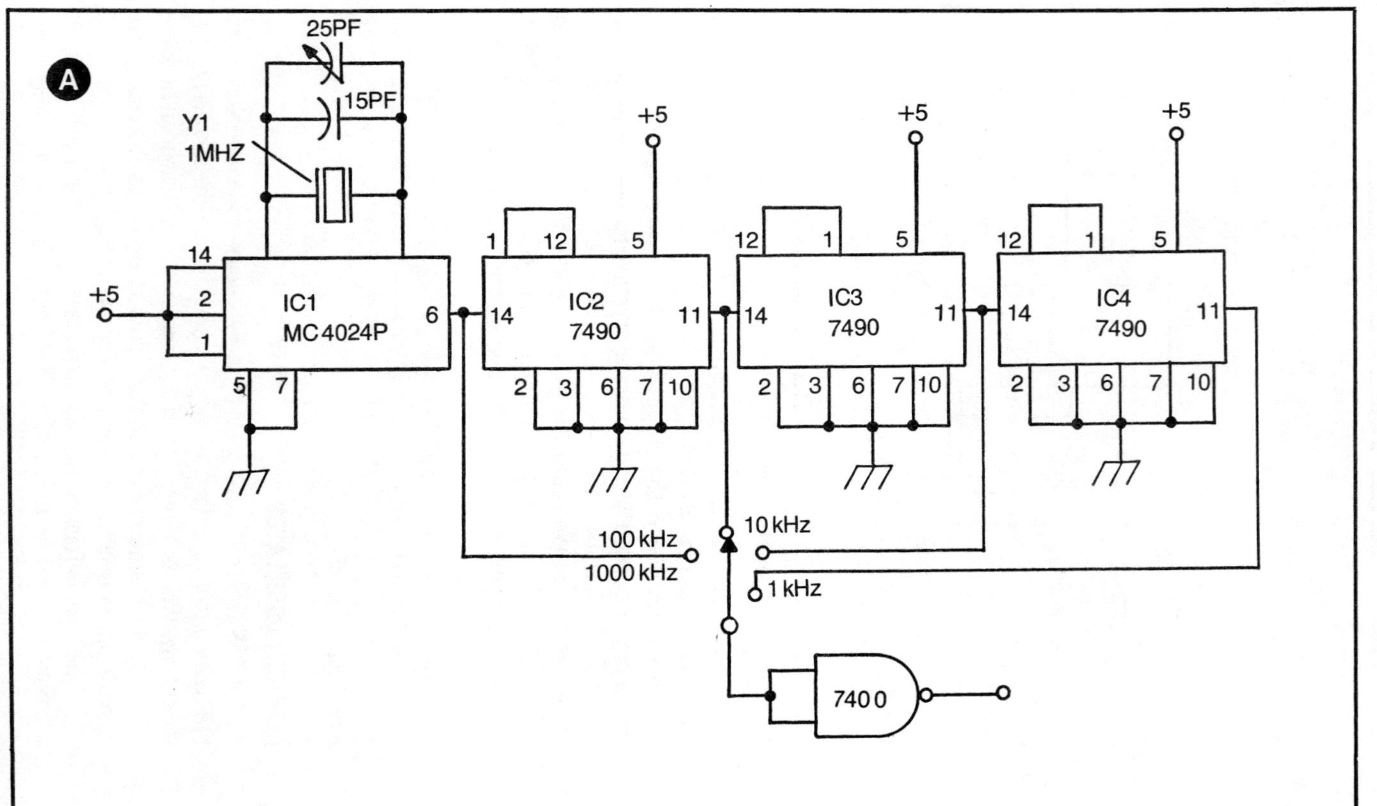
A
25PF
15PF
Y1
1MHZ
+5
IC1
MC4024P
IC2
7490
IC3
7490
IC4
7490
100 kHz
1000 kHz
10 kHz
1 kHz
7400

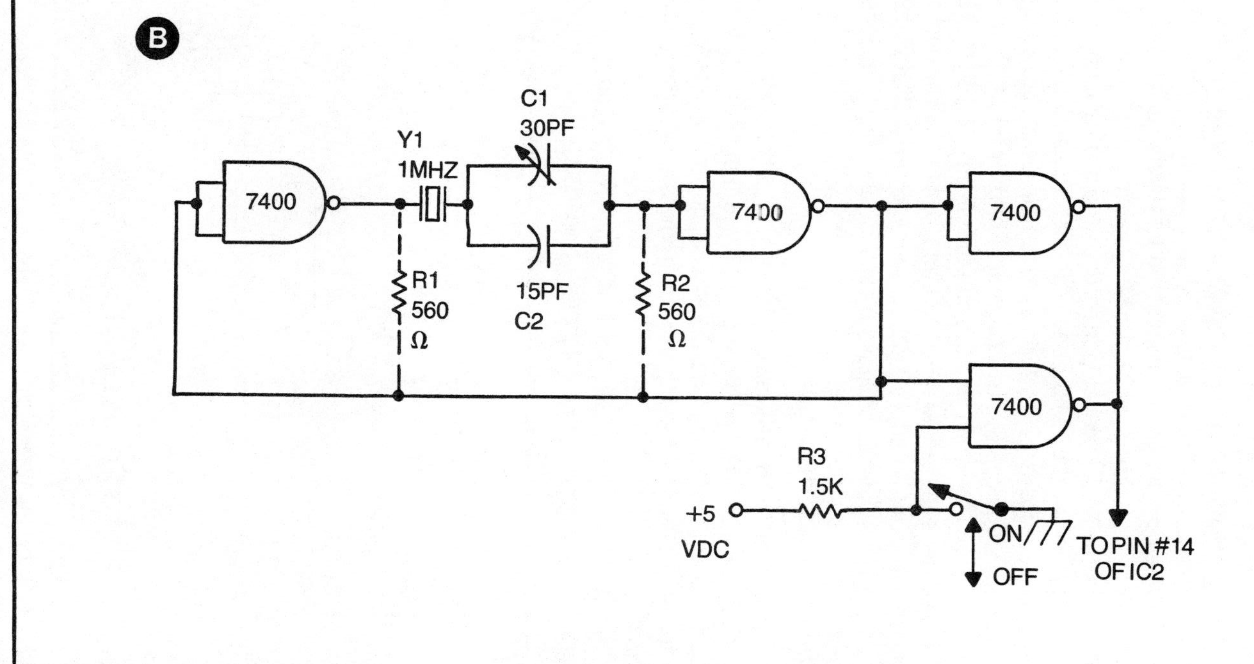

Fig. 17-2. Wide-range crystal marker oscillator.

operators prefer as the ideal listening tone). The oscillator will run only when the reset terminal (pin 4) of the 555 is HIGH. A LOW on this terminal sets the output to zero (LOW) so the oscillator will not run.

The HIGH on the reset terminal is provided by the actual transmitted signal. This signal is picked up on a small whip antenna, which can be one of the small telescoping whips used as replacements for those on AM-FM transistor portable radio receivers. The signal is developed across a 2.5 mH RF choke, and is rectified to DC by the 1N60 diode. The rectified signal is filtered by capacitor C3. Resistor R4 serves to load the diode (providing a return path) and keeps the reset terminal of the 555 LOW when there is no DC present at the output of the rectifier. The loudspeaker is connected to the output (pin 3) of the 555 through an electrolytic capacitor and a 10 kohm volume control. The control will allow you to adjust the volume to a comfortable earphone level.

VARIABLE PULSE GENERATOR

Pulse generators are used to test a variety of electronics circuits and devices. The pulse output of these signal sources is not a square wave but is a squared pulse with an on-time that is usually less than the off-time. We can use timer circuits to make a pulse generator.

Figure 17-4 shows the block diagram for a simple pulse generator made from circuits in this book. The actual pulse is

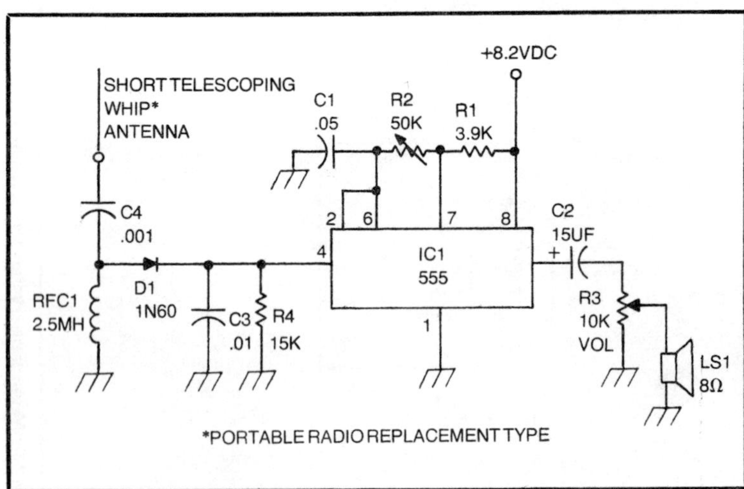

Fig. 17-3. CW sidetone monitor.

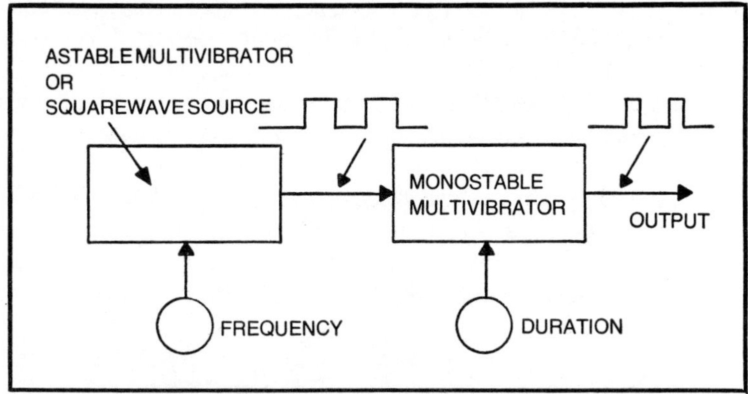

Fig. 17-4. Simple pulse generator.

formed by a monostable multivibrator. This stage must have a variable duration control so that we can adjust the pulse width of the output signal. This control can be a linear-taper potentiometer in some cases. But if we are making a high-frequency (very short duration) pulse generator then we might want to make the duration control a variable capacitor, such as used to tune radio receivers.

The repetition rate of the pulse generator is set by an astable multivibrator. The output pulses from this stage are used to trigger the input of the monostable multivibrator that follows. We can make the repetition rate variable by either a potentiometer or a variable capacitor in the RC timing network that sets the frequency of the astable multivibrator.

WARBLE SIREN

A warble siren (there is another project in an earlier chapter) is the two-pitch tone heard from many speeding police cars and other emergency vehicles. You are probably familiar with the peculiar beeeeeee-booooooo sound of this electronic siren. The earlier circuit used an electronic switching system to change the frequency of oscillation of an astable multivibrator. The version in Fig. 17-5A, however, uses a different tactic and is more nearly like the professional circuits seen in some real electronic sirens.

In this circuit, we use two separate astable multivibrators (not shown) as the tone generators. The astable circuits used, incidentally, can be made from 555 IC timers but must be operated at +5 volts DC to be TTL compatible. We can vary the tones from those shown to suit taste and preference. One tone is fed to one of the TTL NAND gates, while the other tone is fed to an input of the other TTL NAND gate. A 7404 TTL inverter (or another section of the 7400 NAND gate wired for inverter operation) connects together the opposite inputs of the NAND gates. This circuit tactic causes the two NAND gates to be driven out of phase with each other. When line A is HIGH, line B is LOW. A LOW applied to the input of NAND gate IC1a will cause it to turn off (i.e., the output is forced HIGH). On the half cycle when line A is LOW, IC1b is turned off and IC1a is operating. Each NAND gate will pass the oscillator signal to the output when the control input is HIGH. When the control input is LOW, on the other hand, the oscillator signal is blocked.

This warble siren works because each NAND gate output alternates with the other (see Fig. 17-5b), so the tone will warble back and forth between the two frequencies.

The signals from the two NAND gate outputs are combined at the top end of the volume control and are coupled from the NAND gate through a resistor and DC blocking capacitors (C1 and C2). The rate at which the tones bounce back and forth is set by astable multivibrator IC3. The frequency of this stage is less than 1 Hz, but this can be changed to suit your needs (see the section on astable operation of the 555 IC timer).

TOUCH-PLATE SWITCHES

Figure 17-6 shows two forms of touch-plate switch circuit. The circuit in Fig. 17-6A is the ordinary sort of circuit, while that of Fig. 17-6B is a latching type that must be reset prior to the next operation.

The common circuit of Fig. 17-6A is a 555 monostable multivibrator in which the trigger circuit (pin 2) is a simple touch plate. This circuit depends upon the 60 Hz AC field generated by the AC power mains, which causes 60 Hz currents in your body. It is possible to produce a rather high voltage from these currents in your body. If you do not believe this statement, then try a little experiment: grab the input lead of an AC voltmeter or oscilloscope vertical channel. It is likely that the display will show several volts.

When a person touches the plate, the 60 Hz signal will trigger the one-shot multivibrator (IC1). The output terminal will snap HIGH for a period of approximately 5 seconds. This period can be changed to meet any need that you have within the range of the 555 device. Consult the chapter on monostable operation of the 555.

The latching type of touch-plate switch, also a 555 circuit, is shown in Fig. 17-6B. Note the slightly unusual configuration of this

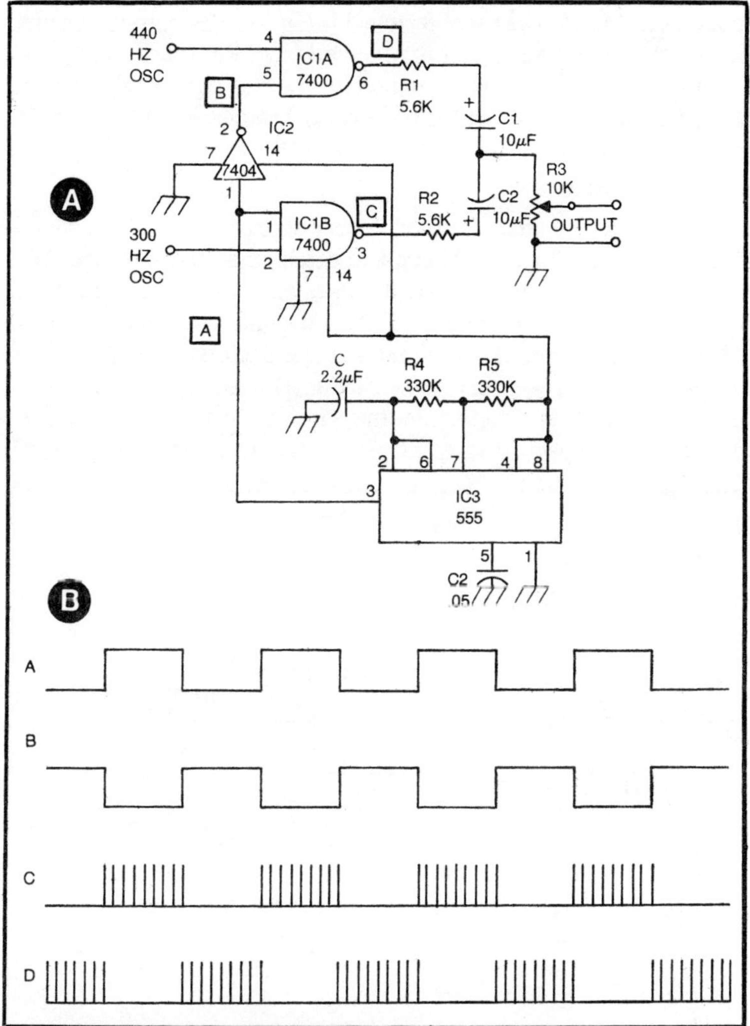

Fig. 17-5. Warble alarm (another type).

circuit. Pins 6 and 7 are connected together in the usual manner, but there is no resistor from the V+ power supply to pins 6 and 7 (this is what gives the circuit its latching capability). Also, note that in this circuit the reset terminal (pin 4) is used. In most of the previous circuits, the reset terminal is connected to the V+ power supply, or wired HIGH through a pull-up resistor. In this circuit we use a 10 kohm pull-up resistor (R1) to the V+ supply and a grounding reset switch. The device will operate in the normal manner as long as the reset terminal is HIGH. The output remains LOW until a trigger signal is received from the touch plate, and then it snaps HIGH and remains HIGH. The output will remain HIGH until the circuit reset terminal is brought LOW by closing S1.

PULSE CATCHER

Many digital circuits produce very short duration, high speed transient pulses that are difficult to catch on an oscilloscope. We sometimes try to catch transient pulses on a storage oscilloscope or with an oscilloscope camera. But some pulses are simply too fast for the writing speeds of either storage oscilloscope or oscilloscope cameras. The circuit shown in Fig. 17-7 is a *pulse catcher*.

The pulse is coupled to the circuit through a low-value disc-ceramic capacitor (C1, 33 pF) to the input of a 4050 non-

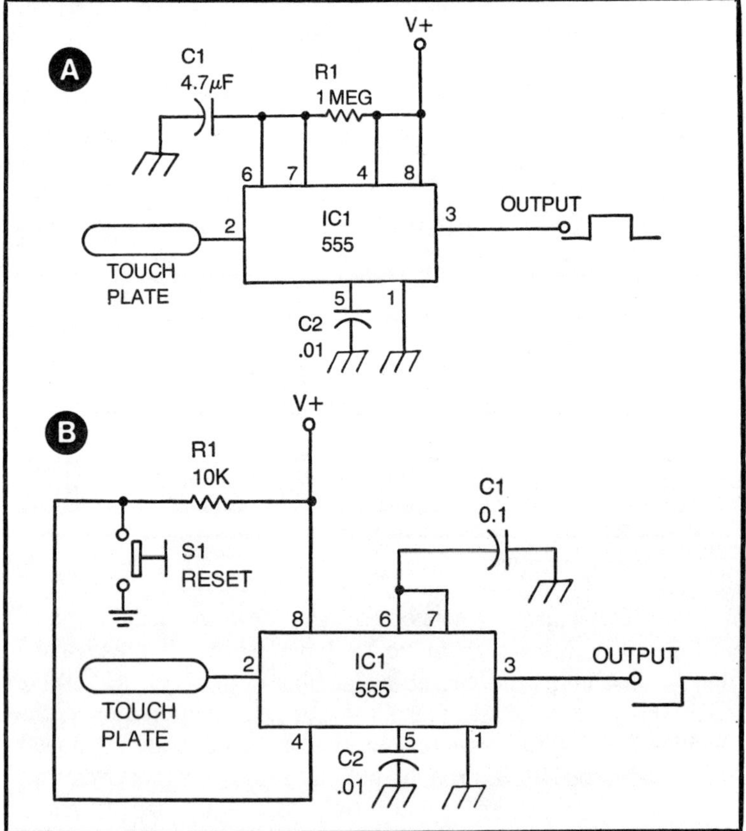

Fig. 17-6. Touch-plate switches.

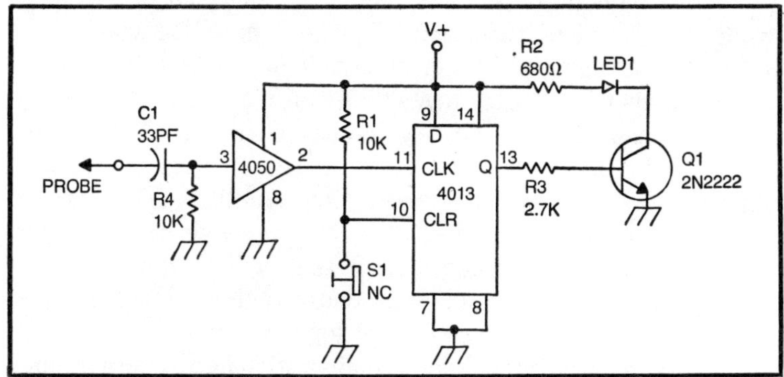

Fig. 17-7. Pulse catcher.

inverting buffer (CMOS). The output of the 4050 is applied to the clock input of a 4013 Type-D flip-flop. Since the D input of this stage is wired permanently HIGH, the Q output will go HIGH when the pulse is received from the 4050. A reset switch (S1) will cause the normally LOW clear input to go HIGH when we want to reset the circuit.

This circuit is used in testing digital circuits. Press the reset switch, clearing the 4013. When a pulse is detected, the output of the 4013 will go HIGH causing transistor Q1 to be forward biased, turning on the LED, and giving an indication that the pulse existed.

DIGITAL LOGIC PROBE

The digital logic probe has become a primary piece of test equipment for technicians and others who troubleshoot digital electronic devices. The hand-held probe will signal the user whether the logic level is HIGH or LOW.

The circuit shown in Fig. 17-8A is based on the 555 IC timer. When the signal level at the probe is LOW, then the output will go HIGH. It remains HIGH (note the lack of a V+ pull-up resistor from pins 6 and 2) until the reset switch (S1) is closed, bringing pin 4 LOW.

If, on the other hand, the logic level is HIGH, then the output will be LOW. We can, therefore, tell which logic level is present at the input by the condition of the output of the 555. If the output is LOW (indicating a HIGH on the input), then LED1 is grounded and will light up. In this case, transistor Q1 is turned off so LED2 will be dark. But, when the output of the 555 is HIGH, indicating a LOW on the input of the probe, then transistor Q1 is forward biased and LED2 lights up.

The circuit must be reset after each use, and this feature (while good for some purposes) keeps us from telling whether or not a clock is running. The clock circuit in a digital device will produce a chain of squarewaves. We can make the logic probe sensitive to pulsing by connecting the circuit of Fig. 17-8B in series with the input of the logic probe circuit. The inverter is used as a buffer amplifier, and drives a coupling capacitor. This circuit differentiates the input pulse, producing positive and negative spikes (differentiated squarewaves) that will trigger the 555 device. Since the capacitor only responds to pulses, it will indicate that the pulse train is present. The logic probe will have to be modified to the ordinary one-shot configuration by placing a timing resistor from pins 6 and 2 to the V+ line.

OSCILLOSCOPE CALIBRATOR

An oscilloscope calibrator must provide two pieces of data: frequency and voltage. The frequency is used to calibrate the

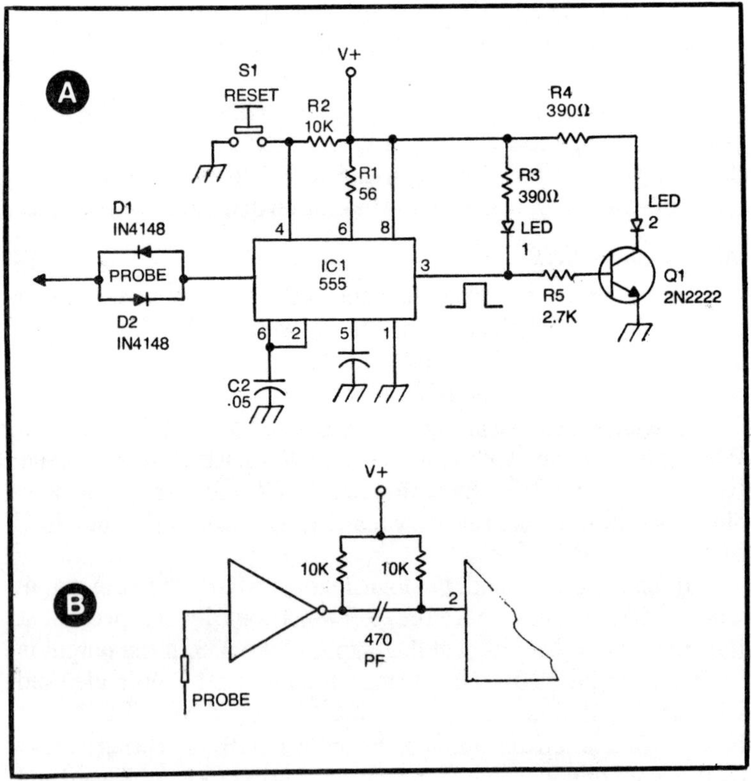

Fig. 17-8. Digital logic probe.

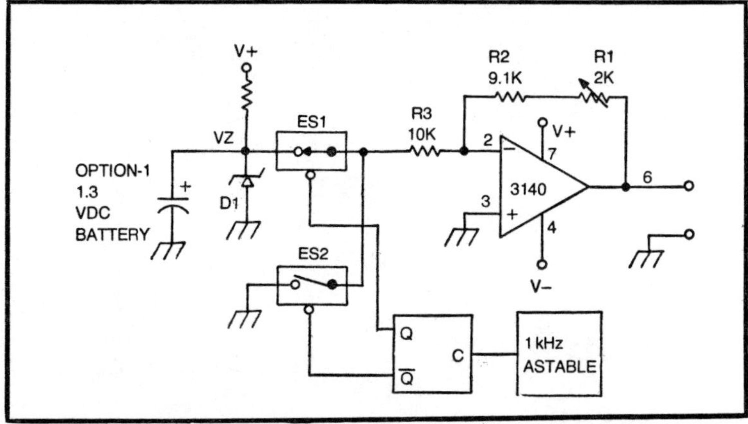

Fig. 17-9. Oscilloscope calibrator.

timebase section of the oscilloscope. The voltage level is used to calibrate the vertical amplifier attenuator. In many cases, there are separate voltage and frequency calibrators. The voltage calibrator will be a precision voltage source that will cause a precise deflection of the beam when the voltage is connected to the vertical input. The frequency calibrator is usually a 1000-Hz oscillator producing squarewaves.

The circuit in Fig. 17-9 provides both types of calibration signal. The operational amplifier (RCA CA3140) is used as a precision voltage source. The output voltage is set by the zener diode voltage (D1) and the ratio (R1+R2)/R3. The polarity of the output voltage will be opposite of the Zener potential, which in this case is negative.

We use a 2000-Hz oscillator to provide the timebase calibration signal. The output of the astable multivibrator is applied to the clock input of a J-K flip-flop. The complementary Q and NOT-Q outputs of the flip-flop are applied to the control lines of CMOS electronic switches ES1 and ES2. Since the J-K FF outputs are complementary, these switches are opened and closed opposite each other. To the amplifier input, this means that the input signal is alternately V_z or zero volts (one side of ES2 is grounded). The output signal, then, will be a square wave of one-half of the astable frequency, with a potential equal to the potential produced by the amplifier. The precise output amplitude will be set by potentiometer R1. We can use the regular operational amplifier offset control (not shown) to adjust the ground level at the output of the amplifier, should this be needed.

Appendix A

Universal Experimenter's Power Supply

Most of the experiments and projects in this book require a DC power supply. Most well-equipped experimenters will have such power supplies or will build one when needed. The circuit in Fig. A-1 is a universal experimenter's power supply that provides the following voltage and current levels:

- ☐ +12 volts, 1 ampere
- ☐ –12 volts, 1 ampere
- ☐ +5 volts, 3 or 5 amperes

These power supplies are well regulated, meaning that they can be used in almost any application. There will be less than 15 mV of 120-Hz power-supply ripple on any of the three outputs. In addition, there are output capacitors provided to improve the transient response of the circuit (i.e., answer a heavy drain on the output that lasts only a few dozen milliseconds). A certain amount of noise immunity is also built into the power supply circuit.

+ 5 VOLT SECTION

The +5-volt section is designed to power a considerable number of TTL integrated circuit devices. It will also power IC timers, such as the 555, 556, or XR-2240, that will operate down to +4.5 volts.

Transformer T2 is a 6.3 volt AC filament transformer that produces at least 3 amperes. If you want the supply to run cool, or to provide up to 5 amperes output, then look for a transformer with

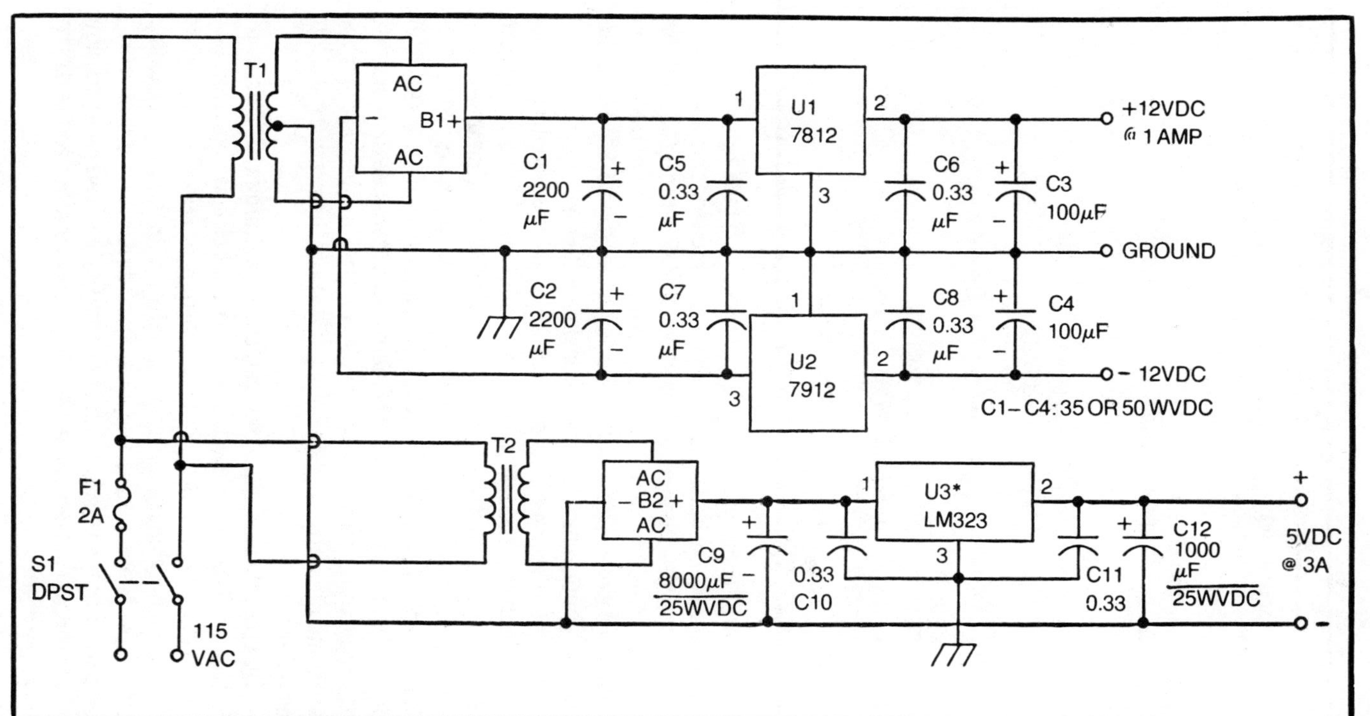

Fig. A-1. Universal experimenter's power supply.

Parts list for Fig. A-1

Parts List	
Transformers	
T1	25.2 volt, center-tapped, @ 2 amps(or more)
T2	6.3 volt, 5 amp (or more)
Semiconductors	
B1	50 PIV @ 1 ampere bridge rectifier
B2	50 PIV @ 6 ampere (25 ampere OK) bridge rectifier
Capacitors	
C1,C2	2200 μF/35 or 50 WVDC
C3,C4	100 μF/35 or 50 WVDC
C5,C6, C7,C8, C10,C11	0.33 μF/50 WVDC mylar or tantalum
C9	8000 to 10,000 μF/15 WVDC (or more)
C12	1000 μF/25 WVDC
Miscellaneous	
F1	2 ampere fuse with fuse holder
S1	DPST AC switch

the appropriate current rating. Triad makes a handy little filament transformer that produces 8 amperes. It is nice for this application because it will produce either 6.3 or 7.5 volts RMS on the secondary winding, depending upon which portions of the primary winding are used. The 7.5 VAC works better than 6.3 VAC during brownout conditions (i.e., low AC mains voltage).

The rectifier is a bridge stack (although you could use four ordinary rectifier diodes connected as a bridge) with a rating of at least 6 amperes. Motorola sells some with this rating under their HEP brand. It is also likely, however, that most experimenters will use a 25 ampere bridge stack because they are considerably more available than the lower current types. Most dealers seem to stock only the 1 amp and 25 amp sizes (sigh). They are not too costly, and the extra current rating will allow the rectifier stack to run cooler.

By now, you may have noticed that I like things to "run cool." It is true in all areas of electronics that cool circuits burn out less than circuits that run hot. The transformer or rectifier that is too hot to touch is aching and may well "belly up" on you sooner than is normal. On heavy power supplies, i.e., those over 3 amperes, I prefer to place the regulator and the rectifier on heatsinks and blow air over them. Conservative, yes, but my regulators and rectifiers rarely short out!

The filter capacitor is selected according to the more conservative of the two methods normally given, i.e., 2000 μF/ampere. In this case, therefore, we need 8000 μF, or more, to do the job. This filter should have a working voltage rating of at least 15 volts, and 25 volts is nearer the mark.

The regulator for the 3 ampere power supply is the LM323 device. This three-terminal voltage regulator comes in the T0-3 transistor package in which the "base" terminal is pin 1, the "emitter" terminal is pin 2, and the case is pin 3. This diamond shaped package should be mounted on an ample heatsink, using silicone heat transfer grease between the case of the regulator and the heatsink.

If you want to provide more current, then use the Lambda Electronics 5-volt, 5-ampere regulator, the LAS1905. This device looks exactly like the LM323 device, but will produce more output current at the rated voltage—without burning out.

Capacitors C10 and C11 (0.33 μF) are used to improve the noise immunity of the circuit. Fast acting, tight regulation requires internal reference amplifiers (i.e., inside U3) that have a high frequency response. If a noise pulse comes along, with its inherent high frequency spectrum, then the noise will either pass through or interrupt the regulator. The noise-suppression capacitors should be mounted as close as possible to the body of the regulator. In most cases, this means mount them right on the package (i.e., connect C10 from pin 1 of the regulator to the case, and connect C11 from pin 2 of the regulator to the case).

The output capacitor (C12) is optional but highly recommended. This capacitor forms a current reservoir for the output in cases where there is a brief, transient increase in the load demanded of the regulator. If this brief load is sufficient to invoke the current limiter circuit inside the regulator, then a "glitch" will occur. The purpose of this capacitor is to smooth out those glitches. The value of C12 is calculated from 100 μF/ampere, so a 100 μF unit will provide protection up to 10 amperes. +5 volt power supplies that deliver more than 10 amperes sometimes require complex circuits (there are only a few over-10 ampere regulators on the market, see the Lambda Electronics catalog).

Note that this circuit keeps the ground of the +5 volt supply together with the ground of the dual polarity low voltage power supply. This is usually OK, but can sometimes lead to problems. In many cases, the ± 12 volt supply is used for the analog portion of a circuit, and the +5 volts is for the digital. The noise and other problems present due to the operation of the digital logic devices will interfere with the operation of the analog circuitry. As a result of this and the inherent problem of ground loops in high-current power supplies, it is sometimes advantageous to separate the two grounds and then strap them together only at *one point* in the circuit that you are powering. Single-point grounding seems to take care of a host of problems.

12 VOLT POWER SUPPLY SECTION

The dual-polarity low-voltage section is powered from a 25.2-volt (or more) AC filament transformer. This type of transformer is relatively common and is usually available at Radio Shack stores. Be absolutely sure that the transformer that you buy is *center-tapped,* the circuit will not work without the C.T.! Note that some of the Radio Shack models are not center tapped, so they are useless for this project. If this type of transformer is not available, then substitute two appropriately connected 12.6-volt transformers, each rated at 1 ampere. The primaries of these transformers are connected in parallel, while the secondary windings are connected in series. If the output voltage seems to be zero, or very low, then suspect a phasing problem in the connection of the transformers. This problem is corrected by reversing the primary leads on *one* (not both!) of the 12.6-volt transformers.

Two regulators are used in this circuit. The positive power-supply regulator (U1) is the 7812, while the negative supply regulator is the matching 7912 device. Note well that there is a pinout difference between these devices! The numbers are the same as on the LM323 discussed above, but, on the 7912, the meanings of the pins are different! Failure to note this can result in destruction of the device.

I have a personal preference for the T0-3 versions of the 7812/7912 devices. These seem to handle 1 ampere of current better, especially when heatsinked. The T0-220 plastic-case versions are rated for only 750 mA in free air, and I suspect the 1 ampere heatsinked rating (too many failed on me over the years). If you cannot obtain the T0-3 style, or have some other reason why the T0-220 must be used, make sure that the device is properly heatsinked (with silicone grease), and that the mounting screw is tightened to the rated torque.

The functions of the capacitors, and their ratings, are the same as in the previous discussion on the +5-volt supply, so the discussion will not be repeated.

Appendix B
Interfacing Timer ICs with High Power Loads

The timer IC is usually a special purpose device (in either bipolar or MOS technology) or some configuration of CMOS and/or TTL devices. No matter which devices are used, they are essentially low power, low voltage devices that will not easily tolerate the high voltages of the AC power line, or the high DC potentials found in some other applications (i.e., controls or radio transmitters). The solution seems to be a proper interface that will allow the timer IC to be connected to such sources without problems. Some of the circuits for this purpose have been seen earlier in this book, but this appendix is a way of getting them together in one place. You may be especially interested in these circuits if you plan to use digital circuitry in order to control your home, or perform some similar task (as is popular with energy conscious people who also know how to program microcomputers!).

The most basic form of interface is the electromechanical *relay*. These devices have a set of switch contacts that are actuated by an electromagnetic coil. The magnetic field set up when a current flows in the coil will attract the armature that is connected to the "toggle" arm of the switch contacts. I can already hear some of my sophisticated electronut friends bemoaning the fact that "Carr told 'em to stick a *#!! magnet in the circuit!" They are right in that electromechanical relays have been around since the "year one," but that does not disqualify them for use in modern projects. All those years of experience allow relay makers to turn out a highly reliable product that still has validity. Besides, if you look into a lot of modern equipment, there are tons of anachronistic relays in use, especially where high AC powers are encountered.

Figure B-1 shows how to connect a relay into the circuit with a 555 timer. This relay is wired so that it is normally energized; when the 555 output is LOW, current flows in the relay coil. Since most relays have SPDT or DPDT contacts, this is of little consequence. The relay must be a low-voltage type so that it will operate with the potentials that are compatible with the 555 device. Also, the relay coil resistance must be high enough to limit the coil current to 200 milliamperes or less. The 555 output will sink up to 200 mA when LOW, so the coil current must be limited to this value. Select a coil resistance that is greater than (V+)/0.200.

Figure B-1B shows a method for slowing down the relay pull-in time. This is sometimes necessary in order to avoid relay races where two or more relays are operated together. I can recall an X-ray machine that burned up because two relays in the three-phase AC power supply (440 VOLTS!) were out of sync. The protocol called for them to make at the same time, but in reality it was necessary to ensure that one of them made before the other—but certainly no later than the other! The solution proposed by the X-ray manufacturer was to place a 5 to 20 uF capacitor across the coil of one relay in order to slow it down, proportional to the time constant of the capacitor and the resistance of relay coil.

We can also connect the relay coil from the 555 output to ground because the 555 output will source current as well as sink it. The situation in this case is exactly reverse of the previous situation, i.e., the relay is energized when the 555 output is HIGH. Make sure, on other styles of timer, that the output will source sufficient current or the relay may be starved. The 555, and certain others, will easily handle small relay coils in this manner.

Note in Fig. B-1A that there is a diode shunted across the relay coil, and that the diode *seems* to be reverse biased. What earthly use is a reverse biased diode in this application? That diode is used to suppress the inductive kick of the relay coil when the current is interrupted. The voltage across an inductor is proportional to the inductance L, and the time rate of change of the current ($V = LdI/dt$). Since the break current has a very rapid fall time, the time rate of change is quite high—thus the voltage is also quite high. The spike from breaking the current can easily rise to 1000 volts or more, and will have a polarity opposite that of the V+ power supply. This is the reason for using a reverse biased diode in

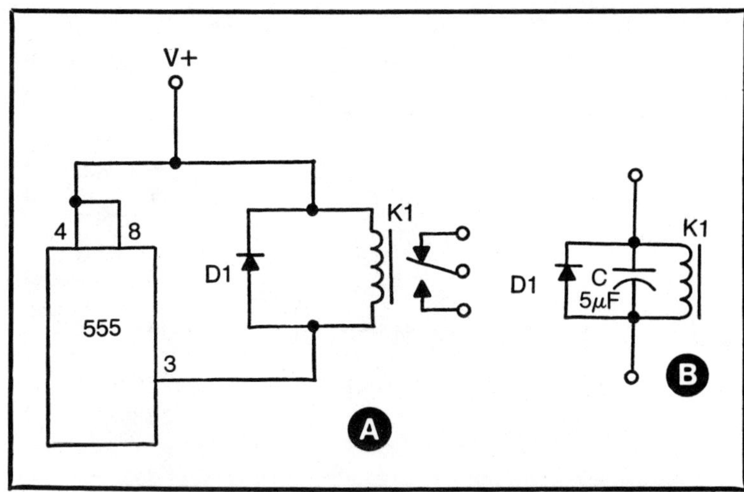

Fig. B-1. Relay interface: A) Relay connection; B) Delayed pull-in.

shunt with the relay coil. It will clip this high voltage transient to 0.7 volts, thereby preventing any damage to the integrated circuit(s). The correct diode to use in this application is a 1000 PIV, 1 ampere model such as the 1N4007 (or TV repairmen's replacement types, if you don't mind paying the bucks).

Some timers, such as the 555 by some manufacturers, are sensitive to spikes applied to the output terminal. The 555 will sometimes trigger when hit by such spikes. This is ordinarily not a problem, except in cases such as we are considering in this appendix. When a high-power inductive load is being handled, or when a relay is being used, then the inductive spike, even when clipped, can raise havoc with the 555 output. The solution is to place diodes in series with the output line in order to isolate the output from the load. Diodes D1 through D3 in Fig. B-2 are used for this purpose. In this case, we have used three type 1N4148 (or 1N914) devices. The fast rise time of the diodes will not cause a deterioration of the output pulse rise time and will produce an additional 2.1 volts of immunity to noise (negative pulsed) and a lot more isolation to positive pulses.

Figure B-2 also shows the use of a relay control transistor (Q1) to actually energize the relay. This tactic allows us to keep the relay turned off when the 555 output is LOW, and turn it on only when the 555 output is HIGH. A HIGH on the timer output line will forward bias the transistor, causing it to conduct. When transistor Q1 is turned on, a current will flow in the relay coil causing it to

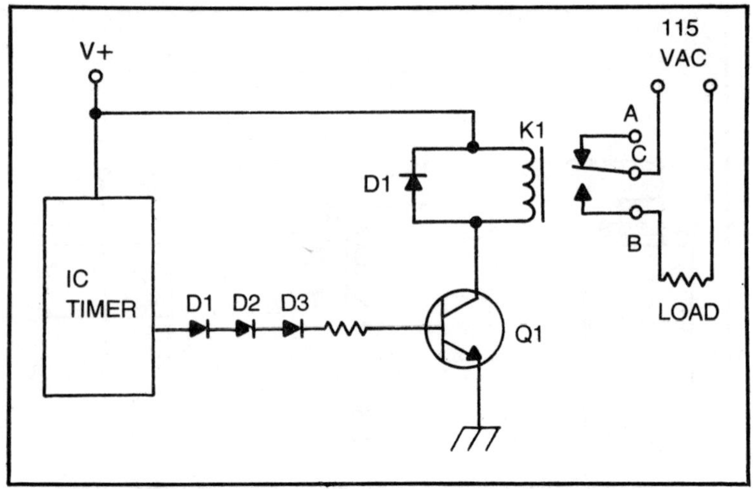

Fig. B-2. Transistor relay driver.

energize. The same diode across the coil is needed in this circuit because of the transistor's sensitivity to high voltage spikes.

The type of transistor selected for Q1 will depend upon the relay coil current. In this circuit, we are no longer constrained by the 200 mA limit of the 555 output (at least, not *directly*). Select a transistor with a collector current rating that exceeds the relay coil current.

The circuit of Fig. B-2 also allows us to split the relay power supply from the timer power supply, as long as they both share a common ground. The timer power supply is usually limited to +18 volts, or less. But relays are very often available in 28 volt DC coils. We could connect the V+ end of relay K1 to a +28 volt source, and not hurt the timer. Two additional constraints are to be sure that the collecter-emitter potential of the transistor is greater than the relay V+ potential; and that the product of the relay V+ potential and the relay coil current do not exceed the collector power dissipation specification.

A self-latching relay circuit is shown in Fig. B-3. This circuit uses the number 1 set of contacts (A1, B1, and C1) to control the AC power (as in the previous case), and the number 2 contacts (A2, B2, and C2) to latch the circuit. Contact C2 is grounded. When the transistor becomes forward biased, it will energize the relay. This will close contact pair B2-C2 causing the relay coil current to flow also in this path. When the one-shot timer expires, the relay will hold because of the current in these contacts, even though Q1 is now turned off again. The relay will remain energized until reset switch S1 is opened, causing the contacts to separate when the coil current ceases. It normally requires a lot less current to *hold* a relay armature than it did to *pull* it in. Because of this, some designers will place a resistance in series with contact B2. The lower hold current will reduce the current consumption of the unit. *Delco Electronics,* in their *Wonder Bar* signal seeker radios for GMC cars, used this method in order to turn off the 1 ampere seek relay.

There are sometimes cases where we will want to use two transistors in cascade to drive the relay coil. Examples are shown in Fig. B-4. This may be used in cases where we wish to invert the action of the relay in response to the timer output or have to handle a heavy-current relay coil (sometimes seen in automotive applications).

In the first instance, we might want to invert the action of the relay in response to the timer output. We might, for example, wish to gain the speed advantage inherent in some relays when they drop from the energized position. It seems that it is sometimes faster to let the relay drop from this position than to allow it to overcome the inertia of being deenergized. Figure B-4A shows a circuit in which the relay action is reversed. Normally, the timer output is LOW, and this keeps a zero bias on transistor Q1. Since the transistor is effectively turned off in this condition, it will pass

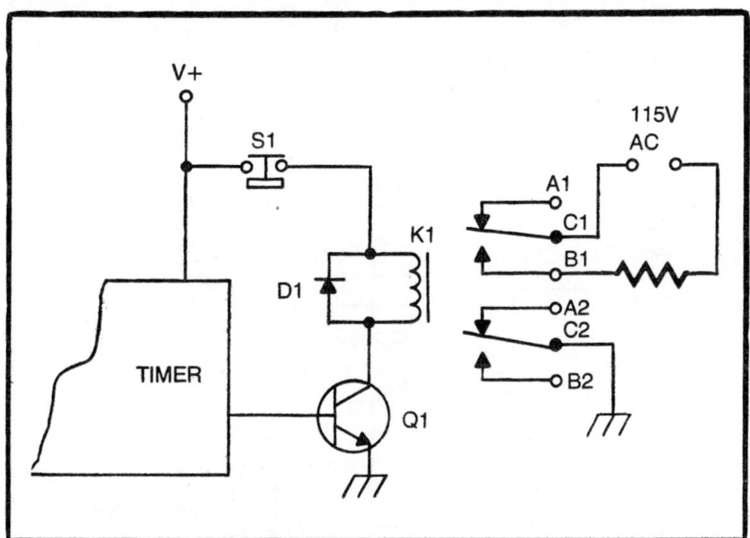

Fig. B-3. Self-latching relay.

no collector current. The collector voltage, therefore, is maximum and will approximate V+. A HIGH on the collector of Q1 will cause Q2 to be forward biased (through resistor R2). We can conclude, therefore, that relay K1 will be turned on when the output of the timer is LOW.

When the timer output snaps HIGH, however, the situation reverses. A HIGH on the timer output will forward bias transistor Q1 sufficiently to cause saturation. This condition causes the collector voltage to drop to near zero, thereby removing the bias from transistor Q2. The relay will no longer be energized, and will drop out. We would usually wire the SPDT contacts of such a relay

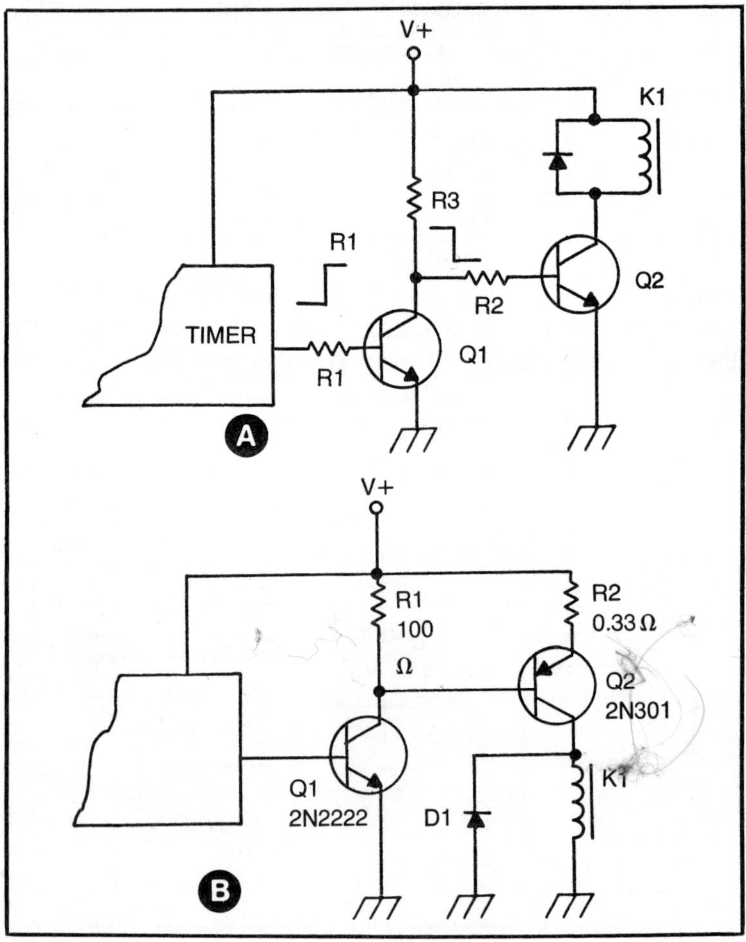

Fig. B-4. Inverting relay drivers.

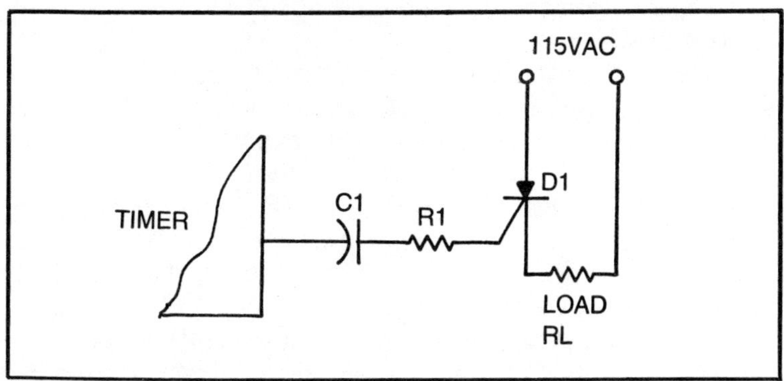

Fig. B-5. SCR interface.

in the opposite manner than would be done in the case of a single transistor relay amplifier (Figs. B-2 and B-3).

The circuit of Fig. B-4A used an NPN transistor to control the relay. The transistor might be an ordinary 2N2222 device, or a 2N3057 if a medium to heavy relay were used. If, on the other hand, a heavy current relay is used, then transistor Q2 must be of the 2N3055 variety (which infers that a 2N3057-class device must be used for Q1).

In any event, we must sometimes be concerned with the beta of the transistors used and the available driver current. In the case of a 555 device, for example, we can drive a base current of 200 mA. The beta of the transistor relates the collector current to the base current (beta $= I_c/I_b$), so we can figure out the minimum beta needed in order to support any given relay current. A 25-ampere relay, for example, requires a minimum beta of 25/0.2, or 125. This figure is a little high for the ordinary power transistor, so a cascade chain must be used as shown in Figs. B-4A and B-4B.

The use of a Germanium PNP transistor such as the old-fashioned 2N301 device is shown in Fig. B-4B. The circuit is essentially the same as in the previous example, except that a PNP device is used. Note that the roles of the emitter and collector are reversed in this case, as befits the inverted polarity of the PNP device over the NPN.

The use of a silicon controlled rectifier (SCR) is shown in Fig. B-5. The SCR operates like a gated rectifier. When the device is dormant, the resistance between the anode and the cathode is very high. But, once a gate pulse is received, the resistance in the anode-cathode path drops markedly and the device operates like an ordinary rectifier diode. SCRs are available in current ranges from milliamperes to hundreds of amperes, at voltages from 25 PIV to kilovolts PIV.

The SCR requires a certain minimum current in order to sustain conduction between cathode and anode. If the anode-cathode current drops below the holding current value, then the device will turn off. In some cases, the turn off will consist of a series connected normally-closed SPST switch that interrupts the current to the anode. In other cases, a pulse is applied to either the cathode or the anode circuit with a polarity and amplitude sufficient to turn off the device.

In Fig. B-5 we are connecting the output of the timer to the gate of the SCR through a series differentiating circuit. The leading edge of the timer output pulse produces a positive spike at the output of the differentiator. This spike is used to turn on the gate of the SCR. The negative-going edge of the timer output pulse will produce a negative spike which most SCRs will ignore. In some cases, however, we will want to place a series rectifier diode between R1 and the gate terminal of the SCR. The polarity of the rectifier is arranged to pass the positive going spike and block the negative-going spike.

The timer output terminal must be capable of sourcing sufficient current to turn on the SCR. If not enough current is available, then the timer might as well be whistling in the wind for all the good it does.

We sometimes must operate the timer from a high voltage supply. We can do this by referencing the timer potentials to one line of the AC from the SCR, resulting in a very low (relatively) potential for the timer. It is important to make sure that the common in the timer circuit is ***not grounded***, but operates above ground. Alternatively, we would rectify the 115 volt AC potential, and then feed it to the 555 (or other timer) through a magnificent dropping resistor.

The modern method for isolating the timer from the nasty 115 volt (or higher) line is to use an optoisolator (Figs. B-6A and B-6B). The optoisolator is a small IC package that contains a light emitting diode (LED) and a phototransistor. The LED is positioned inside of the plastic case in such a way that its light falls on the light sensitive base of the phototransistor. The package of the optoisolator is opaque, so outside light does not affect the operation.

Figure B-6A shows the use of an active-LOW output with the LED. When the output is HIGH (dormant period), the potentials on each side of the LED are equal so the LED will not be lighted. But

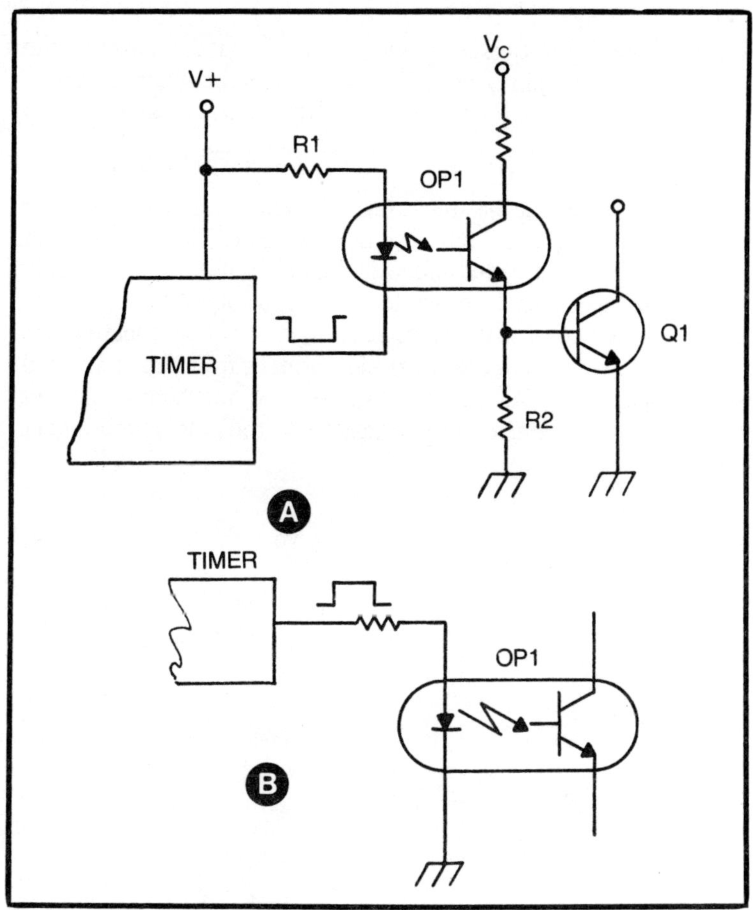

Fig. B-6. Optoisolator interface.

when the timer is triggered and the output line drops LOW, the LED sees a potential proportional to V+ and the drop across series limiting resistor R1. The value of this resistor, incidentally, depends upon V+ and the LED current. When the LED is turned on, it illuminates the base of the phototransistor, causing that transistor to conduct. A conduction voltage appears across R2, and that will turn on control transistor Q1. The rest of the circuit will depend upon the task being performed, but will most likely be similar to some of those from previous examples.

The circuit in Fig. B-6B is used when the timer output will source a current. The operation, however, is essentially the same as in Fig. B-6A. When the timer output is HIGH, the current through the LED causes a light to shine on the base of the phototransistor. This turns on the phototransistor which, in turn . .

The use of the optoisolators provides a certain elegance that is difficult to overlook. The isolation is tremendous, up to 10^{12} ohms has been claimed. The device is also less sensitive to noise potentials. It seems that most high current loads will produce a certain amount of transient noise pulse energy. These pulses can create all kinds of dreaded affects in digital circuits, so they should be shunned as much as possible. The optoisolator can be a low frequency device (if you plan correctly), so it will attenuate those pulses that come from the timer side. Pulses from the controlled side simply have no place to go and no mechanism to get there—how can a pulse in the transistor of the optoisolator affect the LED?

Appendix C
Resistors and Capacitors

Most of the timer circuits in this book depend upon the RC network as the principle timing element. Some use the RC network as part of the frequency-determining network of an astable multivibrator that is used as a clock. In other cases, the timer will be a monostable multivibrator, and the RC network determines the output-active duration. In either event, proper timer design sometimes depends upon knowing the simple RC network and how it works.

We also see some voltage dividers used in timers. The 555 and XR-2240 devices, for example, use resistor voltage dividers in the timing network. The advantage of these circuits is that they are all but insensitive to small variations in power-supply voltage occurring during the timer period. This is in distinct contrast to the situation in some bipolar or unijunction transistor timer circuits.

In this appendix, we will consider the practical operation of resistor voltage dividers and of resistor-capacitor networks. These circuits are used sufficiently often that you should be aware of them. There will be some arithmetic (I hesitate to call it *math*) involved, but it is of the variety that a simple scientific calculator (i.e., we need Log, Ln, and e^x functions) will work the problems almost by themselves (well, not *quite* by themselves). In any event, we will work a few practical examples to show the non-math freaks how they can be worked in a minimum time.

Figure C-1 shows a simple two-resistor voltage divider. The output voltage V_o will be a fraction of the input voltage V. Ohm's

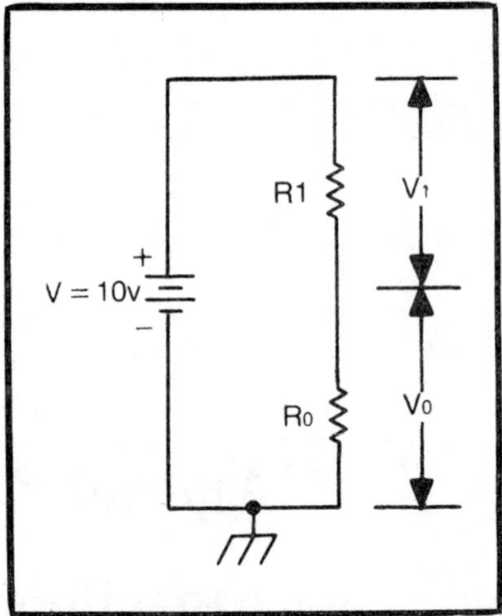

Fig. C-1. Resistor voltage divider.

law tells us that the voltage drop is proportional to the resistance. We can further invoke a great and mighty mind of electronics-past by calling on Dr. Kirchhoff. In his famous Kirchhoff's voltage law (KVL) he told us minions that the algebraic sum of all voltage drops and rises (?? ? what the. . .) around a closed circuit is zero. What the dickens does this mean? What is a voltage drop and a voltage rise? Well, a voltage drop is the voltage appearing across a resistance, and it is proportional to the resistance and the current flowing in that resistance ($V = IR$). A voltage rise is the opposite of a voltage drop, but let's leave it at that, saying only that a battery or power supply gives a voltage rise, while resistors give voltage drops. In the circuit of Fig. C-1, the voltage rise is the battery V, while the voltage drops are V1 and V_o. The Essence of KVL is that:

$$V - V1 - V_o = 0$$

Which is a dumb way of saying:

$$V = V1 + V_o$$

The proportion of voltage that appears across resistor R_o is a fraction of the total voltage V:

$$V_o = \frac{R_o}{R1 + R_o} \times V$$

☐ **Example:**

Suppose that a 1 kohm and a 1.5 kohm resistor are in series across a 10-volt power supply. Find the voltage drop across the 1.5 kohm resistor.

Solution:

$$V_o = \frac{1.5\,\text{kohm}}{(1+1.5)\text{kohm}} \times (10\,\text{V})$$

$$= \frac{1.5}{2.5} \times (10\text{V})$$

$$= (0.6)\,(10\,\text{V}) = 6.0\,\text{volts}$$

The procedure is a little more complicated in multiple-resistor circuits. We still place the resistor (across which the drop is calculated) in the numerator of the fraction, and the sum of all resistances in the denominator. Such a circuit, involving three resistors, is shown in Fig. C-2. Let's assume that R1 = 500 ohms, R2 = 1000 ohms, and R3 = 1200 ohms. Calculate voltage V_2 (i.e., the potential across resistor R2). We could, if we wanted to waste some time, calculate the total resistance (R1 + R2 + R3), and then calculate the current flowing in the circuit. This current could then be plugged into Ohm's law with the value of R2, etc. But, it is more elegant to use the voltage divider equation:

$$V_2 = \frac{R2}{R1+R2+R3} \times V$$

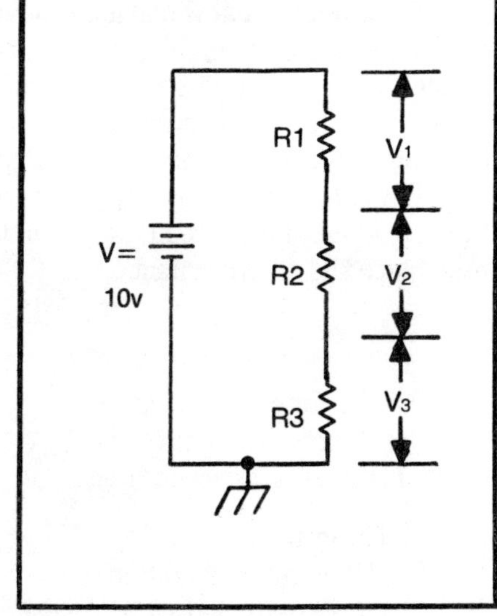

Fig. C-2. Resistor voltage divider (three resistors).

□ **Example:**

Calculate V2 with the values given in the text above.

Solution:

$$V_2 = \frac{1000 \text{ ohms}}{(500 + 1000 + 1200) \text{ ohms}} \times (10\text{ V})$$

$$= \frac{1000}{2700} \times (10\text{ V})$$

$$= (0.37)(10\text{ V}) = 3.7 \text{ volts}$$

Capacitors and resistors are used as the timing networks in a large variety of timer circuits. The capacitor is a "storage tank" for electrical charge. The voltage that appears across a capacitor is proportional to the amount of charge stored in the capacitor. The energy in a capacitor is stored in an electrostatic field in a dielectric (insulator) that is placed between two metallic plates.

Consider the circuit in Fig. C-3A. Here, we have a capacitor in series with a resistor, and we are interested in the voltage appearing across the capacitor. Initially, the voltage is zero because the switch connects the RC network to ground. If the switch is toggled so that the battery V_s is applied to the RC network, the capacitor will begin to change. The voltage waveform across the network at the instant the switch is closed is shown in Fig. C-3B. This potential rises almost immediately. But note the voltage waveform across the capacitor ("A" in Fig. C-3C). It does not rise instantly, but increases exponentially instead. This is in keeping with the idea that a capacitor opposes changes in voltage.

The rate at which the capacitor voltage rises is set by the RC time constant of the circuit, which is defined as the product of the resistance and the capacitance:

$$T = RC$$

Where:

T is the time in seconds
R is the resistance in ohms
C is the capacitance in farads

□ **Example:**

Find the time constant, in seconds, of a series combination of a 15,000 ohm resistor and a 0.01 μF capacitor.

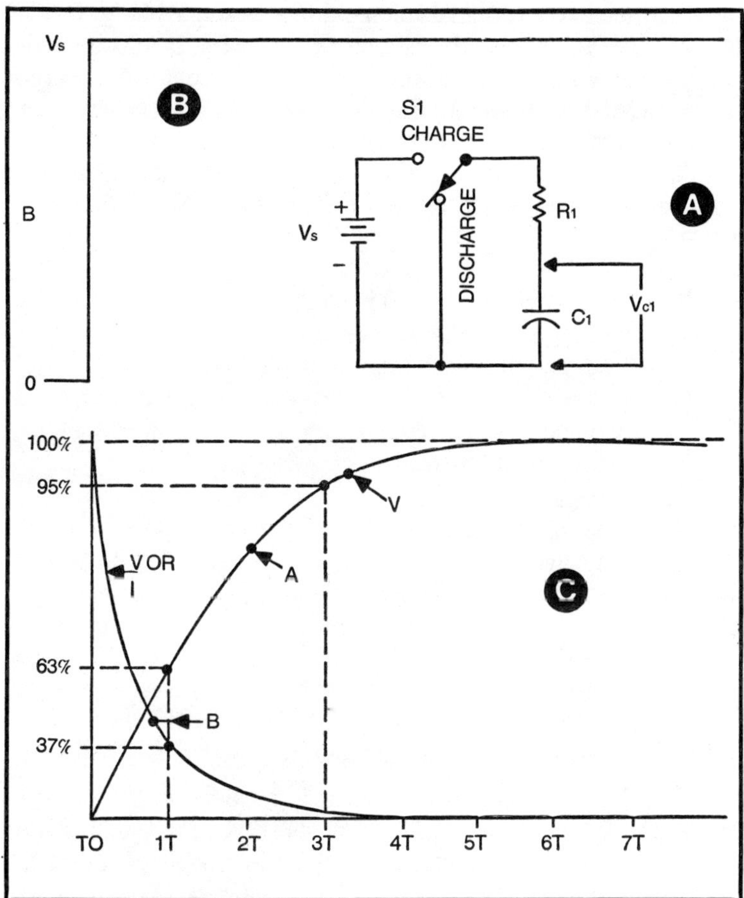

Fig. C-3. RC charging patterns.

Solution:

$$T = RC$$
$$= (1.5 \times 10^4 \text{ ohms})(0.01 \times 10^{-6} \text{ farads})$$
$$= 1.5 \times 10^{-4} \text{ seconds} = 150 \text{ microseconds}$$

The time constant is also defined as the time required for the voltage to rise from zero to 63 percent of its total voltage. Alternatively, it is also the time required for a fully charged capacitor to discharge to 37 percent of the fully charged voltage. These points are shown in the capacitor charging waveform of Fig. C-3C. The capacitor is considered to be fully charged after five (5) RC time constants. In actuality, the capacitor approaches the fully charged voltage asymptotically, so will theoretically never become fully charged to the terminal voltage. In the example of Fig. C-3C, the voltage will approach, but never quite reach 100 percent charge. The voltage will, however, reach 95 percent of full charge by the end of three time constants, and over 99 percent charge after five time constants.

The time constant is used in the calculation of the instantaneous voltage across a capacitor that is charging. This equation is:

$V_{C1} = V_S (1 - e^{-t/RC})$

Where:

V_{C1} is the potential across C1 in volts
V_s is the supply potential in volts
e indicates that the natural logarithms are used
t is the time, in seconds, after the capacitor began to charge
R is the resistance in ohms
C is the capacitance in farads

☐ **Example**

An RC network consists of a 0.1 μF capacitor and a 100 kohm resistor. Find the voltage across the capacitor 13 milliseconds after the charging begins if the supply voltage is 20 volts.

Solution:

$$V_{C1} = V_s (1 - e^{-t/RC})$$
$$= (20\text{ V})(1 - e^{-(0.013\text{ s})/(0.1 \times 10\text{-}6\text{F})(10^5\text{ ohms})})$$
$$= (20\text{ V})(1 - e^{-1.3}) = (20\text{ V})(0.727)$$
$$= (20\text{ V})(0.727) = 14.54\text{ volts}$$

Clearly, when the time element is included in the voltage expression for an RC network, we can use this voltage to indicate the passage of time. The typical RC timer is capable of operating from tenths or hundredths of a second to microseconds.

The discharge time from some voltage V_s is expressed by the equation

$$V_{C1} = V_s (e^{-t/RC})$$

Where all terms are as previously defined.

☐ **Example:**

Consider the RC network in the previous example. Find the voltage 8 milliseconds after the switch is set to the discharge position.

Solution:

$$V_{C1} = V_s (e^{-t/RC})$$
$$= (20\text{ V})(e^{(.008)/(0.1 \times 10^{-6}\text{ farads})(10^5\text{ ohms})})$$
$$= (20\text{ V})(e^{-0.8})$$
$$= (20\text{ V})(0.45) = 9\text{ volts}$$

The accuracy and the stability of the RC network timer is dependent upon the quality of the components used. The resistors should be precision types (1 percent, or better tolerance). These resistors normally have a low temperature coefficient, so will maintain the resistance value. Carbon composition resistors have the nasty habit of changing resistance drastically with changes in temperature. Also, the values of carbon composition resistors tends to be 5 percent (at best) and may be as bad as 20 percent.

The capacitors used should be either silver mica or polyethylene types. These capacitors have low temperature coefficients and are generally available in low-tolerance values.

Index

Edited by: Jack Chadwick/Ray Collins